Through-Silicon Vias
for 3D Integration

About the Author

John H. Lau, Ph.D., has been a fellow of the Industrial Technology Research Institute (ITRI) in Taiwan since January 2010. Prior to that, he was a visiting professor at Hong Kong University of Science and Technology (HKUST) for 1 year; the director of the Microsystems, Modules, and Components (MMC) Laboratory at the Institute of Microelectronics (IME) in Singapore for 2 years; and a senior scientist/MTS at HPL/Agilent in California for more than 25 years.

Dr. Lau's professional competences are design, analysis, materials, process, manufacturing, qualification, reliability, testing, and thermal management of electronic, optoelectronic, light-emitting diode, and microelectromechanical system (MEMS) components and systems (which include devices, substrates, packages, and printed circuit boards), with emphases on surface-mount technology (SMT), flip-chip wafer-level package (WLP), through-silicon via (TSV), and other key-enabling technologies for three-dimensional (3D) integrated circuit (IC) integration and system-in-package.

With more than 36 years of research and development and manufacturing experience, he has authored or coauthored more than 310 peer-reviewed technical publications and more than 120 book chapters, has invented more than 30 issued and pending patents, and has given more than 270 lectures/workshops/ keynotes worldwide. He has authored or coauthored 17 textbooks on TSVs, 3D MEMS packaging, reliability of 3D IC integrations, advanced packaging, ball-grid-array, chip-scale package, tape automated bonding, flip-chip WLP, high-density interconnects, chip-on-board, direct chip attach, SMT, and lead-free materials, soldering, manufacturing, and reliability.

Dr. Lau earned a Ph.D. degree in theoretical and applied mechanics from the University of Illinois at Urbana–Champaign, an M.S. degree in structural engineering from the University of British Columbia, a second M.S. degree in engineering physics from the University of Wisconsin at Madison, and a third M.S. degree in management science from Fairleigh Dickinson University. He also has a B.E. degree in civil engineering from National Taiwan University.

He was a member of the editorial boards of *ASME Transactions; Journal of Electronic Packaging; IEEE Transactions on Component, Packaging, and Manufacturing Technology; Circuit World; Soldering and Surface Mount Technology;* and others. He was the general chair and program chair of the IEEE Electronic Components and Technology Conference (ECTC) from 1990 to 1995 and the International Electronic Manufacturing Technology Symposium from 1987 to 1992, Solder Mechanics Symposium organizer of the ASME Winter Annual Meeting from 1987 to 2002, and 3D IC Integration Symposium organizer of the ASME IMECE 2010. Also, he was the publications chair of the IEEE ECTC from 1995 to 2006. He served on the board of governors of IEEE/CPMT and has been one of its distinguished lecturers every year for the past 11 years.

Dr. Lau has received many awards from the American Society of Mechanical Engineers (ASME), Institute of Electrical and Electronics Engineers (IEEE), Society of Manufacturing Engineers (SME), and other societies. He is an elected ASME fellow and has been an IEEE fellow since 1994.

Through-Silicon Vias for 3D Integration

John H. Lau, Ph.D.

New York   Chicago   San Francisco
Lisbon   London   Madrid   Mexico City
Milan   New Delhi   San Juan
Seoul   Singapore   Sydney   Toronto

1 2 3 4 5 6 7 8 9 0   CTP/CTP   1 8 7 6 5 4 3 2

ISBN 978-0-07-178514-3
MHID 0-07-178514-0

Sponsoring Editor	**Copy Editor**
Michael Penn	James K. Madru
Editorial Supervisor	**Proofreader**
Stephen M. Smith	Upendra Prasad,
	Cenveo Publisher Services
Production Supervisor	
Pamela A. Pelton	**Indexer**
	Robert Swanson
Acquisitions Coordinator	
Bridget L. Thoreson	**Art Director, Cover**
	Jeff Weeks
Project Manager	
Sapna Rastogi,	**Composition**
Cenveo Publisher Services	Cenveo Publisher Services

McGraw-Hill books are available at special quantity discounts to use as premiums and sales promotions, or for use in corporate training programs. To contact a representative, please e-mail us at bulksales@mcgraw-hill.com.

This book is printed on acid-free paper.

Contents

Foreword

Three-dimensional (3D) integration, especially 3D integrated circuit (IC) integration, is taking the semiconductor industry by storm. It has been (1) affecting the chip suppliers, "fabless" design houses, foundries, integrated device manufacturers, outsourced semiconductor assembly and test, packaging substrate, electronics manufacturing services, original design manufacturers, original equipment manufacturers, universities, research institutes, and materials and equipments suppliers; (2) attracting researchers and engineers from all over the world to go to conferences, lectures, workshops, forums, and meetings to present their findings, exchange information, look for solutions, learn the latest technologies, and plan for their future; and (3) pushing the industry to build the standards, ecosystems, and infrastructures for 3D integration. I've never seen anything like this before!

This is a perfect storm! People and companies think that Moore's law is going to take a bow soon, and 3D integration is the next one. In order to prepare for their future and have a competitive edge, people have been investing heavily in both human and physical resources for 3D IC integration. 3D IC integration is defined as the stacking of thin chips/interposers with through-silicon vias (TSVs) and solder microbumps. Thus TSV, thin-wafer/chip handling, microbumps, bonding, and thermal management are the most important key-enabling technologies for 3D IC integration.

Unfortunately, for most practicing engineers and managers, as well as scientists and researchers, TSVs, thin-wafer strength measurement and handling, solder microbumping, chip-to-chip (C2C) bonding, chip-to-wafer (C2W) bonding, wafer-to-wafer (W2W) bonding, and the reliability of 3D integration for electronics and optoelectronics packaging and interconnect systems are not well understood. Thus there is an urgent need, both in the industry and in research institutes, to create a comprehensive book on the current state of knowledge of these key-enabling technologies. This book should be written so that readers can quickly learn about the basics of problem-solving methods and understand the tradeoffs inherent in making system-level decisions.

To meet these needs, John H. Lau at the Electronics and Optoelectronics Research Laboratories has collected a great deal of useful information from the latest technical publications and produced *Through-Silicon Vias for 3D Integration*. This is an excellent book for industry and research institutes as well as universities. It is equally appropriate as an introduction to 3D IC integration, 3D Si integration, and 3D IC packaging for those just entering the field and as an up-to-date reference for those already engaged in interconnect design and process developments for 3D integration.

This book has 11 chapters that cover the whole spectrum of 3D integration from the fundamentals to the latest developments in this field. Chapter 1 briefly discusses nanotechnology and 3D integration for the semiconductor industry. Chapter 2 provides the six key tasks (via forming, dielectric layer deposition, barrier- and seed-layer deposition, via filling, chemical-mechanical polishing, and copper revealing) in making TSVs. Chapter 3 shows the mechanical, thermal, and electrical behaviors of a TSV. Chapters 4 and 5 present, respectively, thin-wafer strength measurements and thin-wafer handling during semiconductor and packaging assembly. Chapter 6 provides the bumping, assembly, and reliability of microbumps and joints. Microbump electromigration is examined in Chap. 7. Chapter 8 provides C2C, C2W, and W2W transient liquid-phase bonding. The thermal management of 3D IC integration system-in-package (SiP) is presented in Chap. 9. Finally, the competitive technologies that keep 3D integration technologies away from high-volume production such as the 3D IC packaging technology are presented in Chap. 10.

This comprehensive guide provides cutting-edge information on 3D Si integration, 3D IC integration, and 3D IC packaging and their applications to high-density, high-performance, low-power-consumption, wide-bandwidth, lightweight, low-profile, and green products. It is essential for electronics and optoelectronics manufacturing and packaging professionals who wish to master TSV, thin-wafer strengthening and handling, microbumping, assembly and reliability, and thermal management problem-solving methods and those demanding cost-effective designs and high-yield environmental benign manufacturing processes. This valuable reference book covers all aspects of this fast-growing field.

Will 3D integration (like Moore's law) provide the characteristics necessary to allow the world to depend on it in the future? This book may not be able to answer this question, but it will help all participants in the electronics and optoelectronics design and assembly world better understand what needs to be done and how to answer this question, plan for the future, and make it happen.

Dr. Ian Yi-Jen Chan, Vice President
Director of the Electronics and Optoelectronics Research Laboratories
Industrial Technology Research Institute
Hsinchu, Taiwan

Preface

Through-silicon vias (TSVs) are the heart and most important key-enabling technology of three-dimensional (3D) silicon (Si) integration and 3D integrated circuit (IC) integration. TSVs provide the opportunity for the shortest chip-to-chip interconnects and smallest pad size and pitch of interconnects. Compared with other interconnection technologies, such as wire bonding, the advantages of TSVs include (1) better electrical performance, (2) lower power consumption, (3) wider data width and thus bandwidth, (4) higher density, (5) smaller form factor, and (6) lighter weight.

TSV is a disruptive technology. As with all disruptive technologies, the questions to ask are "What is it displacing?" and "What's the cost?" Unfortunately, TSV technology is attempting to displace wire-bonding technology, which is the most mature, high-yield, and low-cost technology. On the other hand, there are six key steps in making a TSV, namely, vias by deep reactive ion etching (DRIE), dielectric layers by plasma-enhanced chemical-vapor deposition (PECVD), barrier and seed layers by physical vapor deposition (PVD), via filling by (e.g., copper) electroplating, overburden copper removal by chemical-mechanical polishing (CMP), and TSV copper revealing. Thus, compared with wire bonding, TSV technology is very expensive! However, just as with solder-bumped flip-chip technology, because of their unique advantages, TSVs are here to stay for high-performance, high-density, low-power, and high-bandwidth applications.

Besides TSV technology, the other key-enabling technologies for 3D Si integration and 3D IC integration are thin-wafer strength measurement and handling and thermal management. For 3D IC integration, solder wafer microbumping and assembly and electromigration of solder microjoints are additional key-enabling technologies. In this book, all these key-enabling technologies will be presented and discussed.

There are seven major subjects in this book, namely, (1) nanotechnology and 3D integration for the semiconductor industry (Chap. 1); (2) TSV technology and the mechanical, thermal, and electrical behaviors of TSVs (Chaps. 2 and 3); (3) thin-wafer strength measurements and handling (Chaps. 4 and 5); (4) solder wafer microbumping, assembly, and reliability and electromigration of solder microjoints (Chaps. 6 and 7); (5) transient liquid-phase bonding (Chap. 8); (6) thermal management (Chap. 9); and (7) 3D IC packaging (Chap. 10). The book concludes with a simple thought (Chap. 11) about future (up to 2020) efforts warranted in the field.

Chapter 1 briefly discusses the origin and outlook of nanotechnology for the semiconductor industry. The recent advances, challenges, and trends of 3D Si integration and 3D IC integration are also provided. Furthermore, some embedded 3D IC integration examples are presented. Finally, this chapter concludes with a list of TSV patents.

Chapter 2 details the six key steps in making a TSV. Also, the TSV processes for 3D Si integration (via before and after bonding) and 3D IC integration (via-first, via-middle, and via-last) are discussed briefly. The mechanical, thermal, and electrical behaviors of a TSV are presented and examined in Chap. 3.

Chapter 4 presents the design, fabrication, and calibration of a piezoresistive stress sensor for thin-wafer strength measurement. The effects of wafer back-grinding on the mechanical behavior of Cu–low-k chips are also provided. Some thin-wafer handling issues such as temporary bonding and debonding of a carrier and potential solutions are given in Chap. 5. Thin-wafer handling with carrierless technology is also presented and discussed.

Chapter 6 presents solder wafer microbumping, characterization, assembly, and reliability assessments of lead-free fine-pitch solder microjoints. Ultrafine-pitch solder microbumps are also explored. Chapter 7 presents the electromigration of solder microjoints with larger solder volume and pitch as well as smaller solder volume and pitch. Failure mechanisms of the multiphase solder-joint interconnect are also provided.

Chapter 8 presents chip-to-chip (C2C), chip-to-wafer (C2W), and wafer-to-wafer (W2W) low-temperature (transient liquid-phase) bonding. Scanning electron microscopy (SEM), focused ion beam (FIB), transmission electron microscopy (TEM), x-ray diffraction (XDR), differential scanning calorimetry (DSC), and C-mode scanning acoustic microscopy (C-SAM) of intermetallic compounds (IMCs) are presented and discussed.

Chapter 9 presents the effects of TSV chip/interposers on the thermal performance of 3D IC integration systems-in-package (SiPs). The thermal performance of 3D memory-chip stacking is also provided. Furthermore, the effect of thickness of the TSV chip on its hot spot temperature is presented. Finally, a thermal management system with TSVs and microchannels for 3D integration SiPs is provided and discussed.

Chapter 10 presents the recent advances in 3D IC packaging that keep TSVs away from volume production. The first example is wire bonding of stacked dies on Cu–low-*k* chips. The second example is a bare C2C with solder bumps and face-to-face interconnects. The last example is a fan-out-embedded WLP-to-chip with solder bumps and face-to-face interconnects. A note on the reliability of wire bonding is also presented.

For whom is this book intended? Undoubtedly it will be of great interest to three groups of specialists: (1) those who are active or intend to become active in research and development of TSVs, thin-wafer handling, solder wafer microbumping, assembly and electromigration, thermal management, and low-temperature C2C, C2W, and W2W bonding; (2) those who have encountered practical TSV and other key-enabling technology problems and wish to understand and learn more methods for solving such problems; and (3) those who have to choose a reliable, creative, high-performance, high-density, low-power-consumption, wide-bandwidth, and cost-effective 3D integration technique for their products. This book also can be used as a text for college and graduate students who have the potential to become our future leaders, scientists, and engineers in the electronics and optoelectronics industries.

I hope that this book will serve as a valuable reference source for all those faced with the challenging problems created by the ever-increasing interest in TSVs for 3D IC integration and 3D Si integration. I also hope that it will aid in stimulating further research and development on TSVs and other key-enabling technologies and more sound applications to 3D integration products.

The organizations that learn how to design and manufacture TSV interconnects in their 3D integration systems have the potential to make major advances in the electronics and optoelectronics industry and to gain great benefits in performance, functionality, density, power, bandwidth, quality, size, and weight. It is my hope that the information presented in this book may assist in removing roadblocks, avoiding unnecessary false starts, and accelerating design, materials, and process development of TSVs and other key-enabling technologies.

John H. Lau, Ph.D.

Acknowledgments

Development and preparation of *Through-Silicon Vias for 3D Integration* was facilitated by the efforts of a number of dedicated people. I would like to thank them all, with special mention of Bridget Thoreson, Pamela Pelton, and Stephen Smith of McGraw-Hill, Sapna Rastogi of Cenveo Publisher Services, and James K. Madru for their unswerving support and advocacy. My special thanks go to Michael Penn and Steve Chapman, who made my dream of this book come true by approving and effectively sponsoring the project, patiently listening to my excuses for delay, and solving many problems that arose during the book's preparation. It has been a great pleasure and fruitful experience to work with all of them in transferring my messy manuscript into a very attractive printed book.

The material in this book clearly has been derived from many sources, including individuals, companies, and organizations, and I have attempted to acknowledge by citations in the appropriate parts of the book the assistance that I have been given. It would be quite impossible for me to express my thanks to everyone concerned for their cooperation in producing this book, but I would like to extend due gratitude. Also, I would like to thank several professional societies and publishers for permitting me to reproduce some of their illustrations and information in this book, including the American Society of Mechanical Engineers (ASME) conference proceedings (e.g., *International Intersociety Electronic Packaging Conference*) and transactions (e.g., *Journal of Electronic Packaging*), the Institute of Electrical and Electronics Engineers (IEEE) conference proceedings (e.g., *Electronic Components and Technology Conference* and *Electronics Packaging and Technology Conference*) and transactions (e.g., *Advanced Packaging, Components and Packaging Technologies,* and *Manufacturing Technology*), and the International Microelectronics and Packaging Society (IMAPS) conference proceedings (e.g., *International Symposium on Microelectronics*) and transactions (e.g., *International Journal of Microcircuits & Electronic Packaging*).

I would like to thank my former employers, the Hong Kong University of Science and Technology (HKUST), the Institute of Microelectronics (IME), Agilent, and HPL, for providing me with excellent working environments that have nurtured me as a human being, fulfilled my need for job satisfaction, and enhanced my professional reputation. Also, I would like to thank Dr. Don Rice (HPL), Dr. Steve Erasmus (Agilent), Professor Dim-Lee Kwong (IME), and Professor Ricky Lee (HKUST) for their kindness and friendship while I was at their organizations. Furthermore, I would like to thank Dr. Ian Yi-Jen Chan, vice president and director of the Electronics and Optoelectronics Research Laboratories (EOL), for his trust, respect, and support of my work at the Industrial Technology Research Institute (ITRI) in Taiwan. Finally, I would like to thank the following colleagues for their stimulating discussions and significant contributions to this book: Dr. Zhang Xiaowu, Mr. C. S. Premachandran, Mr. Vincent Lee, Dr. V. N. Sekhar, Mr. D. Pinjala, Dr. Tang Gongyue, Dr. Ricky Lee, Dr. M. S. Zhang, Dr. Y. S. Chan, Ms. Sharon Lim Pei-Siang, Mr. Vempati Srinivasa Rao, Dr. Vaidyanathan Kripesh, Mr. Juan Milla, Mr. Andy Fenner, Dr. Daquan Yu, Dr. Aibin Yu, Mr. Navas Khan, Dr. Li Ling Yan, Dr. Won Kyoung Chou, Dr. Seung Wock Yoon, Ms. Cheryl Selvanayagam, Dr. Y. S. Chan, Mr. Chai Tai Chong, Mr. Shiguo Liu, Mr. Charles Vath, Mr. John Doricko, Ms. Germaine Yen, Dr. Wen Hui Zhu, Dr. Jui-Chin Chen, Dr. Ching-Kuan Lee, Dr. Tao-Chih Chang, Dr. Yu-Min Lin, Dr. Chau-Jie Zhan, Dr. Pei-Jer Tzeng, Dr. Cha-Hsin Lin, Dr. Shin-Yi Huang, Mr. Chun-Hsien Chien, Mr. Chien-Ying Wu, Ms. Yu-Chen Hsin, Mr. Shang-Chun Chen, Mr. Chien-Chou Chen, Mr. Hsiang-Hung Chang, Mr. Jing-Ye Juang, Mr. Wen-Li Tsai, Ms. Chia-Wen Chiang, Mr. Cheng-Ta Ko, Dr. Ra-Min Tain, Dr. Heng-Chieh Chien, Mr. Sheng-Tsai Wu, Mr. Ming-Ji Dai, Mr. Yu-Lin Chao, Mr. Shyh-Shyuan Sheu, Mr. Zhe-Hui Lin, Mr. Jui-Feng Hung, Mr. Shih-Hsien Wu, Mr. Shinn-Juh Lai, Mr. Peng-Shu Chen, Dr. Li Li, Dr. Yu-Hua Chen, Mr. Tai-Hung Chen, Mr. Chih-Sheng Lin, Dr. Tzu-Kun Ku, Dr. Wei-Chung Lo, and Dr. Ming-Jer Kao. Definitely, I would like to thank my eminent colleagues (the enumeration of whom would not be practical here) at ITRI, HKUST, IME, Agilent, EPS, HPL, and Sandia National Lab, and throughout the electronics industry, for their useful help, strong support, and stimulating discussions. Working and socializing with them has been a privilege and an adventure. I learned a lot about life and advanced integrated circuit (IC) packaging and three-dimensional (3D) IC integration technologies from them.

Last, I would like to thank my daughter, Judy, and my wife, Teresa, for their love, consideration, and patience by allowing me to work peacefully on this book. Their simple belief that I am making my small contribution to the electronics industry was a strong motivation for me. Thinking that Judy, Teresa, and I are in good health, I want to thank God for His generous blessings.

CHAPTER 1

Nanotechnology and 3D Integration for the Semiconductor Industry

1.1 Introduction

Some important milestones of nanotechnology will be mentioned briefly in this chapter, and emphasis will be placed on the applications and outlook of nanotechnology in electronics. "Computing Machines in the Future" was the title of the Nishina Memorial Lecture at Gakushuin University in Tokyo given by the 1965 Nobel Physics Laureate, Richard Feynman, on August 9, 1985. During the lecture, Feynman not only told us to go for three-dimensional (3D) integration, but he also taught us how to do it! Three-dimensional integration will be the focus of this book, and emphasis will be placed on 3D integrated circuit (IC) integration and recent advances and new trends. 3D silicon (Si) integration will be discussed briefly.

1.2 Nanotechnology

1.2.1 Origin of Nanotechnology

On the evening of December 29, 1959, Richard Feynman gave a lecture entitled, "There's Plenty of Room at the Bottom," to the American Physical Society at Caltech. His lecture is generally considered to be the very first speech about nanotechnology, even though Professor Norio Taniguchi of the University of Tokyo coined the term *nanotechnology* in his paper, "On the Basic Concept of Nano-Technology," published in the *Proceedings of the International Conference on Production Engineering* in 1974. In that paper, Taniguchi defined nanotechnology for fabrication methods below 1 μm to describe semiconductor processes such as thin-film deposition and ion-beam milling that exhibited characteristic control on the order of a nanometer. However, today the industry defines nanotechnology as ≤0.1 µm, or 100 nm.

1.2.2 Important Milestones of Nanotechnology

Feynman was amazingly prophetic, and his insights were really pretty visionary. Basically, in his 1959 speech he told us to make things "small and smaller," as suggested in Moore's law, proposed by Gordon Moore in 1965 [1]. (Both of them are from Caltech, where Feynman was a professor and Moore earned a Ph.D. degree.) Among many visionary things, Feynman said the following in his 1959 speech: "What would happen if we could arrange the atoms one by one the way we want them?" Since then, many scientists around the world have been doing exactly that, and some significant results are listed in the following.

1. 1974: First molecular electronic device patent.

2. 1981: IBM invented the scanning probe microscope (SPM) to measure and identify structures on the nanoscale. The SPM has the ability to move individual atoms and molecules on a surface. (Fig. 1.1a).

3. 1985: Curl, Kroto, and Smalley discovered buckyballs (1996 Nobel Prize in Chemistry)—stable molecules that contain 50 to 500 carbon atoms in a ball—using laser vaporized carbon (Fig. 1.1b).

4. 1989: IBM Almaden Research Center wrote IBM with 35 xenon atoms.

5. 1991: Discovery of carbon nanotubes (CNTs) by Sumio Iijima at the NEC Research Labs (Fig. 1.1c).

6. 2004: Intel launched the Pentium IV PRESCOFT processor based on 90-nm process technology. Even this is not arranging any atoms, but it is a very important milestone of Moore's law (the first high-volume product) by nanotechnology in the electronics industry!

7. 2004: Geim and Novoselov (2010 Nobel Prize in Physics) discovered a simple and stable method for isolating single atomic layers of graphite, known as *graphene* (Fig. 1.1d).

1.2.3 Why Graphene Is So Exciting and Could Be Very Important for the Electronics Industry

- Andre Geim and Konstantin Novoselov (at the University of Manchester) discovered how to make stable graphene, a honeycomb sheet of carbon atoms just one atom thick, and

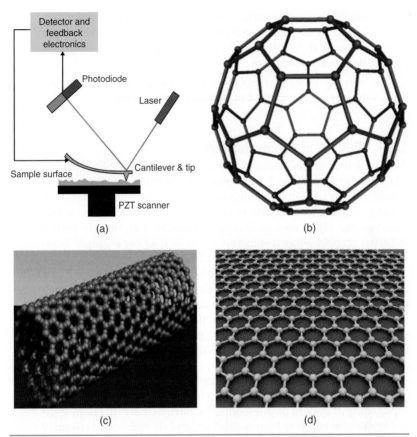

FIGURE 1.1   Some important milestones of nanotechnology. (a) Scanning probe microscope by IBM; (b) Buckyballs by Curl, Kroto, and Smalley; (c) CNTs by Sumio Iijima; and (d) graphene by Geim and Novoselov.

published their results in the journal *Science* 306, 666–669, October 2004.

- Since then, more than 4000 research papers have cited their paper.

- Since then, more than 2000 papers with *graphene* in the title have appeared in *Physical Review Letters,* the world's most prestigious physics journal (only 20 before 2004).

- In 6 years, Geim and Novoselov won the Nobel Prize for Physics (October 2010).

- Graphene (which is stronger than steel, more electrically conductive than copper, and able to transmit electrical signals with amazing rapidity) could be a candidate for the material to make faster and more powerful electronics. Sure there are challenges and issues to bring it into high-volume production, but fortunately, carbon-based electronics is an active area of research at, for example, IBM, Samsung, and other electronic device manufacturers.

- Already more than 500 papers about graphene-based electronics have appeared in *Applied Physics Letters.*

1.2.4   Outlook of Nanotechnology

As mentioned in his National Science Foundation (NSF) report in 2006, "Nano Hype: The Truth Behind the Nanotechnology," David Berube told us to stop all the hype; otherwise, the "nano-bubble" is coming. The industry needs to develop and commercialize a few high-volume killer nanotechnology products! For the electronics industry, graphene-based substrate could be one! For example, Samsung is planning to use it to make flexible displays.

In order to make high-volume nanotechnology products, it is a must to perform research and development to systemically, cost-effectively, and reliably package the nanotechnology devices to the next level of interconnects. (Most scientists working on nanotechnologies, especially those in academics and research institutes, tend to forget about these priorities, and thus their research results can never enter volume production.)

The nanotechnology device is not an isolated island. It must communicate with other IC/opto chips in a circuit through an input/output system of interconnects. Furthermore, the nanotechnology device needs to be powered up, and its embedded circuitry and elements are delicate, requiring the package to both carry and protect the device. Consequently, the major functions of nanotechnology packaging are (1) to provide a path for the electric current that powers the device circuit, (2) to distribute the signals onto and off the nanotechnology device, (3) to remove the heat generated by the circuit, and (4) to support and protect the nanotechnology device from hostile environments. These are the key reasons why the semiconductor industry has a key segment called *packaging assembly and testing* or *outsourced semiconductor assembly and test* (OSAT) that enables high-volume production of semiconductors with nanotechnology.

Table 1.1 shows the latest world ranking of these five key segments (i.e., semiconductor, foundry, OSAT, substrate/printed circuit board [PCB], and electronics manufacturing services [EMS]) of the semiconductor industry (please note the table footnotes). The world ranking of buyers of semiconductors is shown in Table 1.2, and they are called *original equipment manufacturers* (OEMs). The world ranking of fabrication-less (fabless) design houses is (1) Qualcomm,

Semiconductor	Foundry	OSAT	Substrate/PCB	EMS
Intel	TSMC	ASE	Unimicron	Foxconn
Samsung	UMC	Amkor	Ibiden	Flextronics
Texas Instruments	Globalfoundries	SPIL	Samsung	Jabil
Toshiba	SMIC	STATSChipPAC	Nippon Mektron	Celestica
Renesas	Dongbu HiTek	Powertech	CMK	Sanmina-SCI
Qualcomm	TowerJazz	UTAC	Nanya	Cal-Comp
STMicroelectronics	Vanguard (VIS)	ChipMOS	Shinko	Benchmark
Hynix	IBM	KYEC	KB Group	Elcoteq
Micron	MagnaChip	CARSEM	Compeq	Venture
Broadcom	Samsung	Unisem	Multek	Plexus

1. Semiconductor (excluding Foundry) mixing IDM and Fabless together.
2. Integrated device manufacturers (IDM) design, manufacture, and sell their chips. For example, Intel and Samsung.
3. Fabless companies design and sell chips but outsource manufacturing to foundry. For example, Qualcomm, Broadcom, and NVIDIA.
4. Foundry companies manufacture chips designed and sold by their customers especially from fabless companies. For example, TSMC and UMC.
5. Outsourced Semiconductor Assembly & Test (OSAT). For example, ASE, Amkor, and SPIL.
6. Electronics manufacturing service (EMS). For example, Foxconn (HH) and Flextronics.

TABLE 1.1 Ranking of the Five key Segments in Semiconductor Industry

Rank 2010	Rank 2011	Company	2010*	2011*	Growth (%)	Share (%)
3	1	Apple	12,819	17,257	34.6	5.7
2	2	Samsung Electronics	15,272	16,681	9.2	5.5
1	3	HP	17,585	16,618	−5.5	5.5
5	4	Dell	10,497	9,792	−6.7	3.2
4	5	Nokia	11,318	9,042	−20.1	3.0
6	6	Sony	9,020	8,210	−9.0	2.7
7	7	Toshiba	7,768	7,589	−2.3	2.5
10	8	Lenovo	6,091	7,537	23.7	2.5
8	9	LG Electronics	6,738	6,645	−1.4	2.2
9	10	Panasonic	6,704	6,267	−6.5	2.1
		Others	195,552	196,413	0.4	65.0
		Total	**299,364**	**302,051**	**0.9**	**100.0**

*Data are in millions of dollars.
Source: Gartner (January 2012)

TABLE 1.2   Top 10 Buyers of Semiconductors in 2011

(2) Broadcom, (3) Nvidia, (4) SanDisk, (5) Marvell, (6) LSI Logic, (7) Xilinx, (8) MediaTek, (9) Altera, and (10) Conexant.

1.2.5   Moore's Law: Nanotechnology for the Electronics Industry

Meanwhile, let's look at Moore's law [1], as shown in Fig. 1.2. Today, 32-nm process technology is in high-volume production. TSMC announced in September 2011 that it had totally qualified its 28-nm process technology. Samsung announced on September 23, 2011 that some of its (2-Gb) dynamic random access memories (DRAMs) have been shipping with the 20-nm technology. Toshiba and SanDisk Flash Alliance will soon implement the 19-nm process technology for their NAND Flash chips. On February 18, 2011, Intel announced that it will spend $5 billion to build a 14-nm plant (Fab 42) near Chandler, AZ, and 10 nm is in planning. The CEO (Morris Chang) of TSMC said (September 2, 2011) that Moore's law will have 10 more years of glory. Moore's law (for costs and innovations) arguably has been the most powerful driver for the semiconductor industry in the past 46 years. This law emphasizes lithography scaling and integration of all functions on a two-dimensional (2D) surface of a single chip, perhaps through system-on-chip (SoC).

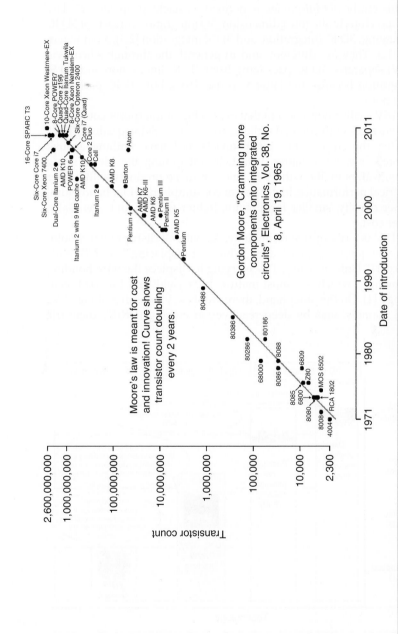

FIGURE 1.2 Moore's law for costs and innovations.

1.3   Three-Dimensional Integration

1.3.1   Through-Silicon Via Technology

In this study, *3D integration* is defined as stacking up Moore's law wafers/chips in the third dimension. 3D integration consists of 3D IC packaging, 3D IC integration, and 3D Si integration [2,3], as shown in Fig. 1.3. They are different, and in general, the through-silicon via (TSV) separates 3D IC packaging from 3D IC integration and 3D Si integration because the latter two use TSVs, but 3D IC packaging does not.

TSV (with a new concept that every chip or interposer could have two surfaces with circuits) is the heart of 3D IC/Si integrations. TSVs shorten the chip-to-chip interconnects and have nothing against Moore's law. As a matter of fact, by integrating the Moore's law chips/wafers in 3D with TSVs, one can make products with higher electrical performance, lower power consumption, wider bandwidth, higher density, smaller form factor, lighter weight, and eventually, lower cost.

TSV is a disruptive technology. As with all disruptive technologies, the questions to ask are: "What is it displacing?" and "What's the cost?" Unfortunately, TSV is trying to displace the wire-bonding technology, which is a most mature, high-yield, and low-cost technology [4]. On the other hand, there are six key steps in making TSVs [2,3], namely, vias by deep reactive-ion etching (DRIE), dielectric

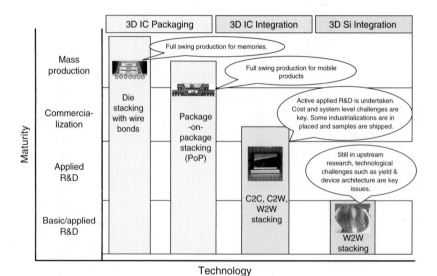

Figure 1.3   3D integration technologies.

layer by plasma-enhanced chemical vapor deposition (PECVD), barrier and seed layers by physical vapor deposition (PVD), via filling by (e.g., Cu) electroplating, overburden Cu removal by chemical-mechanical polishing (CMP), and TSV Cu revealing. Thus, compared with wire bonding, TSV technology is very expensive! However, just like solder-bumped flip-chip technology [5,6], because of their unique advantages, TSVs will be here to stay and for a very long time for high-performance and high-density applications. In this chapter, the origin of 3D integration is presented. Also, the evolution, challenges, and outlook of 3D IC/Si integrations are discussed. Finally, a few generic, low-cost, and thermal-enhanced 3D IC integration system-in-packages (SiPs) with various passive TSV interposers are proposed.

1.3.2 Origin of 3D Integration

Again, since the focus of this book is TSV, 3D IC packaging will not be discussed until at the end of this book (Chap. 10). 3D integration is a very old idea [7] that consists of two or more layers of active electronic components that are integrated vertically through TSV (it used to be called *vertical interconnection*) into a single circuit. It was trigged by the advance of the silicon-on-insulator (SOI) technology first reported by Gat and colleagues more than 30 years ago [8], when semiconductor people thought Moore's law could be hitting the wall by the 1990s. Of course, the facts showed otherwise.

In the early 1980s, there were two schools of thought [7]. One was to stack the chips up with TSVs and flip-chip solder bumps (3D IC integration), as shown in the right-hand side (RHS) of Fig. 1.4, and the other was to stack up the wafers/chips with TSVs alone, that is, bumpless (3D Si integration), as shown in the left-hand side (LHS) of Fig. 1.4. The advantages of 3D Si integration over 3D IC integration are (1) better electrical performance, (2) less power consumption, (3) lower profile, (4) less weight, and (5) higher throughput. In general, the industry favored the 3D Si integration.

The most powerful pusher on 3D integration was the 1965 Nobel Physics Laureate Richard Feynman. At the Yoshio Nishina Memorial Lecture, titled "Computing Machines in the Future," at Gakushuin University (Tokyo) on August 9, 1985, Feynman said: "Another direction of improvement [of computing power] is to make physical machines three dimensional instead of all on a surface of a chip. That can be done in stages instead of all at once—you can have several layers and then add many more layers as time goes on." Wow, Feynman not only told us to go for 3D, but he also taught us how to do it more than 26 years ago. Feynman went on to say: "Another important device would be a way of detecting automatically defective elements on a chip; then this chip itself automatically rewiring itself so as to avoid the defective elements." Wow again, Feynman

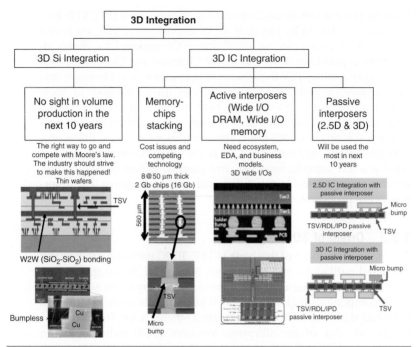

FIGURE 1.4   3D integration (excluding 3D IC packaging).

told us to do build-in self-test (BIST) and build-in self-repair (BISR) more than 26 years ago. He fully understood the importance of known-good die (KGD) for 3D integration, and that was the first thing on his mind. Figure 1.5 shows the visionary contributions of Richard Feynman to nanotechnology and 3D integration for the electronics industry.

1.4   Challenges and Outlook of 3D Si Integration

1.4.1   3D Si Integration

3D Si integration stacks up the TSV wafers/chips without any solder microbumps, as shown on the LHS of Fig. 1.4. Basically, wafer-to-wafer (W2W) is the logical way to perform the bumpless bonding operation, and yield is a big issue (e.g., some bad chips are forced to bond on the good chips). In addition, the absence of a gap between wafers/chips and thermal management are huge problems. Furthermore, the requirements of the bonding conditions, such as the surface cleanness, surface flatness, and class of clean room for 3D Si integration, are very high [9–12].

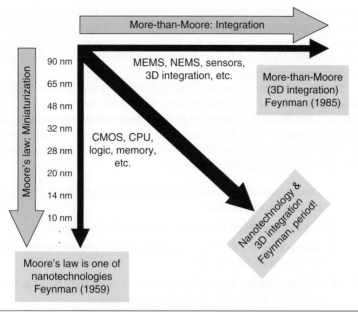

FIGURE 1.5   The contributions of Richard Feynman to nanotechnology and 3D integration for the semiconductor industry.

1.4.2   3D Si Integration Bonding Assembly

There are at least two different W2W bumpless bonding methods for 3D Si integration, namely, Cu-to-Cu bonding and oxide-to-oxide bonding, as shown on the LHS of Fig. 1.4. It should be noted and emphasized that for 3D Si integration [9–12], (1) it is bumpless, (2) TSVs before bonding for Cu-to-Cu bonding, (3) TSVs after bonding for SiO_2-to-SiO_2 bonding, (4) the TSV diameter is very small (usually ≤ 1 µm) and thus the aspect ratios are very high, and (5) the TSVs are usually filled with tungsten or copper by PVD/CVD. Also, for heterogeneous systems, owing to different chip size and pin-out, W2W bonding is very difficult. In this case, chip-to-wafer (C2W) bonding is one way to go, and it improves yield. Furthermore, owing to alignment issues, hybrid (Cu-to-SiO_2) bonding is required.

1.4.3   3D Si Integration Challenges

Tremendous amounts of work still must be done before products can be manufactured using the 3D Si integration technology [9–12]. Besides thermal management, via formation and filling, and thin-wafer handling, more research and development (R&D) efforts also should be placed on such areas as cost reduction, design and process parameter optimization, bonding environment, W2W bonding alignment, wafer distortion, wafer bow (warpage), inspection and testing, contact performance, contact integrity, contact reliability, and manufacturing yield issues. In addition, packaging the 3D Si integration module systematically, cost-effectively, and reliably to the next level of interconnect pose another great challenge.

Besides technology issues, electronic design automation (EDA), which is the soul of 3D Si integration, is far from ready. Urgently, the industry needs an ecosystem (e.g., standard and infrastructure) for 3D Si integration. Then EDA can write the design, simulation, analysis, and verification; manufacturing preparation; and test software with the following guidelines [9–12]:

1. Design automation from high-level description to layout generation/optimization
2. Verification of all dedicated and tuned 3D integration
3. Addressing the third dimension not like a packaging bumping (it should be bumpless)
4. Addressing the true third dimension, with partitioning, floor planning, automatic placing, and routing
5. Full extraction with the third dimension, full 3D design rule checks (DRC), 3D layout versus schematic (LVS) with all tiers together in the same database
6. The 3D Si integration then has to be seen as a whole system distributed in several tiers and not just a stack of predefined chips

1.4.4   3D Si Integration Outlook

In the next few years, the industry will be hard pressed to make products with the 3D Si integration technology, except for very niche applications. However, it should be noted and emphasized that 3D Si integration is the right way to go and compete with Moore's law. The industry should *strive* to make this happen [9–12].

1.5   Potential Applications and Challenges of 3D IC Integration

1.5.1   Definition of 3D IC Integration

Unlike 3D Si integration, 3D IC integration (as shown on the RHS of Fig. 1.4) stacks up whatever Moore law's IC chips in the third dimension with TSVs, thin chips/interposers, and microbumps [1,2,9–12] to achieve high performance, low power, wide bandwidth, small form factor, and eventually, low cost. Today, unlike approximately 26 years ago, most people favor 3D IC integration, and the most likely applications are shown in Fig. 1.6.

The data width is limited by IC packaging technology. With TSV technology, which provides very small via size (5- to 10-µm sizes are common) and pitch (20- to 40-µm pitches are common), a much wider I/O data path (width), such as 512-bit data width, is more than possible. On the other hand, with wire-bonding technology (which has pad sizes and pitches that are many times larger than those of TSV), in order to achieve 512-bit data width, the chip size (and thus the cost) has to be increased substantially. This is why TSV is so attractive for memory bandwidth. Let's say that if we have TSVs run through a 4-DRAM stacking with a ×512-bit data path, we could have the same DDR3-1600 chip with a total memory bandwidth of $512 \times 1600 = 819.2$ Gb/s = 102.4 GB/s. Of course, this DRAM stack has to interconnect to the logic/SoC in order to get this bandwidth.

1.5.5 Memory-Chip Stacking

The drawing and sample on the far left in Fig. 1.6 show the simplest example of memory-chip stacking published by Samsung in 2006 (Fig. 1.8). These chips could be DRAM or NADA Flash with fewer than 100 I/Os (78 to be exact). Even with eight-chip stacking, their total thickness (560 µm) is still less than that of an ordinary chip (this is significant). Unfortunately, owing to cost and competing wire-bonding technology such as the one used by Apple's iPhone4 (http://www.infoneedle.com/posting/99670?snc=20641), memory-chip stacking with TSVs is not in volume production today for consumer products. Currently, Samsung is shooting for the next-generation server products, most likely with the DDR4 (double data rate, type 4).

Known-good die (KGD) is one of the most important issues for 3D integration because it affects the 3D integration yield (and translates into cost). For homogeneous structures such as memory-chip stacking, the yield y is given by $y = Y^N$, where Y is chip yield and N is the number of chips in the stack. For example, if we stack up eight homogeneous chips ($N = 8$), and the chip yield Y is 80 percent, then the 3D chip-stack yield is 17 percent. However, if the chip yield is increased from 80 to 99 percent, the 3D chip-stack yield increases dramatically to 92 percent. Of course, for KGD ($Y = 100$ percent), the 3D chip-stack yield is 100 percent (assuming that the assembly yield is 100 percent, and the test detects all the possible faults whenever they are present). It can be seen that "good" or "high" chip yield is very important for 3D integration.

1.5.6 Wide I/O Memory

The second LHS of Fig. 1.6 shows a wide I/O memory, which consists of a low-power and wide-bandwidth memory usually with more than thousands of interface pins. This memory is supported by a CPU/logic or SoC with TSVs [13]. Since mobile products want it, samples have been made and published by, for example, Samsung, as

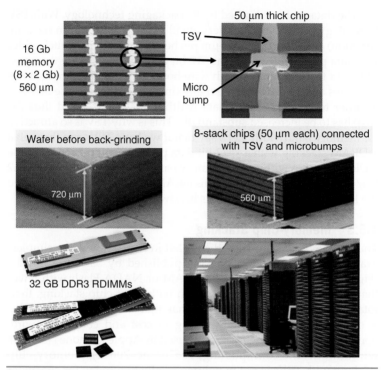

Figure 1.8 Samsung's eight-memory-chip stack with a thickness that is less than that of an ordinary chip. (Potential application is the next generation servers.)

shown in Fig. 1.9. Unfortunately, the readiness of "fabless" supply-chain infrastructure, including industry standards, supply-chain business models, and competitive pricing, takes time and is not here yet. Also, much cheaper solutions such as package-on-package (PoP) have been keeping TSV wide I/O memory away from being used for consumer products such as smart phones and tablets.

1.5.7 Wide I/O DRAM

The second RHS of Fig. 1.6 shows a wide I/O DRAM for mobile applications. Samsung has published many papers on this topic over the last 3 years, for example, [14], and finally, in 2011 at the IEEE International Solid-State Circuits Conference (ISSCC) in San Francisco (Fig. 1.10), the company showed a sample of two DRAMs on a master logic chip with TSVs. For this DRAM, the number of interface pins was slightly more than a thousand. The Joint Electron Device Engineering Council (JEDEC) standard put it to 1200 pins in four channels (http://www.jedec.org/). Again, owing to infrastructure, competitive pricing, and competing technology such as embedded PoP, there are no mobile consumer products in high-volume production yet.

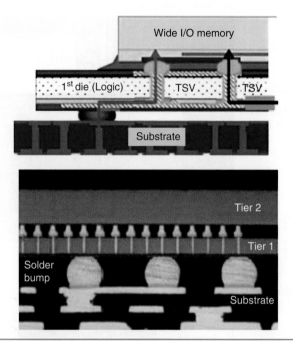

FIGURE 1.9   Samsung's wide I/O memory for mobile application. (*3D Architectures for Semiconductor Integration and Packaging, Burlingame, CA, December 2010.*)

For heterogeneous structures, the 3D integration yield y is given by $y = (X^R)(Y^S)(Z^T)$. . . , where X, Y, Z, . . . are chip yield for different chip types and R, S, T, . . . are the number of different types of chips. For example, if we stack up a wide I/O DRAM, which consists of one logic and four DRAMs, their respective chip yields are 66 and 68 percent. The wide I/O DRAM stack yield y is $y = (0.66)^1(0.68)^4 = 21$ percent. When the chips are mature, for example, when the chip yield for the logic is 90 percent and the DRAM is 95 percent, then the yield of the wide I/O DRAM is $y = (0.90)(0.95)^4 = 73$ percent. Thus BIST and BISR are desperately needed for widespread use of 3D Si and 3D IC integrations. The electronics industry is currently working very hard in this area.

Recently, the Hybrid Memory Cube (HMC) Consortium, which includes such companies as Micron, Intel, Altera, Samsung, Open Silicon, Xilinx, and IBM, created an entirely new technology (http://www.micron.com/innovations/hmc.html). The end result could be a high-bandwidth (15× more than DDR3), low-power (70 percent less energy per bit than DDR3), and small-form-factor (90 percent less space than RDIMMS) product, as shown schematically in Fig. 1.11*a*, and this product is planned for fabricated with one of IBM's via-middle TSV technologies, as shown in Fig. 1.11*b* [15]. The cross section is

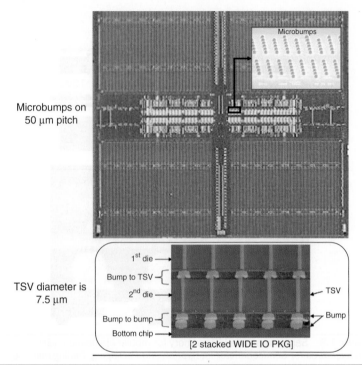

Microbumps on
50 µm pitch

TSV diameter is
7.5 µm

FIGURE 1.10   Samsung's wide I/O DRAM for mobile applications. *(IEEE/ISSCC, February 2011.)*

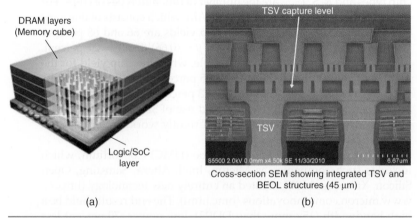

(a)   (b)

FIGURE 1.11   (a) Micron's hybrid memory cube (HMC). (b) IBM's via-middle TSV technology.

shown in Fig. 1.12a, and the logic-memory interface (LMI) follows the JEDEC wide I/O SDR (JESD229) standard (http://www.jedec.org/), which is highlighted in Fig. 1.12b.

Figure 1.12b shows that the wide I/O defines four memory channels on the LMI, each with 128 bits of data width, thus with 512 data bits in total. Each channel has 300 connections (6 rows by 50 columns), thus with a total of 1200 connections. The dimensions of the total LMI area are 5.25×0.52 mm and located at the center. The bottom chip of the memory cube and the logic (SoC) chip are face-to-face interconnected, and the pin/pad locations are symmetric between channels. The pad pitch is 50 μm (in the 5.25-mm direction) and 40 μm (in the 0.52-mm direction). Each channel includes all control, power, and ground for the channel, and power connections are shared between channels. However, each channel is controlled independently on the control, clock, and data. The DRAM has 1.2-V complementary metal-oxide semiconductor (CMOS) signal levels and no termination. The total memory bandwidth (for a data rate of 266 Mb/s, SDR) is 17 GB/s or 4.26 GB/s per channel.

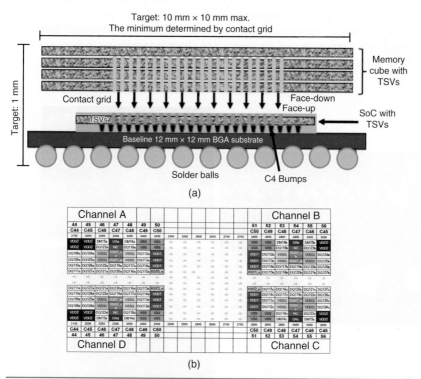

FIGURE 1.12   (a) Schematic of the cross section of an HMC assembly. (Dan Skinner, Micron.) (b) JEDEC standard. (JESD229.)

1.5.8 Wide I/O Interface

The far RHS of Fig. 1.6 shows a wide I/O interface for CPU/router/
telecommunications/etc. applications. The Moore's law chips (without
any TSV) have memory/ASIC/CPU/etc., and the I/O could be between
hundreds and thousands, which are supported by a piece of dummy
silicon with TSVs and redistribution layers (RDLs). The sample in
Fig. 1.13 is presented by Xilinx [16–18], where the field-programmable
gate array (FPGA) is fabricated by TSMC's 28-nm process technology,
and the interposer is by 65 nm. There are four RDLs on top of the
interposer that let those four FPGAs communicate with each other at
very short distances. The TSVs in the interposer are for powers,
grounds, and some signals. The bandwidth improvement is $100 \times$ [16].
The shipment of samples for Xilinx's customers will begin in 2012.

1.5.9 2.5D and 3D IC Integration (Passive and Active Interposers)

In Fig. 1.6, the memory-chip stacking, wide I/O memory, and wide
I/O DRAM are the examples of 3D IC integration because the active
chips are stacked in 3D with at least one active chip with TSVs. The
wide I/O interface is an example of 2.5D IC integration because the
TSVs are in a piece of dummy silicon (so-called TSV passive interposer or simply interposer) and not in any of the active Moore's law
chips, which are on top of the interposer. Since the active (with devices)
CPU/Logic/SoC chip with TSVs is supporting the memory (wide I/O

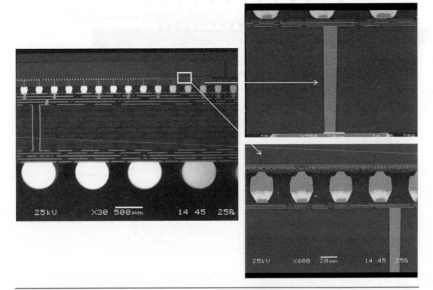

FIGURE 1.13   Xilinx/TSMC's wide I/O interface. *(IEEE/ECTC, May 2011.)*

memory) and the master logic active (with devices) chip with TSVs is supporting the memory cube (wide I/O DRAM), they are called the *TSV active interposers* (or simply *active interposers*) [9–12].

1.6 Recent Advances of 2.5D IC Integration (Interposers)

For 2.5D IC integration, the TSV passive interposer (or simply interposer) can be used as [9–12]: (1) an intermediate substrate, (2) a stress-relief (reliability) buffer, (3) a carrier, and (4) a thermal management tool. Some recent advances involving interposers are discussed in the following subsections.

1.6.1 Interposer Used as Intermediate Substrate

Today the most well-known interposer is in Xilinx's FPGA wide I/O interface [16–18], which has been mentioned in Sec. 1.5.8. Figure 1.14 shows a cross section of ITRI's 3D IC integration SiP test vehicle [19], which consists of an interposer supporting four memory chips, one thermal chip, and one mechanical chip. It is overmolded for pick-and-place purpose, as well as for protecting the chips from harsh environments. There are RDLs on both the top and bottom sides of the interposer. Also, stress sensors are implanted on the top side, and integrated passive devices (IPDs) are fabricated through the thickness (100 μm) of the interposer (12.3 × 12.2 mm). Figure 1.15 shows the sample, and Fig. 1.16 shows

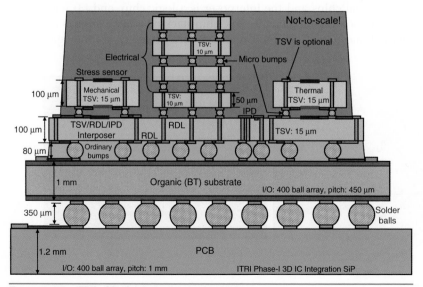

FIGURE 1.14 ITRI's 3D IC integration test vehicle.

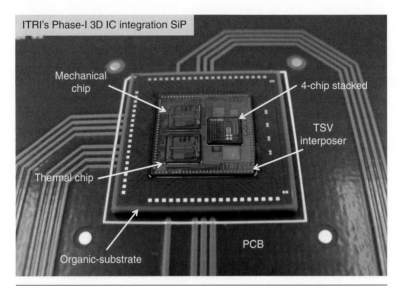

FIGURE 1.15   Sample of ITRI's 3D IC integration test vehicle.

a scanning electron microscope (SEM) image of a cross section and an x-ray image of the SiP [20]. Even though the assembly is not perfect yet, it is not so bad for the first time.

This test vehicle can be degenerated to the case of (1) wide I/O memory if there is neither memory-chip stacking nor TSVs in the mechanical/thermal chips and the interposer is either a logic, microprocessor, or SoC, (2) wide I/O DRAM if there are no mechanical and thermal chips and the interposer is a logic chip, and (3) wide I/O interface if there is no memory-chip stacking and no TSVs in the thermal/mechanical chips. Thus the core enabling technologies

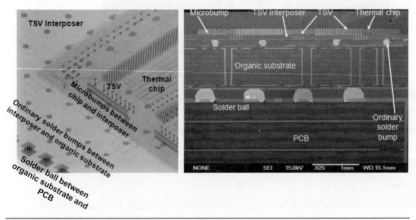

FIGURE 1.16   X-ray and SEM images of ITRI's 3D IC integration test vehicle.

(e.g., via etching, dielectric, barrier and seed layer deposition, via filling, CMP, thin-wafer handling, electrical and thermal design and test of TSVs, wafer bumping of ultrafine-pitch lead-free microbumps, fluxless C2W bonding, electronmigration of microbumps, and reliability of microbump assemblies) developed [19–36] with this test vehicle are very useful and can have very broad applications.

Figure 1.17 shows the largest interposer, developed in 2007 [37,38] to support a very large active chip (21 × 21 mm) with 11,000 I/Os fabricated by Charter Semiconductor (called Globalfoundries today) in 2006 by the 65-nm Cu–low-k technology. Simulation results show [37] that owing to the very larger local thermal expansion coefficient (TEC) mismatch between the Si ($2.5 \times 10^{-6}/°C$) and the Cu ($17.5 \times 10^{-6}/°C$) in the vertical direction, during heating, the TSV Cu tends to pump out from the surrounding Si, as shown in Fig. 1.18. If the TSV Cu is partially covered by the SiO_2, during heating, the local thermal expansion mismatch with Cu may (push out to) crack the SiO_2, as shown in Fig. 1.19 (the SEM is provided by Tezzaron), which is called *Cu pumping*. Reference 37 shows that, in general, above an aspect ratio (thickness/diameter) of 5, there is little dependence of stress and strain on the aspect ratio of the TSV.

The wide I/O memory shown in Fig. 1.6 also can be redesigned in such a way that the maximum-bandwidth and low-power-consumption

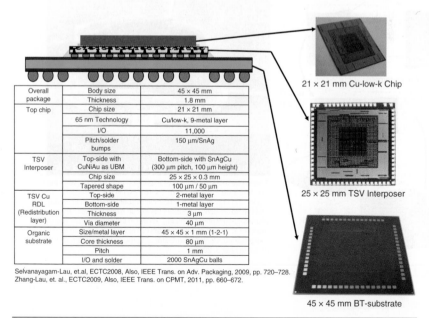

Overall package	Body size	45 × 45 mm
	Thickness	1.8 mm
Top chip	Chip size	21 × 21 mm
	65 nm Technology	Cu/low-k, 9-metal layer
	I/O	11,000
	Pitch/solder bumps	150 µm/SnAg
TSV Interposer	Top-side with CuNiAu as UBM	Bottom-side with SnAgCu (300 µm pitch, 100 µm height)
	Chip size	25 × 25 × 0.3 mm
	Tapered shape	100 µm / 50 µm
TSV Cu RDL (Redistribution layer)	Top-side	2-metal layer
	Bottom-side	1-metal layer
	Thickness	3 µm
	Via diameter	40 µm
Organic substrate	Size/metal layer	45 × 45 × 1 mm (1-2-1)
	Core thickness	80 µm
	Pitch	1 mm
	I/O and solder	2000 SnAgCu balls

Selvanayagam-Lau, et.al, ECTC2008, Also, IEEE Trans. on Adv. Packaging, 2009, pp. 720–728.
Zhang-Lau, et. al., ECTC2009, Also, IEEE Trans. on CPMT, 2011, pp. 660–672.

21 × 21 mm Cu-low-k Chip

25 × 25 mm TSV Interposer

45 × 45 mm BT-substrate

FIGURE 1.17   IME's 25- × 25-mm interposer (2007) for a 65-nm Cu–low-k chip with 11,000 I/Os.

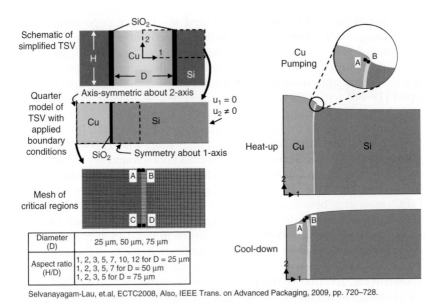

Diameter (D) : 25 μm, 50 μm, 75 μm

Diameter (D)	25 μm, 50 μm, 75 μm
Aspect ratio (H/D)	1, 2, 3, 5, 7, 10, 12 for D = 25 μm 1, 2, 3, 5, 7 for D = 50 μm 1, 2, 3, 5 for D = 75 μm

Selvanayagam-Lau, et.al, ECTC2008, Also, IEEE Trans. on Advanced Packaging, 2009, pp. 720–728.

FIGURE 1.18   TSV Cu pumping owing to local thermal expansion mismatch.

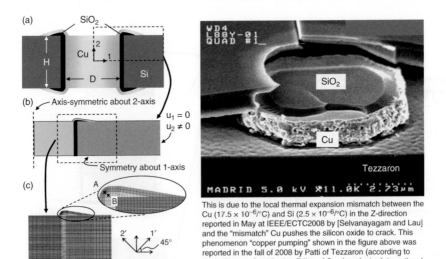

This is due to the local thermal expansion mismatch between the Cu ($17.5 \times 10^{-6}/°C$) and Si ($2.5 \times 10^{-6}/°C$) in the Z-direction reported in May at IEEE/ECTC2008 by [Selvanayagam and Lau] and the "mismatch" Cu pushes the silicon oxide to crack. This phenomenon "copper pumping" shown in the figure above was reported in the fall of 2008 by Patti of Tezzaron (according to Philip Garrou, Contributing Editor of Semiconductor International on December 3, 2009.)

FIGURE 1.19 SiO_2 cracking owing to TSV Cu pumping.

memory chip is side by side with the SoC on top of a TSV/RDL/IPD interposer. In this case, there is no need to make TSVs in the active chip.

1.6.2   Interposer Used as a Stress-Relief (Reliability) Buffer

Figure 1.20 shows a Moore's law chip that is attached (1) on top of a Cu-filled TSV interposer and then the interposer is attached on an organic (e.g., bismaleimide triazene (BT)) substrate (*top drawing*) and (2) on top of the organic substrate directly (*bottom drawing*). It has been shown in Refs. 2, 37, and 38 that the Cu-filled TSV interposer acts like a stress-relief (reliability) buffer and reduces the stress (from 250 to 125 MPa), acting at the Cu–low-k pads of the Moore's law chip. This is more important for smaller-feature-size devices because the allowable stress on their Cu–low-k pads is smaller. If a special (with a very small filler size) underfill is added between the chip and the interposer, the stress acting at the Cu–low-k pads of the chip is further reduced to 42 MPa.

Why? Because the TEC of the BT substrate is $15 \times 10^{-6}/°C$. The effective TEC of Cu-filled TSV interposer, depending on the number of vias, is 8 to $10 \times 10^{-6}/°C$. Thus, with Cu-filled TSV interposers, the global thermal expansion mismatch between the Moore's law chip and the interposer is smaller than that between the chip and the BT substrate.

1.6.3   Interposer Used as Carrier

Figure 1.21 shows a 3D module consisting of two stacks assembled one over the other with three chips [39]. The module size is 12×12 mm, with 1.3-mm thickness. The silicon carrier is $12 \times 12 \times 0.2$ mm with 168 peripherally populated vias. The bottom carrier (carrier 1) is assembled with a 5- $\times$ 5-mm flip-chip. The top carrier (carrier 2) is assembled with a 5- $\times$ 5-mm flip-chip and two stacked 3- $\times$ 6-mm wire-bonded chips. Carrier 2 is overmolded (using the transfer molding process) to protect the wire-bond chips. The silicon carriers have been fabricated with two metal layers with SiO_2 as the dielectric/passivation layer. Electrical connections through the carrier are formed by TSVs. Carrier 1 is mounted on an FR4 PCB using SAC305 of 250-μm diameter. Carrier 1 assembly is underfilled (cured at 165°C for three hours). More technical information is given in Ref. 39.

1.6.4   Interposer Used as a Thermal Management Tool

Figure 1.22 shows two identical silicon interposers (carriers) with microchannels for thermal management and TSVs for electrical feedthrough [40]. The size of the whole system is about one-quarter of a fist. Each interposer is supporting a chip (10×10 mm) with 100 W of (heat) power dissipation. This interposer is fabricated by wafer-to-wafer bonding two silicon wafers together, and an optimized liquid cooling channel structure (fabricated by DRIE) is embedded in between the

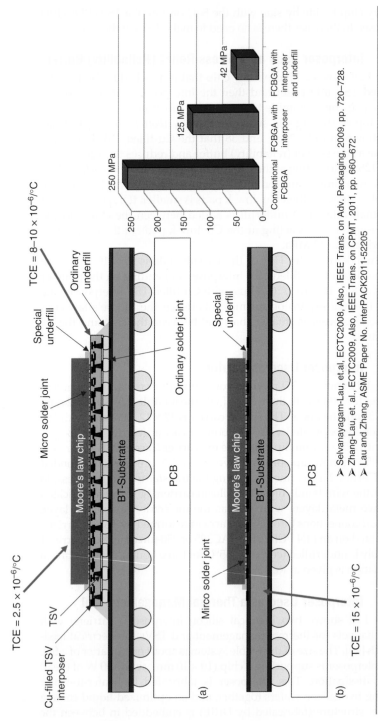

Figure 1.20 Interposer can be used as a stress-relief (reliability) buffer for the Cu-low-k pads of a Moore's law chip.

➤ Selvanayagam-Lau, et.al, ECTC2008, Also, IEEE Trans. on Adv. Packaging, 2009, pp. 720–728.
➤ Zhang-Lau, et. al., ECTC2009, Also, IEEE Trans. on CPMT, 2011, pp. 660–672.
➤ Lau and Zhang, ASME Paper No. InterPACK2011-52205

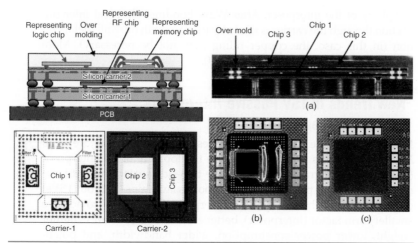

FIGURE 1.21 IME's interposers used as carriers.

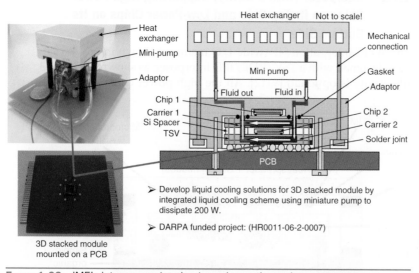

➤ Develop liquid cooling solutions for 3D stacked module by integrated liquid cooling scheme using miniature pump to dissipate 200 W.

➤ DARPA funded project: (HR0011-06-2-0007)

FIGURE 1.22 IME's interposers (carriers) used as a thermal management tool.

interposer. A minipump is used to pump the cold water from the heat exchanger to the inlet of the interposer, through the microchannels, out from the outlet of the interposer, and then back into the heat exchanger, as shown in Fig. 1.22. Silicon is chosen as the interposer material because it is a suitable material for the integration of both electrical (TSV) and fluidic structures (microchannel) in the same substrate with DRIE. The difference between these two interposers is that the bottom Si interposer does not have any outlets. TSVs can be designed along the periphery of the interposer. After W2W bonding, electrical interconnection through the carrier is made by the TSV with on-wall metallization (in this case, the copper filling may not be necessary). More technical details can be found in Ref. 40.

1.7   New Trends in TSV Passive Interposers for 3D IC Integration

Most TSV passive interposer samples published so far are with Moore's law chips (without TSVs) on the top side of the interposer (i.e., 2.5D IC integration). Just like the two-side surface-mount technology (SMT), it is possible to stack Moore's law chips (without TSVs) on both sides of the TSV passive interposer [9–12,41,42] to achieve smaller form factor (interposer), better electrical performance, lighter weight, lower power consumption, wider bandwidth, and potentially lower cost.

1.7.1   Interposer (with a Cavity) Supporting High-Power Chips on Its Top Side and Low-Power Chips on Its Bottom Side

The 3D IC integration SiP (wide I/O memory) shown in Fig. 1.6 can be redesigned with a TSV/RDL/IPD interposer, as shown in Fig. 1.23 (the cavity is optional) [9–12]. All the high-power chips such as the

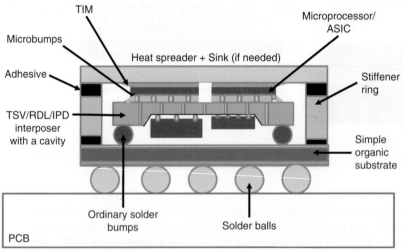

Special underfills are needed between the Cu-filled interposer and all the chips.
Ordinary underfills are needed between the interposer and the organic substrate.

FIGURE 1.23   ITRI's interposers with cavity (optional) supporting high-power chips on its top side and low-power chips on its bottom side.

SoC, microprocessor unit (MPU), graphic processor unit (GPU), application-specific IC (ASIC), digital signal processor (DSP), microcontroller unit (MCU), and radiofrequency (RF) chips (with pads that have different pitches, sizes, and locations) are mounted on the top side of the interposer in a flip-chip format [4,5] so that the backside of these chips can be attached to a heat spreader via a thermal interface material (TIM). Most of the heat from the high-power chips can be dissipated through the heat spreader (with a heat sink if necessary). All low-power chips, for example, MEMS, OMEMS, CMOS image sensors, and low-power-dissipation memory are mounted on the bottom side of the interposer.

A simple organic substrate with standard (in size and pitch) solder balls for PCB assembly supports the passive interposer. A ring stiffener connecting the organic substrate and the heat spreader provides adequate standoff for 3D IC integration with the passive interposer and to support the heat spreader with or without the heat sink. Underfill encapsulants are needed between the TSV interposer and the high- and low-power flip-chips and between the TSV interposer and the organic substrate. However, underfill is not needed between the 3D IC integration SiP organic substrate and the PCB. In this case, (1) no new EDA is needed, (2) there are no TSVs in the active chips, (3) the heat from the high-power chip can be removed from its backside, and (4) solder joint reliability is not an issue because it is a standard package. The outlook (package) of this proposed 3D IC integration SiP is very attractive to integrated device manufactures (IDMs), original equipment manufacturers (OEMs), outsourced semiconductor assembly and test (OSAT), and electronics manufacturing services (EMS) because it is a standard plastic ball grid array (PBGA) package that has been used by the electronics industry for more than 17 years [43].

If the memory chips are too thick to fit in the space between the interposer and the organic substrate and backgrinding is too expensive for the active chips, then we can make a cavity in the bottom side of the interposer, which can be fabricated easily by either lasers or wet anisotropic etches, for example, by using the potassium hydroxide (KOH) solution.

1.7.2   Interposer (on an Organic Substrate with a Cavity) Supporting High-Power Chips on Its Top Side and Low-Power Chip Stacking on Its Bottom Side

Figure 1.24 shows a new design, which is exactly the same as the one shown in Fig. 1.23 except that the current organic substrate has a cavity at the center. The advantages of this new design are (1) there is no need to etch the interposer, (2) no need to thin down the

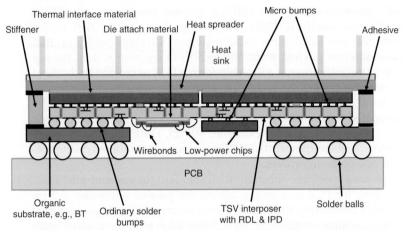

Special underfill between the TSV interposer and the high-and low-power flip chips. Ordinary underfill between the TSV interposer and the organic substrate. Encapsulant for wirebonding memory stack.

FIGURE **1.24** Interposer (on an organic substrate with a cavity) supporting high-power chips on its top side and low-power chips at its bottom side.

Moore's law chips, and (3) to allow higher number of memory-chip stacking [9–12].

1.7.3   A Simple Design Example

Figure 1.25 shows a design example for detailed analysis of a 3D IC integration SiP that consists of four identical high-power chips (e.g., microprocessors) uniformly distributed on the top side of a TSV/RDL/IPD interposer and 16 identical low-power chips (e.g., memories) uniformly distributed at the bottom side. The dimensions of the high-power chips are 10 mm × 10 mm × 200 μm, the low-power chips are 5 mm × 5 mm × 200 μm, and the interposer's dimensions are 35 mm × 35 mm × 200 μm. There are 1600 identical TSVs with 20-μm diameter and 850-μm pitch. The backsides of the high-power chips are attached to a heat spreader through a 100-μm TIM. An aluminum heat sink with 21 uniformly distributed fins is attached to the back of the heat spreader.

The 3D IC integration SiP is attached (with ordinary solder bumps) to a BT substrate (44 × 44 × 0.8 mm with a cavity of 33 × 33 × 0.8 mm). The substrate is also connected to the heat spreader with the heat sink through an aluminum ring stiffener. Then the substrate is lead-free (SnAgCu) soldered onto a FR-4 PCB (50 × 50 × 2.5 mm). Nonlinear finite-element modeling and analyses of the temperature distributions and creep strain energy density per temperature cycle have been reported and show that with proper selection of the heat sink and underfills, the micro-bumps and solder joints are reliable for most of the operating conditions [44].

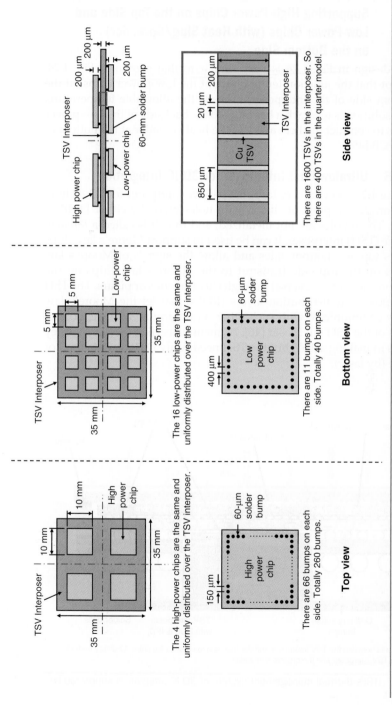

FIGURE 1.25 Interposer (as a cost-effective 3D IC integrator) supporting four CPUs on its top side and 16 memories on its bottom side.

1.7.4   Interposer (on an Organic Substrate with a Cavity) Supporting High-Power Chips on the Top Side and Low-Power Chips (with Heat Slug/Spreader) on the Bottom Side

The design in Fig. 1.26 is very similar to that shown in Fig. 1.24, except that the junction temperature at the low-power chips (at the bottom side of the interposer) exceeds the allowable temperature. One solution is to attach a heat slug to the backside of the low-power chips to conduct the heat away to the heat spreader at the bottom of the PCB [45].

1.7.5   Ultralow-Cost Interposer for 3D IC Integration

Figure 1.27 shows a very low-cost passive interposer for 3D IC integration. It is a piece of silicon with holes made by either DRIE or laser. These holes are not metalized, and thus it is called a *through-Si hole* (TSH) interposer. It can be used to support Moore's law chips on its top and bottom sides and allow the signals of Moore's law chips on the top side transmit to the Moore's law chips on the bottom side (or vise verse) through Cu/Au wires or pillars. The TSH interposer's redistribution layers (RDLs) can let the Moore's law chips communicate with each other on the top and/or the bottom sides of the TSH interposer [46]. As mention earlier, there are six key steps in making a TSV, but the TSH needs only DRIE (by etching) or none (by laser).

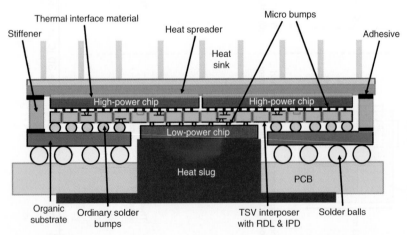

Special underfill between the TSV interposer and the high and low-power flip chips. Ordinary underfill between the TSV interposer and the organic substrate.

Figure 1.26   ITRI's thermal management system of 3D IC integration supported by an interposer.

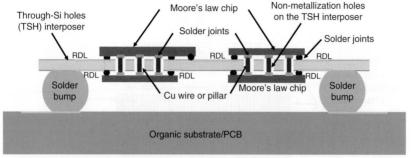

1. Underfills are optional between the Moore's law chips and the interpose when they are subjected to thermal loading! However, for shock and vibration loads, and depending on chip size, underfills may be needed! Underfills between the TSH and the organic substrate/PCB are necessary!

2. There are six key steps in making a TSV with semiconductor equipments: Via forming (DRIE); Dielectric layer (PECVD); Barrier/seed layers (PVD); Via filling (Cu plating); CMP; Cu revealing. The TSH needs only DRIE (by etching) or none (by laser).

FIGURE 1.27   ITRI's ultralow-cost TSH (through-Si hole) interposer for 3D IC integration.

1.7.6   Interposer Used as a Thermal Management Tool for 3D IC Integration

The 3D IC integration SiP shown in Figs. 1.23 through 1.27 cannot be stacked into 3D because of the heat spreader/sink. Also, there may not be the luxury to separate all the high-power chips (on top) from the low-power chips (at the bottom) of the interposer. Figures 1.28 and 1.29 show a new 3D IC integration SiP design that is able to stack up any chips randomly on top or bottom of the interposer.

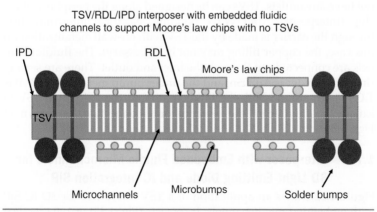

FIGURE 1.28   ITRI's TSV/RDL/IPD interposer with embedded microchannels for all kinds of Moore's law chips randomly distributed on its top and bottom sides.

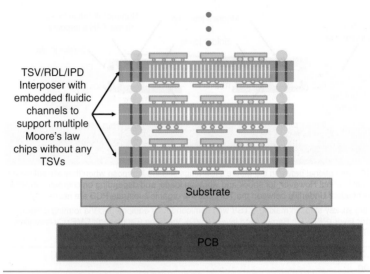

TSV/RDL/IPD
Interposer with
embedded fluidic
channels to
support multiple
Moore's law
chips without any
TSVs

Substrate

PCB

Figure 1.29   ITRI's 3D IC integration SiP consists of a series of interposers with embedded microchannels to support multiple Moore's law chips without any TSVs.

The basic unit is shown in Fig. 1.28, which consists of the TSV/RDL/IPD interposer with embedded fluidic channels. This interposer is fabricated by bonding two silicon wafers together by W2W bonding, and an optimized liquid-cooling channel structure is embedded in between the chips, as shown in Figs. 1.30 and 1.31. As mentioned in Sec. 1.6.4, silicon is chosen as interposer material because it is a suitable material for the integration of both electric and fluidic structures in the same substrate with microfabrication process. The difference between these two chips (no devices) is that the bottom Si chip does not have any outlets. TSVs can be designed along the periphery of the chip (interposer). After W2W bonding, electrical interconnection through the carrier is made by the TSV with on-wall metallization (in this case, the copper filling may not be necessary). The fluidic channels are connected out through the inlet and outlet. There are sealing rings (the solder is Au20Sn, and the under bump metallogy (UBM) is TiCuNiAu) around both the fluidic path and the individual TSV to isolate the fluid from the electrical interconnection, as shown in Figs. 1.30 and 1.31 [40].

1.7.7   Interposer with Embedded Fluidic Microchannels for 3D Light-Emitting Diode and IC Integration SiP

Figure 1.32 shows an application of a TSV interposer for 3D IC SiP with embedded microchannels. It consists (Fig. 1.33) of 100 identical light-emitting diodes (LEDs) uniformly distributed on the top side of

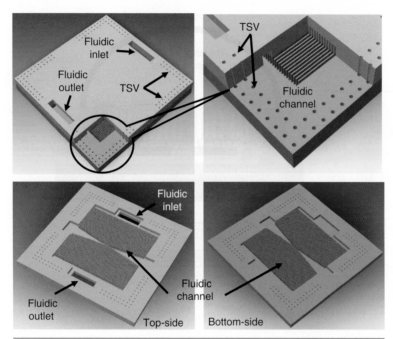

FIGURE 1.30   Interposer with TSVs for electrical feed through and embedded fluidic channels for thermal management.

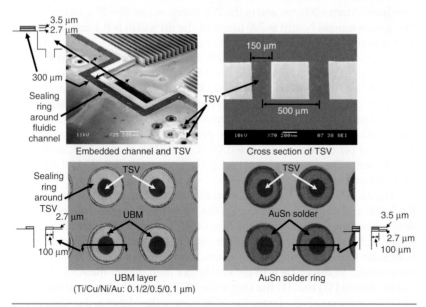

FIGURE 1.31   Details of the TSV and microchannel interposer (carrier).

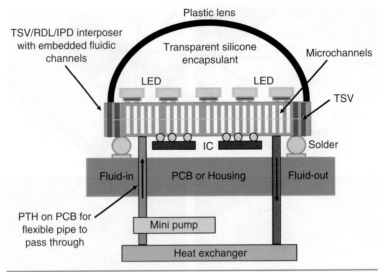

FIGURE 1.32   ITRI's 3D LED and IC integration SiP with interposer with embedded microchannels.

the TSV interposer and four identical logics (e.g., application-specific integrated circuit (ASIC)) uniformly distributed on the bottom side. The dimensions of the LEDs are 1 mm × 1 mm × 300 µm, the IC chips are 6 mm × 6 mm × 300 µm, and of the TSV interposer are 25 × 25 × 1.4 mm. The microchannel height is 700 µm, and the fin width is 0.1 mm on a 1.25-mm pitch. The fluidic inlet/outlet openings are 20 × 1.5 mm.

The material properties of the 3D IC and LED SiP are (1) for water: $\rho = 989$ kg/m^3; $C_p = 4177$ J/kg-k; $k = 0.6367$ W/m-k; and viscosity = 5.77×10^{-4} kg/m-s; and (2) for Si: $\rho = 2330$ kg/m^3; $C_p = 660$ J/kg-k; and $k = 148$ W/m-k. The boundary conditions are (1) all exposed surfaces are adiabatic (the worst case for thermal responses), (2) the powers of each LED are 1, 1.5, and 2 W, (3) the powers of each ASIC are 5 and 10 W, and (4) the flow rates are 0.18, 0.36, 0.54, 0.72, 0.9, 1.08, and 1.26 L/min. The finite-volume codes (ANSYS and ICEPAK 12.1.6.) are used for the analyses.

Figure 1.34 shows the simulation results: (1) The temperature of the LEDs and chips near the inlet is the lowest and near the outlet is the highest (the water carriers the heat out), (2) the temperature of the LEDs and chips near the interposer center is larger than that near both sides (parallel to the fluid direction) of the interposer, and (3) the temperature in the LEDs is in general larger than that in the ASIC. It can be shown [47] that (1) the larger the LED power, the higher is the LED temperature, (2) the larger the ASIC power, the higher is the ASIC temperature, (3) the larger the SiP chip power, the higher the LED, ASIC, and water outlet temperatures, and (4) the rate of change of the

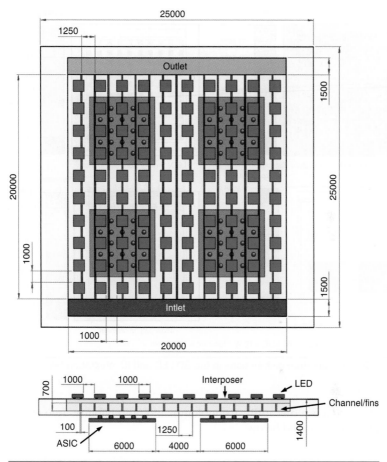

Figure 1.33 TSV interposer with embedded microchannels for multiple LEDs and driver chips.

average temperature of the LEDs is larger than that of the ASIC and water outlet. Also, it should be noted that with only a 0.54-L/min flow rate and inlet water temperature of 20°C, it can take away 240 W of power consumption from the interposer, and the maximum temperature in the LEDs will be 88.7°C and in the ASICs will be 59.2°C. This shows that the current design has very high cooling ability.

1.8   Embedded 3D IC Integration

Embedding devices in the substrate/PCB is the best way to make low-profile products. The basic requirements of embedded structures are (1) reworkability, (2) thermal management, (3) minimum disruption

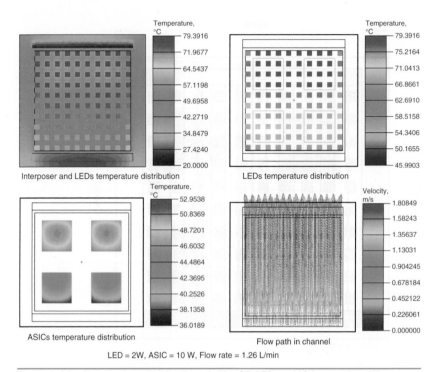

FIGURE 1.34 Temperature distributions in the 3D LED and IC integration SiP.

(because it is a disruptive technology), and (4) reliability. Two examples are discussed below.

1.8.1 Semiembedded Interposer with Stress-Relief Gap

Figure 1.35 shows a semiembedded interposer (with a stress-relief gap) supporting Moore's law chips on its top surface. The advantages of this design are (1) low profile, (2) free to use of any Moore's law chips without TSVs, (3) short design cycle, (4) low manufacturing cost, (5) RDLs allow chips to talk with each other at short distance, (6) lots TSVs can be used for powers, grounds, and some signals, (7) reworkability (test the chip-interposer module on substrate/PCB before underfilling), (8) the heat can be removed from chips' backside through a heat spreader/sink and/or a heat slug with a spreader (not shown) conducting the heat from the solder bumps to the bottom side of the substrate/PCB, (9) reliability (because the stress-relief gap reduces the global thermal expansion mismatch between the embedded interposer (6 to $8 \times 10^{-6}/°C$) and the organic substrate/PCB (15 to $18.5 \times 10^{-6}/°C$), and (10) potentially low system cost [45].

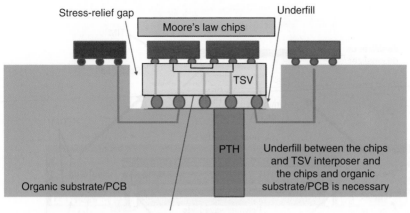

The advantages of this design are:
1. Low profile and low cost
2. Free to use any Moore's law chips without TSVs
3. RDLs allow chip-to-chip short interconnect
4. TSVs can be used for powers, grounds, and some signals
5. Very reliable (because the stress-relief gap reduces the thermal expansion mismatch between the embedded TSV interposer and the organic substrate/PCB)

FIGURE 1.35 Semiembedded TSV interposer with stress-relief gap.

1.8.2 Embedded 3D Hybrid IC Integration for Optoelectronic Interconnects

Figure 1.36 shows a low-cost (with bare chips) and high-performance (optical, electrical, thermal, and mechanical) optoelectronic system embedded into a PCB or an organic laminated substrate. This system consists of a rigid PCB (or a substrate) with an embedded optical polymer waveguide, vertical cavity surface-emitted laser (VCSEL), driver chip, serializer, photodiode detector, transimpedance amplifier (TIA), and deserializer. The bare VCSEL, driver chip, and serializer chip are 3D stacked and then attached on one end of the embedded optical polymer waveguide in the PCB. Similarly, the bare photodiode detector, TIA chip, and deserializer chip are 3D stacked and then attached on the other end of the embedded optical polymer waveguide in the PCB. The backside of the driver or serializer and the TIA or deserializer chips is attached to a heat slug (with a heat spreader if necessary). This novel structure offers low-profile optoelectronic packaging for chip-to-chip optical interconnects. Optical, thermal management, and mechanical performances have been demonstrated [48] by simulations based on optic theory, heat-transfer theory, and continuum mechanics.

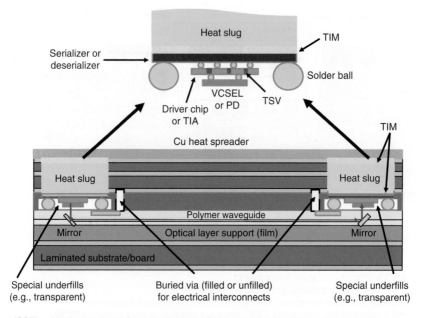

VCSEL = Vertical cavity surface emitted laser (transparent); PD = Photo diode detector (transparent); TIA = Trans-impedance amplifier

FIGURE 1.36   Embedded 3D hybrid IC integration for optoelectronic interconnects.

1.9   Summary and Recommendations

The milestones and outlooks of nanotechnology in the semiconductor industry have been mentioned briefly. Also, the evolution, challenges, recent advances, and new trends in 3D integration have been discussed. Furthermore, a few generic, thermal-enhanced, and potentially low-cost 3D IC integration SiPs with various passive interposers have been presented for small-form-factor, high-performance, high-density, low-power-consumption, and wide-bandwidth applications. Some important results and recommendations are summarized in the following. (At press time for this book, IBM announced two breakthroughs: One is a new storage device at 12 atoms, and the other is a 9-nm CNT transistor that beats silicon. Feynman must have a big smile on his face up there.)

1. For nanotechnology, scientists have been listening to Feynman to arrange atoms, and the most significant results are (with carbon) the zero-dimension buckyballs (1996 Nobel Prize), one-dimensional CNTs, and two-dimension graphene (2010 Nobel Prize). Maybe it is time to arrange other atoms to discover new applications.

2. The 2D graphene (which is stronger than steel, more electrically conductive than copper, and able to transmit electric signals with an amazing rapidity) could be a candidate for the material to make faster and more powerful electronics.

3. Research and development to systemically, cost-effectively, and reliably package a nanotechnology device are necessary for such a device to be in high-volume production. Otherwise, no matter how beautiful it looks under the SEM, it is just hype!

4. Moore's law drove nanotechnology in volume production for the semiconductor industry in 2004 with the Pentium IV and will continue to drive the semiconductor industry for at least another 10 years.

5. Feynman told us to go for 3D in 1985 and showed us how to stack up the wafers/chips. He also pointed out the importance of KGD and told us to obtain it through BIST and BISR.

6. TSV is a disruptive technology. In order for it to be widely used by the industry, the disruption should be as minimum as possible, at least in the early stage. Thus, currently passive interposers (either 2.5D or 3D) are the best candidates. Wait until the standards, infrastructures, and business models are ready; then we can make TSVs in the active chips for heterogeneous structures.

7. Passive interposer is the most cost-effective 3D IC *integrator*. It is not only for intermediate substrates, stress-relief (reliable) buffers, and carriers but also for thermal management tools. Let the passive interposer be the *workhorse* of 3D IC integration SiPs!

8. The passive interposer provides flexible coupling for whatever Moore's law chips are available and/or necessary and enhances the functionality and possibly the routings.

9. With the passive interposer, just about all the thermal problems of 3D IC integration SiPs can be managed. A few examples, such as heat spreaders, heat slugs, heat sinks, and microchannels, have been proposed.

10. Embedded 3D IC integration can reduce product profile and is the best candidate for mobile products. A couple of new designs with TSVs have been proposed.

11. A low-cost TSH interposer has been proposed for 3D IC integration.

12. 3D Si integration (bumpless) is the right way to go and compete with Moore's law. However, it still has a long way to go. Some technology R&D topics have been proposed, and some

guidelines on building the EDA have been recommended. The industry should immediately build an ecosystem incorporating standards and infrastructures so that EDA vendors can create and qualify the software for design, simulation, analysis and verification, manufacturing preparation, and test for 3D Si integration and *strive* to make it happen! It should be noted that even solder is God's gift to the electronics industry [49], but it is "too dirty" to be processed in a semiconductor fab!

1.10   TSV Patents

In the past few years, many patents have been granted for TSV and 3D integration. In this section, some of the patents (since 2005) from the United States have been selected and listed (with their patent number, exact title, inventor name, and issue date) for convenient practice. Failure to include any particular TSV and 3D integration patent indicates only that the author either had not read it or was unaware of it when this book went to press.

8,093,726   Semiconductor packages having interposers, electronic products employing the same, and methods of manufacturing the same; Park, Sung-Yong; January 10, 2012

8,093,722   System-in-package with fan-out WLCSP; Chen, Nan-Cheng; Hsu, Chih-Tai; January 10, 2012

8,093,711   Semiconductor device; Zudock, Frank; Meyer, Thorsten; Brunnbauer, Markus; Wolter, Andreas; January 10, 2012

8,093,705   Dual face package having resin insulating layer; Park, Seung Wook; Kweon, Young Do; Yuan, Jingli; Moon, Seon Hee; Hong, Ju Pyo; Lee, Jae Kwang; January 10, 2012

8,093,703   Semiconductor package having buried post in encapsulant and method of manufacturing the same; Kim, Pyoung-Wan; Lee, Teak-Hoon; Jang, Chul-Yong; January 10, 2012

8,093,696   Semiconductor device; Yoon, Kimyung; Dobritz, Stephan; Ruckmich, Stefan; January 10, 2012

8,093,100   Integrated circuit packaging system having through silicon via with direct interconnects and method of manufacture thereof; Choi, A Leam; Chung, Jae Han; Yang, DeokKyung; Park, HyungSang; January 10, 2012

8,090,250   Imaging device with focus offset compensation; Lusinchi, Jean-Pierre; January 3, 2012

8,076,762   Variable feature interface that induces a balanced stress to prevent thin die warpage; Chandrasekaran, Arvind; Radojcic, Ratibor; December 13, 2011

8,084,866   Microelectronic devices and methods for filling vias in microelectronic devices; Hiatt, William M.; Kirby, Kyle K.; December 27, 2011

8,084,841   Systems and methods for providing high-density capacitors; Pulugurtha, Markondeya Raj; Fenner, Andreas; Malin, Anna; Goud, Dasharatham Janagama; Tummala, Rao; December 27, 2011

8,080,862   Systems and methods for enabling ESD protection on 3D stacked devices; Kaskoun, Kenneth; Gu, Shiqun; Nowak, Matthew; December 20, 2011

8,080,885   Integrated circuit packaging system with multi-level contact and method of manufacture thereof; Chow, Seng Guan (Singapore, SG); Kuan, Heap Hoe (Singapore, SG); Huang, Rui; December 20, 2011

8,080,445   Semiconductor device and method of forming WLP with semiconductor die embedded within penetrable encapsulant between TSV interposers; Pagaila, Reza A.; December 20, 2011

8,076,234   Semiconductor device and method of fabricating the same including a conductive structure is formed through at least one dielectric layer after forming a via structure; Park, Byung-Lyul; Choi, Gil-Heyun; Bang, Suk-Chul; Moon, Kwang-Jin; Lim, Dong-Chan; Jung, Deok-Young; December 7, 2011

8,072,079   Through-hole vias at saw streets including protrusions or recesses for interconnection; Pagaila, Reza A.; Camacho, Zigmund R.; Tay, Lionel Chien Hui; Do, Byung Tai; December 6, 2011

8,072,056   Apparatus for restricting moisture ingress; Mueller, Tyler (Phoenix, AZ); Batchelder, Geoffrey; Danzl, Ralph B.; Gerrish, Paul F.; Malin, Anna J.; Marrott, Trevor D.; Mattes, Michael F.; December 6, 2011

8,063,475   Semiconductor package system with through silicon via interposer; Choi, DaeSik (Seoul, KR); Yang, DeokKyung (Hanam-si, KR); Kim, Seung Won; November 22, 2011

8,063,469   On-chip radiofrequency shield with interconnect metallization; Barth, Hans-Joachim; Koerner, Heinrich; Meyer, Thorsten; Brunnbauer, Markus; November 22, 2011

8,059,441   Memory array on more than one die; Taufique, Mohammed H.; Jallice, Derwin; McCauley, Donald

W.; DeVale, John P.; Brekelbaum, Edward A.; Rupley, Jeffrey P., II; Loh, Gabriel H.; Black, Bryan; November 15, 2011

8,067,816   Techniques for placement of active and passive devices within a chip; Kim, Jonghae; Gu, Shiqun; Henderson, Brian Matthew; Toms, Thomas R.; Nowak, Matthew; November 29, 2011

8,067,308   Semiconductor device and method of forming an interconnect structure with TSV using encapsulant for structural support; Suthiwongsunthorn, Nathapong; Marimuthu, Pandi C.; Ku, Jae Hun; Omandam, Glenn; Goh, Hin Hwa; Heng, Kock Liang; Caparas, Jose A.; November 29, 2011

8,063,975   Positioning wafer lenses on electronic imagers; Butterfield, Andrew; Fabre, Sebastien; Kujanpaa, Karo; November 22, 2011

8,063,496   Semiconductor integrated circuit device and method of fabricating the same; Cheon, Keon-Yong; Oh, Tae-Seok; Choi, Jong-Won; Oh, Su-Young; November 22, 2011

8,063,424   Embedded photodetector apparatus in a 3D CMOS chip stack; Gebara, Fadi H.; Ning, Tak H.; Ouyang, Qiqing C.; Schaub, Jeremy D.; November 22, 2011

8,060,843   Verification of 3D integrated circuits; Wang, Chung-Hsing; Tsai, Chih Sheng; Liu, Ying-Lin; Lin, Kai-Yun; November 15, 2011

8,058,137   Method for fabrication of a semiconductor device and structure; Or-Bach, Zvi; Sekar, Deepak C.; Cronquist, Brian; Beinglass, Israel; de Jong, Jan Lodewijk; November 15, 2011

8,058,102   Package structure and manufacturing method thereof; Lin, Diann-Fang; Hu, Yu-Shan; November 15, 2011

8,053,902   Isolation structure for protecting dielectric layers from degradation; Chen, Ming-Fa; Lin, Sheng-Yuan; Chen, Ming-Fa (Taichung, TW); Lin, Sheng-Yuan; November 8, 2011

8,053,900   Through-substrate vias (TSVs) electrically connected to a bond pad design with reduced dishing effect; Yu, Chen-Hua; Chiou, Wen-Chih; Wu, Weng-Jin; November 8, 2011

8,053,898   Connection for off-chip electrostatic discharge protection; Marcoux, Phil P.; November 8, 2011

8,049,327   Through-silicon via with scalloped sidewalls; Kuo, Chen-Cheng; Chen, Chih-Hua; Chen, Ming-Fa; Chen, Chen-Shien; November 1, 2011

8,049,310	Semiconductor device with an interconnect element and method for manufacture; Wolter, Andreas; Hedler, Harry; Irsigler, Roland; November 1, 2011
8,048,761	Fabricating method for crack stop structure enhancement of integrated circuit seal ring; Yeo, Alfred; Chan, Kai Chong; November 1, 2011
8,048,721	Method for filling multilayer chip-stacked gaps; Hsu, Hung-Hsin; Chien, Wei-Chih; November 1, 2011
8,046,727	IP cores in reconfigurable three dimensional integrated circuits; Solomon, Neal; October 25, 2011
8,043,967	Process for through silicon via filling; Reid, Jonathan D.; Wang, Katie Qun; Wiley, Mark J.; October 25, 2011
8,042,082	Three dimensional memory in a system on a chip; Solomon, Neal; October 18, 2011
8,039,393	Semiconductor structure, method for manufacturing semiconductor structure and semiconductor package; Wang, Meng-Jen; Chen, Chien-Yu; October 18, 2011
8,039,386	Method for forming a through silicon via (TSV); Dao, Thuy B.; Noble, Ross E.; Triyoso, Dina H.; Dao, Thuy B. (Austin, TX); Noble, Ross E. (Austin, TX); Triyoso, Dina H.; October 18, 2011
8,039,356	Through silicon via lithographic alignment and registration; Herrin, Russell T.; Lindgren, Peter J.; Sprogis, Edmund J.; Stamper, Anthony K.; October 18, 2011
8,035,194	Semiconductor device and semiconductor package including the same; Lee, Jun-Ho; Lee, Hyung-Dong; Im, Hyun-Seok; October 11, 2011
8,034,708	Structure and process for the formation of TSVs; Kuo, Chen-Cheng; Ching, Kai-Ming; Chen, Chen-Shien; October 11, 2011
8,033,012	Method for fabricating a semiconductor test probe card space transformer; Hsu, Ming Cheng; Chao, Clinton Chih-Chieh; October 11, 2011
8,031,217	Processes and structures for IC fabrication; Sheats, Jayna; October 4, 2011
8,030,780	Semiconductor substrates with unitary vias and via terminals, and associated systems and methods; Kirby, Kyle K.; Parekh, Kunal R.; October 4, 2011
8,030,761	Mold design and semiconductor package; Kolan, Ravi Kanth; Liu, Hao; Toh, Chin Hock; October 4, 2011
8,030,200	Method for fabricating a semiconductor package; Eom, Yong Sung; Choi, Kwang-Seong; Bae, Hyun-Cheol; Lee, Jong-Hyun; Moon, Jong Tae; October 4, 2011

7,981,798	Method of manufacturing substrate; Taguchi, Yuichi; Shiraishi, Akinori; Sunohara, Masahiro; Murayama, Kei; Sakaguchi, Hideaki; Higashi, Mitsutoshi; July 19, 2011
7,973,415	Manufacturing process and structure of through silicon via; Kawashita, Michihiro; Yoshimura, Yasuhiro; Tanaka, Naotaka; Naito, Takahiro; Akazawa, Takashi; July 5, 2011
7,973,358	Coupler structure; Hanke, Andre; Nagy, Oliver; July 5, 2011
8,030,208	Bonding method for through-silicon-via based 3D wafer stacking; Leung, Chi Kuen Vincent; Sun, Peng; Shi, Xunqing; Chung, Chang Hwa; October 4, 2011
8,030,113	Thermoelectric 3D cooling; Hsu, Louis Lu-Chen; Wang, Ping-Chuan; Wei, Xiaojin; Zhu, Huilong; October 4, 2011
8,026,592	Through-silicon via structures including conductive protective layers; Yoon, Minseung; Kim, Namseog; Kim, Pyoungwan; Ma, Keumhee; Jo, Chajea; September 27, 2011
8,026,567	Thermoelectric cooler for semiconductor devices with TSV; Chang, Shih-Cheng; Pan, Hsin-Yu; September 27, 2011
8,026,521	Semiconductor device and structure; Or-Bach, Zvi; Sekar, Deepak C.; September 27, 2011
8,020,290	Processes for IC fabrication; Sheats, Jayna; September 20, 2011
8,018,065	Wafer-level integrated circuit package with top and bottom side, electrical connections; Lam, Ken; September 13, 2011
8,017,515	Semiconductor device and method of forming compliant polymer layer between UBM and conformal dielectric layer/RDL for stress relief; Marimuthu, Pandi C.; Suthiwongsunthorn, Nathapong; Huang, Shuangwu; September 13, 2011
8,017,497	Method for manufacturing semiconductor; Oi, Hideo; September 13, 2011
8,017,451	Electronic modules and methods for forming the same; Racz, Livia M.; Tepolt, Gary B.; Thompson, Jeffrey C.; Langdo, Thomas A.; Mueller, Andrew J.; September 13, 2011
8,014,166	Stacking integrated circuits containing serializer and deserializer blocks using through silicon via; Yazdani, Farhang; September 6, 2011

8,012,808   Integrated microchannels for 3D through silicon architectures; Shi, Wei; Lu, Daoqiang; Bai, Yiqun; Zhou, Qing A.; He, Jianqqi; September 6, 2011

8,008,764   Bridges for interconnecting interposers in multichip integrated circuits; Joseph, Douglas James; Knickerbocker, John Ulrich; August 30, 2011

8,008,195   Method for manufacturing semiconductor device; Koike, Osamu; Kadogawa, Yutaka; August 30, 2011

8,008,192   Conductive interconnect structures and formation methods using supercritical fluids; Sulfridge, Marc; August 30, 2011

8,005,326   Optical clock signal distribution using through-silicon vias; Chang, Shih-Cheng; Lin, Jin-Lien; Hsu, Kuo-Ching; Ching, Kai-Ming; Wu, Jiun Yi; Chen, Yen-Huei; August 23, 2011

7,998,862   Method for fabricating semiconductor device; Park, Kunsik; Baek, Kyu-Ha; Do, Lee-Mi; Kim, Dong-Pyo; Park, Ji Man; August 16, 2011

7,994,041   Method of manufacturing stacked semiconductor package using improved technique of forming through via; Lim, Kwon-Seob; Kang, Hyun Seo; August 9, 2011

7,990,174   Circuit for calibrating impedance and semiconductor apparatus using the same; Park, Nak Kyu; August 2, 2011

7,989,318   Method for stacking semiconductor dies; Yang, Ku-Feng; Wu, Weng-Jin; Chiou, Wen-Chih; Yu, Chen-Hua; August 2, 2011

7,989,226   Clocking architecture in stacked and bonded dice; Peng, Mark Shane; August 2, 2011

7,986,582   Method of operating a memory apparatus, memory device and memory apparatus; Ruckerbauer, Hermann; Sichert, Christian; July 26, 2011

7,986,042   Method for fabrication of a semiconductor device and structure; Or-Bach, Zvi; Cronquist, Brian; Beinglass, Israel; de Jong, Jan Lodewijk; Sekar, Deepak C.; July 26, 2011

7,977,962   Apparatus and methods for through substrate via test; Hargan, Ebrahim H.; Bunker, Layne; Dimitriu, Dragos; King, Gregory; July 12, 2011

7,977,781   Semiconductor device; Ito, Kiyoto; Saen, Makoto; Kuroda, Yuki; July 12, 2011

7,973,416   Thru silicon enabled die stacking scheme; Chauhan, Satyendra Singh; July 5, 2011

7,973,413	Through-substrate via for semiconductor device; Kuo, Chen-Cheng; Chen-Shien, Chen; Ching, Kai-Ming; Chen, Chih-Hua; July 5, 2011
7,973,411	Microfeature workpieces having conductive interconnect structures formed by chemically reactive processes, and associated systems and methods; Borthakur, Swarnal; July 5, 2011
7,972,969	Method and apparatus for thinning a substrate; Yang, Ku-Feng; Chiou, Wen-Chih; Wu, Weng-Jin; Zuo, Kewei; July 5, 2011
7,972,905	Packaged electronic device having metal comprising self-healing die attach material; Wainerdi, James C.; Tellkamp, John P.; July 5, 2011
7,972,902	Method of manufacturing a wafer including providing electrical conductors isolated from circuitry; Youn, Sunpil; Lee, Seok-Chan; July 5, 2011
7,969,193	Differential sensing and TSV timing control scheme for 3D-IC; Wu, Wei-Cheng; Chen, Yen-Huei; Chang, Meng-Fan; June 28, 2011
7,969,013	Through silicon via with dummy structure and method for forming the same; Chen, Chih-Hua; Chen, Chen-Shien; Kuo, Chen-Cheng; Shen, Wen-Wei; June 28, 2011
7,969,009	Through silicon via bridge interconnect; Chandrasekaran, Arvind; June 28, 2011
7,964,916	Method for fabrication of a semiconductor device and structure; Or-Bach, Zvi; Cronquist, Brian; Beinglass, Israel; de Jong, Jan L.; Sekar, Deepak C.; June 21, 2011
7,964,502	Multilayered through via; Dao, Thuy B.; Vuong, Chanh M.; June 21, 2011
7,960,773	Capacitor device and method for manufacturing the same; Chang, Shu-Ming; Chiang, Chia-Wen; June 14, 2011
7,960,282	Method of manufacture an integrated circuit system with through silicon via; Yelehanka, Pradeep Ramachandramurthy; Tan, Denise; Lek, Chung Meng; Thiam, Thomas; Lam, Jeffrey C.; Hsia, Liang-Choo; June 14, 2011
7,960,242	Method for fabrication of a semiconductor device and structure; Or-Bach, Zvi; Cronquist, Brian; Beinglass, Israel; de Jong, Jan Lodewijk; Sekar, Deepak C.; June 14, 2011

7,958,627	Method of attaching an electronic device to an MLCC having a curved surface; Randall, Michael S.; Wayne, Chris; McConnell, John; June 14, 2011
7,957,209	Method of operating a memory apparatus, memory device and memory apparatus; Ruckerbauer, Hermann;, June 7, 2011
7,957,173	Composite memory having a bridging device for connecting discrete memory devices to a system; Kim, Jin-Ki; June 7, 2011
7,956,443	Through-wafer interconnects for photoimager and memory wafers; Akram, Salman; Watkins, Charles M.; Hiatt, Mark; Hembree, David R.; Wark, James M.; Farnworth, Warren M.; Tuttle, Mark E.; Rigg, Sidney B.; Oliver, Steven D.; Kirby, Kyle K.; Wood, Alan G.; Velicky, Lu; June 7, 2011
7,956,442	Backside connection to TSVs having redistribution lines; Hsu, Kuo-Ching; Chen, Chen-Shien; June 7, 2011
7,955,895	Structure and method for stacked wafer fabrication; Yang, Ku-Feng; Chiou, Wen-Chih; Wu, Weng-Jin; Tu, Hung-Jung; June 7, 2011
7,952,904	Three-dimensional memory-based three-dimensional memory module; Zhang, Guobiao; May 31, 2011
7,952,903	Multimedia three-dimensional memory module (M3DMM) system; Zhang, Guobiao; May 31, 2011
7,952,478	Capacitance-based microchip exploitation detection; Bartley, Gerald K.; Becker, Darryl J.; Dahlen, Paul E.; Germann, Philip R.; Maki, Andrew B.; Maxson, Mark O.; May 31, 2011
7,952,176	Integrated circuit packaging system and method of manufacture thereof; Pagaila, Reza Argenty; Do, Byung Tai; Chua, Linda Pei Ee; May 31, 2011
7,952,171	Die stacking with an annular via having a recessed socket; Pratt, Dave; May 31, 2011
7,951,647	Performing die-to-wafer stacking by filling gaps between dies; Yang, Ku-Feng; Chiou, Wen-Chih; Wu, Weng-Jin; Sung, Ming-Chung; May 31, 2011
7,948,064	System on a chip with on-chip RF shield; Barth, Hans-Joachim (Munich, DE); Hanke, Andre; Jenei, Snezana; Nagy, Oliver; Morinaga, Jiro; Adler, Bernd; Koerner, Heinrich; May 24, 2011
7,944,038	Semiconductor package having an antenna on the molding compound thereof; Chiu, Chi-Tsung; Lee, Pao-Nan; May 17, 2011

7,943,513	Conductive through connection and forming method thereof; Lin, Shian-Jyh; May 17, 2011
7,943,473	Minimum cost method for forming high density passive capacitors for replacement of discrete board capacitors using a minimum cost 3D wafer-to-wafer modular integration scheme; Ellul, Joseph Paul; Tran, Khanh; Bergemont, Albert; May 17, 2011
7,940,074	Data transmission circuit and semiconductor apparatus including the same; Ku, Young Jun; May 10, 2011
7,939,941	Formation of through via before contact processing; Chiou, Wen-Chih; Yu, Chen-Hua; Wu, Weng-Jin; May 10, 2011
7,939,369	3D integration structure and method using bonded metal planes; Farooq, Mukta G.; Iyer, Subramanian S; May 10, 2011
7,936,622	Defective bit scheme for multilayer integrated memory device; Li, Hai; Chen, Yiran; Setiadi, Dadi; Liu, Harry Hongyue; Lee, Brian; May 3, 2011
7,936,052	On-chip RF shields with backside redistribution lines; Barth, Hans-Joachim; Pohl, Jens; Beer, Gottfried; Koerner, Heinrich; May 3, 2011
7,923,370	Method for stacking serially-connected integrated circuits and multichip device made from same; Pyeon, Hong Beom; April 12, 2011
7,923,290	Integrated circuit packaging system having dual-sided connection and method of manufacture thereof; Ko, Chan Hoon (Ichon si, KR); Park, Soo-San (Seoul, KR); Kim, YoungChul; April 12, 2011
7,933,428	Microphone apparatus; Sawada, Tatsuhiro; April 26, 2011
7,932,608	Through-silicon via formed with a post passivation interconnect structure; Tseng, Ming-Hong; Jao, Sheng Huang; April 26, 2011
7,930,664	Programmable through silicon via; Feng, Kai Di; Hsu, Louis Lu-Chen; Wang, Ping-Chuan; Yang, Zhijian; April 19, 2011
7,928,563	3D ICs with microfluidic interconnects and methods of constructing same; Bakir, Muhannad S.; Sekar, Deepak; Dang, Bing; King, Calvin, Jr.; Meindl, James D.; April 19, 2011
7,928,534	Bond pad connection to redistribution lines having tapered profiles; Hsu, Kuo-Ching; Chen, Chen-Shien; Huang, Hon-Lin; April 19, 2011

7,916,511   Semiconductor memory device including plurality of memory chips; Park, Ki-Tae; March 29, 2011

7,915,736   Microfeature workpieces and methods for forming interconnects in microfeature workpieces; Kirby, Kyle K.; Hiatt, William M.; Stocks, Richard L.; March 29, 2011

7,913,000   Stacked semiconductor memory device with compound read buffer; Chung, Hoe-ju; March 22, 2011

7,910,473   Through-silicon via with air gap; Chen, Ming-Fa; March 22, 2011

7,906,857   Molded integrated circuit package and method of forming a molded integrated circuit package; Hoang, Lan H.; Chaware, Raghunandan; Yip, Laurene; March 15, 2011

7,906,431   Semiconductor device fabrication method; Mistuhashi, Toshiro; March 15, 2011

7,904,273   In-line depth measurement for thru silicon via; Liu, Qizhi; Wang, Ping-Chuan; Watson, Kimball M.; Yang, Zhijian J.; March 8, 2011

7,902,674   Three-dimensional die-stacking package structure; Chang, Hsiang-Hung; Chang, Shu-Ming; March 8, 2011

7,902,643   Microfeature workpieces having interconnects and conductive, backplanes, and associated systems and methods; Tuttle, Mark E.; March 8, 2011

7,902,069   Small area, robust silicon via structure and process; Andry, Paul S.; Cotte, John M.; Knickerbocker, John Ulrich; Tsang, Cornelia K; March 8, 2011

7,900,519   Microfluidic measuring tool to measure through-silicon via depth; Chandrasekaran, Arvind; March 8, 2011

7,894,199   Hybrid package; Chang, Li-Tien; February 22, 2011

7,893,529   Thermoelectric 3D cooling; Hsu, Louis Lu-Chen; Wang, Ping-Chuan; Wei, Xiaojin; Zhu, Huilong; February 22, 2011

7,893,526   Semiconductor package apparatus; Mun, Sung-ho (Suwon-si, KR); Kang, Sun-won (Seoul, KR); Baek, Seung-duk; February 22, 2011

7,888,806   Electrical connections for multichip modules; Lee, Seok-Chan; Kim, Min-Woo; February 15, 2011

7,888,668   Phase change memory; Kuo, Chien-Li; Wu, Kuei-Sheng; Lin, Yung-Chang; February 15, 2011

7,884,625	Capacitance structures for defeating microchip tampering; Bartley, Gerald K. (Rochester, MN); Becker, Darryl J. (Rochester, MN); Dahlen, Paul E. (Rochester, MN); Germann, Philip R. (Oronoco, MN); Maki, Andrew B. (Rochester, MN); Maxson, Mark; February 8, 2011
7,884,016	Liner materials and related processes for 3D integration; Sprey, Hessel; Nakano, Akinori; February 8, 2011
7,884,015	Methods for forming interconnects in microelectronic workpieces and microelectronic workpieces formed using such methods; Sulfridge, Marc; February 8, 2011
7,880,494	Accurate capacitance measurement for ultra-large-scale integrated circuits; Doong, Yih-Yuh; Chang, Keh-Jeng; Mii, Yuh-Jier; Liu, Sally; Hung, Lien Jung; Chang, Victor Chih Yuan; February 1, 2011
7,875,948	Backside illuminated image sensor; Hynecek, Jaroslav (Lake Oswego, OR); Forbes, Leonard (Lake Oswego, OR); Haddad, Homayoon (Lake Oswego, OR); Joy, Thomas; January 25, 2011
7,872,332	Interconnect structures for stacked dies, including penetrating structures for through-silicon vias, and associated systems and methods; Fay, Owen R. (Meridian, ID); Farnworth, Warren M. (Nampa, ID); Hembree, David R.; January 18, 2011
7,867,910	Method of accessing semiconductor circuits from the backside using ion-beam and gas-etch; Scrudato, Carmelo F.; Gu, George Y.; Hahn, Loren L.; Herschbein, Steven B.; January 11, 2011
7,867,821	Integrated circuit package system with through semiconductor vias and method of manufacture thereof; Chin, Chee Keong; January 11, 2011
7,863,721	Method and apparatus for wafer level integration using tapered vias; Suthiwongsunthorn, Nathapong; Marimuthu, Pandi Chelvam; January 4, 2011
7,863,187	Microfeature workpieces and methods for forming interconnects in microfeature workpieces; Hiatt, William M.; Dando, Ross S.; January 4, 2011
7,863,106	Silicon interposer testing for three-dimensional chip stack; Christo, Michael Anthony; Maldonado, Julio Alejandro; Weekly, Roger Donell; Zhou, Tingdong; January 4, 2011
7,859,099	Integrated circuit packaging system having through silicon via with direct interconnects and method of manufacture thereof; Choi, A Leam; Chung, Jae Han; Yang, DeokKyung; Park, HyungSang; December 28, 2010

7,858,512   Semiconductor with bottom-side wrap-around flange contact; Marcoux, Phil P.; December 28, 2010

7,851,893   Semiconductor device and method of connecting a shielding layer to ground through conductive vias; Kim, Seung Won; Yang, Dae Wook; December 14, 201

7,851,342   In-situ formation of conductive filling material in through-silicon via; Xu, Dingying; Eitan, Amram; December 14, 2010

7,848,153   High-speed memory architecture; Bruennert, Michael (Munich, DE); Gregorius, Peter; Braun, Georg; Gaertner, Andreas; Ruckerbauer, Hermann; Alexander, George William; Stecker, Johannes; December 7, 2010

7,847,383   Multichip package for reducing parasitic load of pin; So, Byung-Se; Lee, Dong-Ho; December 7, 2010

7,847,379   Lightweight and compact through-silicon via stack package with excellent electrical connections and method for manufacturing the same; Chung; Qwan Ho December 7, 2010

7,843,072   Semiconductor package having through holes; Park, Sung Su; Kim, Jin Young; Jin, Jeong Gi; November 30, 2010

7,843,064   Structure and process for the formation of TSVs; Kuo, Chen-Cheng; Ching, Kai-Ming; Chen-Shien, Chen; November 30, 2010

7,843,052   Semiconductor devices and fabrication methods thereof; Yoo, Min; Lee, Ki Wook; Lee, Min Jae; November 30, 2010

7,842,548   Fixture for P-through silicon via assembly; Lee, Chien-Hsiun; Chen, Chen-Shien; Lii, Mirng-Ji; Karta, Tjandra Winata; November 30, 2010

7,839,163   Programmable through silicon via; Feng, Kai Di; Hsu, Louis Lu-Chen; Wang, Ping-Chuan; Yang, Zhijian; November 23, 2010

7,838,975   Flip-chip package with fan-out WLCSP; Chen, Nan-Cheng; November 23, 2010

7,838,967   Semiconductor chip having TSV (through silicon via) and stacked assembly including the chips; Chen, Ming-Yao; November 23, 2010

7,838,337   Semiconductor device and method of forming an interposer package with through silicon vias; Marimuthu, Pandi Chelvam (Singapore, SG); Suthiwongsunthorn, Nathapong (Singapore, SG); Shim, Il Kwon (Singapore, SG); Heng, Kock Liang; November 23, 2010

7,834,440	Semiconductor device with stacked memory and processor LSIs; Ito, Kiyoto; Saen, Makoto; Kuroda, Yuki; November 16, 2010
7,830,692	Multichip memory device with stacked memory chips, method of stacking memory chips, and method of controlling operation of multichip package memory; Chung, Hoe-ju; Lee, Jung-bae; Kang, Uk-song; November 9, 2010
7,830,018	Partitioned through-layer via and associated systems and methods; Lee, Teck Kheng; November 9, 2010
7,829,976	Microelectronic devices and methods for forming interconnects in microelectronic devices; Kirby, Kyle K.; Akram, Salman; Hembree, David R.; Rigg, Sidney B.; Farnworth, Warren M.; Hiatt, William M.; November 9, 2010
7,825,517	Method for packaging semiconductor dies having through-silicon vias; Su, Chao-Yuan; November 2, 2010
7,825,024	Method of forming through-silicon vias; Lin, Chuan-Yi; Lee, Song-Bor; Huang, Ching-Kun; Lin, Sheng-Yuan; November 2, 2010
7,821,107	Die stacking with an annular via having a recessed socket; Pratt, Dave; October 26, 2010
7,816,945	3D chip-stack with fuse-type through silicon via; Feng, Kai Di; Hsu, Louis Lu-Chen; Wang, Ping-Chuan; Yang, Zhijian; October 19, 2010
7,816,776	Stacked semiconductor device and method of forming serial path thereof; Choi, Young-Don; October 19, 2010
7,816,227	Tapered through-silicon via structure; Chen, Chen Shien; Kuo, Chen Cheng; Ching, Kai-Ming; Chen, Chih-Hua; October 19, 2010.
7,816,181	Method of underfilling semiconductor die in a die stack and semiconductor device formed thereby; Bhagath, Shrikar; Takiar, Hem; October 19, 2010
7,813,043	Lens assembly and method of manufacture; Lusinchi, Jean-Pierre; Kui, Xiao-Yun; October 12, 2010
7,812,459	Three-dimensional integrated circuits with protection layers; Yu, Chen-Hua; Chiou, Wen-Chih; Wu, Weng-Jin; Tu, Hung-Jung; Yang, Ku-Feng; October 12, 2010
7,812,449	Integrated circuit package system with redistribution layer; Kuan, Heap Hoe; Chow, Seng Guan; Huang, Rui; October 12, 2010
7,812,446	Semiconductor device; Kurita, Yoichiro; October 12, 2010
7,812,426	TSV-enabled twisted pair; Peng, Mark Shane; Chao, Clinton; Hsu, Chao-Shun; October 12, 2010

7,808,105	Semiconductor package and fabricating method thereof; Paek, Jong Sik; October 5, 2010
7,803,714	Semiconductor through silicon vias of variable size and method of formation; Ramiah, Chandrasekaram; Sanders, Paul W.; September 28, 2010
7,799,678	Method for forming a through silicon via layout; Kropewnicki, Thomas J.; Chatterjee, Ritwik; Junker, Kurt H.; September 21, 2010
7,799,613	Integrated module for data processing system; Dang, Bing; Knickerbocker, John U.; Tsang, Cornelia K.; September 21, 2010
7,796,446	Memory dies for flexible use and method for configuring memory dies; Ruckerbauer, Hermann; Bruennert, Michael; Menczigar, Ullrich; Mueller, Christian; Tontisirin, Sitt; Braun, Georg; Savignac, Dominique; September 14, 2010
7,795,735	Methods for forming single dies with multilayer interconnect structures and structures formed therefrom; Hsu, Chao-Shun; Tang, Chen-Yao; Chao, Clinton; Peng, Mark Shane; September 14, 2010
7,795,650	Method and apparatus for backside illuminated image sensors using capacitively coupled readout integrated circuits; Eminoglu, Selim; Lauxtermann, Stefan C.; September 14, 2010
7,795,139	Method for manufacturing semiconductor package; Han, Kwon Whan; Park, Chang Jun; Suh, Min Suk; Kim, Seong Cheol; Kim, Sung Min; Yang, Seung Taek; Lee, Seung Hyun; Kim, Jong Hoon; Lee, Ha Na; September 14, 2010
7,795,134	Conductive interconnect structures and formation methods using supercritical fluids; Sulfridge, Marc; September 14, 2010
7,791,919	Semiconductor memory device capable of identifying a plurality of memory chips stacked in the same package; Shimizu, Yuui; September 7, 2010
7,791,175	Method for stacking serially connected integrated circuits and multichip device made from same; Pyeon, Hong Beom; September 7, 2010
7,786,008	Integrated circuit packaging system having through silicon vias with partial depth metal fill regions and method of manufacture thereof; Do, Byung Tai; Chow, Seng Guan; Yoon, Seung Uk; August 31, 2010
7,777,330	High bandwidth cache-to-processing unit communication in a multiple processor/cache system; Pelley, Perry H.; McShane, Michael B.; August 17, 2010

7,776,741	Process for through silicon via filing; Reid, Jonathan D.; Wang, Katie Qun; Willey, Mark J.; August 17, 2010
7,775,119	Media-compatible electrically isolated pressure sensor for high temperature applications; Suminto, James Tjanmeng; Yunus, Mohammad; August 17, 2010
7,772,880	Reprogrammable three-dimensional intelligent system on a chip; Solomon, Neal; August 10, 2010
7,772,868	Accurate capacitance measurement for ultra-large-scale integrated circuits; Doong, Yih-Yuh; Chang, Keh-Jeng; Mii, Yuh-Jier; Liu, Sally; Hung, Lien Jung; Chang, Victor Chih Yuan; August 10, 2010
7,772,124	Method of manufacturing a through-silicon-via on-chip passive MMW bandpass filter; Bavisi, Amit; Ding, Hanyi; Wang, Guoan; Woods, Wayne H., Jr.; Xu; Jiansheng; August 10, 2010
7,772,081	Semiconductor device and method of forming high-frequency circuit structure and method thereof; Lin, Yaojian; Fang, Jianmin; Chen, Kang; Cao, Haijing; August 10, 2010
7,760,144	Antennas integrated in semiconductor chips; Chang, Chung-Long; Lu, David Ding-Chung; Chung, Shine; July 20, 2010
7,759,800	Microelectronics devices, having vias, and packaged microelectronic devices having vias; Rigg, Sidney B.; Watkins, Charles M.; Kirby, Kyle K.; Benson, Peter A.; Akram, Salman; July 20, 2010
7,759,165	Nanospring; Bajaj, Rajeev; July 20, 2010
7,755,173	Series-shunt switch with thermal terminal; Mondi, Anthony Paul; Bukowski, Joseph Gerard; July 13, 2010
7,750,459	Integrated module for data processing system; Dang, Bing; Knickerbocker, John Ulrich; Tsang, Cornelia Kang-I; July 6, 2010
7,749,899	Microelectronic workpieces and methods and systems for forming interconnects in microelectronic workpieces; Clark, Douglas; Oliver, Steven D.; Kirby, Kyle K.; Dando, Ross S.; July 6, 2010
7,741,156	Semiconductor device and method of forming through vias with reflowed conductive material; Pagaila, Reza A.; Chua, Linda Pei Ee; Do, Byung Tai; June 22, 2010
7,741,148	Semiconductor device and method of forming an interconnect, structure for 3D devices using encapsulant for structural support; Marimuthu, Pandi C.;

	Suthiwongsunthorn, Nathapong; Heng, Kock Liang; June 22, 2010
7,738,249	Circuitized substrate with internal cooling structure and electrical assembly utilizing same; Chan, Benson; Egitto, Frank D.; Lin, How T.; Magnuson, Roy H.; Markovich, Voya R.; Thomas, David L.; June 15, 2010
7,701,252	Stacked die network-on-chip for FPGA; Chow, Francis Man-Chit; Patel, Rakesh H.; Pistorius, Erhard Joachim; April 20, 201
7,701,244	False connection for defeating microchip exploitation; Bartley, Gerald K.; Becker, Darryl J.; Dahlen, Paul E.; Germann, Philip R.; Maki, Andrew B.; Maxson, Mark O.; Sheets, John E., II; April 20, 2010
7,698,470	Integrated circuit, chip stack and data processing system; Ruckerbauer, Hermann; Savignac, Dominique; April 13, 2010
7,692,946	Memory array on more than one die; Taufique, Mohammed H.; Jallice, Derwin; McCauley, Donald W.; DeVale, John P.; Brekelbaum, Edward A.; Rupley, Jeffrey P., II; Loh, Gabriel H.; Black, Bryan; April 6, 2010
7,692,448	Reprogrammable three-dimensional field programmable gate arrays; Solomon, Neal; April 6, 2010
7,691,748	Through-silicon via and method for forming the same; Han, Kwon Whan; April 6, 2010
7,692,310	Forming a hybrid device; Park, Chang-Min; Ramanathan, Shriram; April 6, 2010
7,692,278	Stacked-die packages with silicon vias and surface activated bonding; Periaman, Shanggar; Ooi, Kooi Chi; Cheah, Bok Eng; April 6, 2010
7,691,748	Through-silicon via and method for forming the same; Han, Kwon Whan; April 6, 2010
7,683,459	Bonding method for through-silicon-via based 3D wafer stacking; Ma, Wei; Shu, Xunqing; Chung, Chang Hwa; March 23, 2010
7,683,458	Through-wafer interconnects for photoimager and memory wafers; Akram, Salman; Watkins, Charles M.; Hiatt, William M.; Hembree, David R., Wark; James M., Farnworth; Warren M., Tuttle; Mark E., Rigg; Sidney B., Oliver; Steven D., Kirby; Kyle K., Wood; Alan G. Velicky, Lu; March 23, 2010
7,670,950	Copper metallization of through silicon via; Richardson, Thomas B.; Zhang, Yun; Wang, Chen; Paneccasio, Vincent, Jr.; Wang, Cai; Lin, Xuan; Hurtubise, Richard; Abys, Joseph A.; March 2, 2010

7,666,768	Through-die metal vias with a dispersed phase of graphitic structures of carbon for reduced thermal expansion and increased electrical conductance; Raravikar, Nachiket R.; Suh, Daewoong; Arana, Leonel; Matayabas, James C., Jr.; February 23, 2010
7,666,711	Semiconductor device and method of forming double-sided through vias in saw streets; Pagaila, Reza A.; Do, Byung Tai; February 23, 2010
7,656,031	Stackable semiconductor package having metal pin within through hole of package; Chen, Cheng-Chung; Wang, Chia-Chung; Tan, Chin Hock; Lin, Charles W. C.; February 2, 2010
7,638,867	Microelectronic package having solder-filled through-vias; Xu, Dingying; Hackitt, Dale A.; December 29, 2009
7,633,165	Introducing a metal layer between SiN and TiN to improve CBD contact resistance for P-TSV; Hsu, Kuo-Ching; Chen, Chen-Shien; Su, Boe; Huang, Hon-Lin; December 15, 2009
7,629,249	Microfeature workpieces having conductive interconnect structures formed by chemically reactive processes, and associated systems and methods; Borthakur, Swarnal; December 8, 2009
7,622,377	Microfeature workpiece substrates having through-substrate vias, and associated methods of formation; Lee, Teck Kheng; Lim, Andrew Chong Pei; November 24, 2009
7,598,617	Stack package utilizing through vias and re-distribution lines; Lee, Seung Hyun; Suh, Min Suk; October 6, 2009
7,598,523	Test structures for stacking dies having through-silicon vias; Luo, Wen-Liang; Kuo, Yung-Liang; Cheng, Hsu Mi; October 6, 2009
7,592,697	Microelectronic package and method of cooling same; Arana, Leonel R.; Newman, Michael W.; Chang, Je-Young; September 22, 2009
7,589,008	Methods for forming interconnects in microelectronic workpieces and microelectronic workpieces formed using such methods; Kirby, Kyle K.; September 15, 2009
7,576,435	Low-cost and ultrafine integrated circuit packaging technique; Chao, Clinton; August 18, 2009
7,564,115	Tapered through-silicon via structure; Chen, Chen-Shien; Kuo, Chen-Cheng; Ching, Kai-Ming; Chen, Chih-Hua; July 21, 2009

7,541,203	Conductive adhesive for thinned silicon wafers with through silicon vias; Knickerbocker, John U.; June 2, 2009
7,531,453	Microelectronic devices and methods for forming interconnects in microelectronic devices; Kirby, Kyle K.; Akram, Salman; Hembree, David R.; Rigg, Sidney B.; Farnworth, Warren M.; Hiatt, William M.; May 12, 2009
7,528,006	Integrated circuit die containing particle-filled through-silicon metal vias with reduced thermal expansion; Arana, Leonel; Newman, Michael; Natekar, Devendra; May 5, 2009
7,514,116	Horizontal carbon nanotubes by vertical growth and rolling; Natekar, Devendra; Tomita, Yoshihiro; Hwang, Chi-Won; April 7, 2009
7,494,846	Design techniques for stacking identical memory dies; Hsu, Chao-Shun; Liu, Louis; Chao, Clinton; Peng, Mark Shane; February 24, 200
7,446,420	Through silicon via chip stack package capable of facilitating chip selection during device operation; Kim, Jong Hoon; November 4, 2008
7,435,913	Slanted vias for electrical circuits on circuit boards and other substrates; Chong, Chin Hui (Singapore, SG); Lee, Choon Kuan; October 14, 2008
7,432,592	Integrated microchannels for 3D through silicon architectures; Shi, Wei; Lu, Daoqiang; Bai, Yiqun; Zhou, Qing A.; He, Jiangqi; October 7, 2008
7,427,803	Electromagnetic shielding using through-silicon vias; Chao, Clinton; Hsu, Chao-Shun; Peng, Mark Shane; Lu, Szu Wei; Karta, Tjandra Winata; September 23, 2008
7,425,499	Methods for forming interconnects in vias and microelectronic workpieces including such interconnects; Oliver, Steven D.; Kirby, Kyle K.; Hiatt, William M.; September 16, 2008
7,413,979	Methods for forming vias in microelectronic devices, and methods for packaging microelectronic devices; Rigg, Sidney B.; Watkins, Charles M.; Kirby, Kyle K.; Benson, Peter A.; Akram, Salman; August 19, 2008
7,410,884	3D integrated circuits using thick metal for backside connections and offset bumps; Ramanathan, Shriram; Kim, Sarah E.; Morrow, Patrick R.; August 12, 2008
7,400,033	Package on package design to improve functionality and efficiency; Cheah, Bok Eng (Penang, MY); Periaman, Shanggar (Penang, MY); Ooi, Kooi Chi; July 15, 2008

7,317,256   Electronic packaging including die with through silicon via; Williams, Christina K.; Thomas, Rainer E.; January 8, 2008

7,241,675   Attachment of integrated circuit structures and other substrates to substrates with vias; Savastiouk, Sergey; Kao, Sam; July 10, 2007

7,111,149   Method and apparatus for generating a device ID for stacked devices; Eilert, Sean S.; September 19, 2006

7,081,408   Method of creating a tapered via using a receding mask and resulting structure; Lane, Ralph L.; Hill, Charles D.; July 25, 2006

6,924,551   Through silicon via, folded flex microelectronic package; Rumer, Christopher L.; Zarbock, Edward A.; August 2, 2005

1.11   References

[1] Moore, G., "Cramming More Components Onto Integrated Circuits," *Electronics*, Vol. 38, No. 8, April 19, 1965.

[2] Lau, J. H., *Reliability of RoHS-Compliant 2D and 3D IC Interconnects*, McGraw-Hill, New York, 2011.

[3] Lau, J. H., C. K. Lee, C. S. Premachandran, and A. Yu, *Advanced MEMS Packaging*, McGraw-Hill, New York, 2010.

[4] Lau, J. H., S. Wu, and J. M. Lau, "A Note on the Reliability of Wirebonding," *ASME Transactions, Journal of Electronic Packaging* (in press).

[5] Lau, J. H., *Flip Chip Technology*, McGraw-Hill, New York, 1995.

[6] Lau, J. H., *Low-Cost Flip Chip Technologies for WLCSP*, McGraw-Hill, New York, 2000.

[7] Akasaka, Y., "Three-dimensional IC Trends," *Proceedings of the IEEE*, Vol. 74, No. 12, December 1986, pp. 1703–1714.

[8] Gat, A., L. Gerzberg, J. Gibbons, T. Mages, J. Peng, and J. Hong, "CW Laser of Polyerystalline Silicon: Crystalline Structure and Electrical Properties," *Applied Physics Letter*, Vol. 33, No. 8, October 1978, pp. 775–778.

[9] Lau, J. H., "Overview and Outlook of TSV and 3D Integrations," *Journal of Microelectronics International*, Vol. 28, No. 2, 2011, pp. 8–22.

[10] Lau, J. H., "3D Integrations," Plenary Keynote at IEEE/EDAPS, Singapore, December 2010.

[11] Lau, J. H., "Evolution, Challenges, and Outlook of 3D IC/Si Integration," Plenary Keynote at IEEE ICEP, Nara, Japan, April 2011, pp. 1–16.

[12] Lau, J. H., "3D IC Integration and 3D Si Integration," Plenary Keynote at IWLPC, San Jose, CA, October 2011, pp. 1–18.

[13] Yu, A., J. H. Lau, Ho, S., Kumar, A., Yin, H., Ching, J., Kripesh, V., Pinjala, D., Chen, S., Chan, C., Chao, C., Chiu, C., Huang, M., and Chen, C., "Three-Dimensional Interconnects with High Aspect Ratio TSVs and Fine Pitch Solder Microbumps," *IEEE/ECTC Proceedings*, San Diego, CA, May 2009, pp. 350–354. Also, *IEEE Transactions in Advanced Packaging* (in press).

[14] Kang, U., H. Chung, S. Heo, D. Park, H. Lee, J. Kim, S. Ahn, S. Cha, J. Ahn, D. Kwon, J. Lee, H. Joo, W. Kim, D. Jang, N. Kim, J. Choi, T. Chung, J. Yoo, J. Choi, C. Kim, and Y. Jun, "8 Gb 3-D DDR3 DRAM Using Through-Silicon-Via Technology," *IEEE Journal of Solid-State Circuits*, Vol. 45, No. 1, January 2010, pp. 111–119.

[15] Farooq, M. G, T. L. Graves-Abe, W. F. Landers, C. Kothandaraman, B. A. Himmel, P. S. Andry, C. K. Tsang, E. Sprogis, R. P. Volant, K. S. Petrarca, K.

R. Winstel, J. M. Safran, T. D. Sullivan, F. Chen, M. J. Shapiro, R. Hannon, R. Liptak, D. Berger, and S. S. Iyer, "3D Copper TSV Integration, Testing and Reliability," *Proceedings of IEEE/IEDM*, Washington, DC, December 2011, pp. 7.1.1–7.1.4.

[16] Dorsey, P., "Xilinx Stacked Silicon Interconnect Technology Delivers Breakthrough FPGA Capacity, Bandwidth, and Power Efficiency," Xilinx White Paper: Virtex-7 FPGAs, WP380, October 27, 2010, pp. 1–10.

[17] Banijamali, B., S. Ramalingam, K. Nagarajan, and R. Chaware, "Advanced Reliability Study of TSV Interposers and Interconnects for the 28-nm Technology FPGA," *IEEE/ECTC Proceedings*, Orlando, FL, June 2011, pp. 285–290.

[18] Kim, N., D. Wu, D. Kim, A. Rahman, and P. Wu, "Interposer Design Optimization for High Frequency Signal Transmission in Passive and Active Interposer Using Through Silicon Via (TSV)," *IEEE/ECTC Proceedings*, Orlando, FL, June 2011, pp. 1160–1167.

[19] Lau, J. H., M. Dai, Y. Chao, W. Li, S. Wu, J. Hung, M. Hsieh, J. Chien, R. Tain, C. Tzeng, K. Lin, E. Hsin, C. Chen, M. Chen, C. Wu, J. Chen, J. Chien, C. Chiang, Z. Lin, L. Wu, H. Chang, W. Tsai, C. Lee, T. Chang, C. Ko, T. Chen, S. Sheu, S. Wu, Y. Chen, R. Lo, T. Ku, M. Kao, F. Hsieh, and D. Hu, "Feasibility Study of a 3D IC Integration System-in-Packaging (SiP)," *IEEE/ICEP Proceedings*, Nara, Japan, April 13, 2011, pp. 210–216.

[20] Lau, J. H., C-J Zhan, P.-J. Tzeng, C.-K. Lee, M.-J. Dai, H.-C. Chien, Y.-L. Chao, W. Li, S.-T. Wu, J.-F. Hung, R.-M. Tain, C.-H. Lin, Y.-C. Hsin, C.-C. Chen, S.-C. Chen, C.-Y. Wu, J.-C. Chen, C.-H. Chien, C.-W. Chiang, H. Chang, W.-L. Tsai, R.-S. Cheng, S.-Y. Huang, Y.-M. Lin, T.-C. Chang, C.-D. Ko, T.-H. Chen, S.-S. Sheu, S.-H. Wu, Y.-H. Chen, W.-C. Lo, T.-K. Ku, M.-J. Kao, and D.-Q. Hu, "Feasibility Study of a 3D IC Integration System-in-Packaging (SiP) from a 300-mm Multi-Project Wafer (MPW)," *Proceedings of IMAPS International Conference*, Long Beach, CA, October 2011, pp. 446–454.

[21] Zhan, C.-J., J. H. Lau, P.-J. Tzeng, C.-K. Lee, M.-J. Dai, H.-C. Chien, Y.-L. Chao, W. Li, S.-T. Wu, J.-F. Hung, R.-M. Tain, C.-H. Lin, Y.-C. Hsin, C.-C. Chen, S.-C. Chen, C.-Y. Wu, J.-C. Chen, C.-H. Chien, C.-W. Chiang, H. Chang, W.-L. Tsai, R.-S. Cheng, S.-Y. Huang, Y.-M. Lin, T.-C. Chang, C.-D. Ko, T.-H. Chen, S.-S. Sheu, S.-H. Wu, Y.-H. Chen, W.-C. Lo, T.-K. Ku, M.-J. Kao, and D.-Q. Hu, "Assembly Process and Reliability Assessment of TSV/RDL/IPD Interposer with Multi-Chip-Stacking for 3D IC Integration SiP," *IEEE/ECTC Proceedings* (in press).

[22] Sheu, S., Z. Lin, J. Hung', J. H. Lau, P. Chen, S. Wu, K. Su, C. Lin, S. Lai, T. Ku, W. Lo, M. Kao, "An Electrical Testing Method for Blind Through Silicon Vias (TSVs) for 3D IC Integration," *Proceedings of IMAPS International Conference*, Long Beach, CA, October 2011, pp. 208–214.

[23] Wu, C., S. Chen, P. Tzeng, J. H. Lau, Y. Hsu, J. Chen, Y. Hsin, C. Chen, S. Shen, C. Lin, T. Ku, and M. Kao, "Oxide Liner, Barrier and Seed Layers, and Cu-Plating of Blind Through Silicon Vias (TSVs) on 300-mm Wafers for 3D IC Integration," *Proceedings of IMAPS International Conference*, Long Beach, CA, October 2011, pp. 1–7.

[24] Chien, H., J. H. Lau, Y. Chao, R. Tain, M. Dai, S. Wu, W. Lo, and M. Kao, "Thermal Performance of 3D IC Integration with Through-Silicon Via (TSV)," *Proceedings of IMAPS International Conference*, Long Beach, CA, October 2011, pp. 25–32.

[25] Chang, H. H., J. H. Lau, W. L. Tsai, C. H. Chien, P. J. Tzeng, C. J. Zhan, C. K. Lee, M. J. Dai, H. C. Fu, C. W. Chiang, T. Y. Kuo, Y. H. Chen, W. C. Lo, T. K. Ku, and M. J. Kao, "Thin Wafer Handling of 300-mm Wafer for 3D IC Integration," *Proceedings of IMAPS International Conference*, Long Beach, CA, October 2011, pp. 202–207.

[26] Lau, J. H., P.-J. Tzeng, C.-K. Lee, C.-J. Zhan, M.-J. Dai, Li Li, C.-T. Ko, S.-W. Chen, H. Fu, Y. Lee, Z. Hsiao, J. Huang, W. Tsai, P. Chang, S. Chung, Y. Hsu, S.-C. Chen, Y.-H. Chen, T.-H. Chen, W.-C. Lo, T.-K. Ku, M.-J. Kao, J. Xue, and M. Brillhart, "Wafer Bumping and Characterizations of Fine-Pitch Lead-Free Solder

Microbumps on 12-inch (300-mm) wafer for 3D IC Integration," *Proceedings of IMAPS International Conference*, Long Beach, CA, October 2011, pp. 650–656.

[27] Hsin, Y. C., C. Chen, J. H. Lau, P. Tzeng, S. Shen, Y. Hsu, S. Chen, C. Wn, J. Chen, T. Ku, and M. Kao, "Effects of Etch Rate on Scallop of Through-Silicon Vias (TSVs) in 200-mm and 300-mm Wafers," *IEEE/ECTC Proceedings*, Orlando, FL, June 2011, pp. 1130–1135.

[28] Chen, J. C., J. H. Lau, P. J. Tzeng, S. C. Chen, C. Y. Wu, C. C. Chen, C. H. Lin, Y. C. Hsin, T. K. Ku, and M. J. Kao, "Impact of Slurry in Cu CMP (Chemical Mechanical Polishing) on Cu Topography of Through Silicon Vias (TSVs), Re-distributed Layers, and Cu Exposure," *IEEE/ECTC Proceedings*, Orlando, FL, June 2011, pp. 1389–1394. Also, *IEEE Transactions on CPMT* (in press).

[29] Chien, J., Y. Chao, J. H. Lau, M. Dai, R. Tain, M. Dai, P. Tzeng, C. Lin, Y. Hsin, S. Chen, J. Chen, C. Chen, C. Ho, R. Lo, T. Ku, and M. Kao, "A Thermal Performance Measurement Method for Blind Through Silicon Vias (TSVs) in a 300-mm Wafer," *IEEE/ECTC Proceedings*, Orlando, FL, June 2011, pp. 1204–1210.

[30] Tsai, W., H. H. Chang, C. H. Chien, J. H. Lau, H. C. Fu, C. W. Chiang, T. Y. Kuo, Y. H. Chen, R. Lo, and M. J. Kao, "How to Select Adhesive Materials for Temporary Bonding and De-Bonding of Thin-Wafer Handling in 3D IC Integration?" *IEEE/ECTC Proceedings*, Orlando, FL, June 2011, pp. 989–998.

[31] Chang, H., J. Huang, C. Chiang, Z. Hsiao, H. Fu, C. Chien, Y. Chen, W. Lo, and K. Chiang, "Process Integration and Reliability Test for 3D Chip Stacking with Thin Wafer Handling Technology," *IEEE/ECTC Proceedings*, Orlando, FL, June 2011, pp. 304–311.

[32] Lee, C. K., T. C. Chang, Y. Huang, H. Fu, J. H. Huang, Z. Hsiao, J. H. Lau, C. T. Ko, R. Cheng, K. Kao, Y. Lu, R. Lo, and M. J. Kao, "Characterization and Reliability Assessment of Solder Microbumps and Assembly for 3D IC Integration," *IEEE/ECTC Proceedings*, Orlando, FL, June 2011, pp. 1468–1474.

[33] Zhan, C., J. Juang, Y. Lin, Y. Huang, K. Kao, T. Yang, S. Lu, J. H. Lau, T. Chen, R. Lo, and M. J. Kao, "Development of Fluxless Chip-on-Wafer Bonding Process for 3D chip Stacking with 30μm Pitch Lead-Free Solder Micro Bump Interconnection and Reliability Characterization," *IEEE/ECTC Proceedings*, Orlando, FL, June 2011, pp. 14–21.

[34] Cheng, R., K. Kao, J. Chang, Y. Hung, T. Yang, Y. Huang, S. Chen, T. Chang, Q. Hunag, R. Guino, G. Hoang, J. Bai, and K. Becker, "Achievement of low Temperature Chip Stacking by a Pre-Applied Underfill Material," *IEEE/ECTC Proceedings*, Orlando, FL, June 2011, pp. 1858–1863.

[35] Huang, S., T. Chang, R. Cheng, J. Chang, C. Fan, C. Zhan, J. H. Lau, T. Chen, R. Lo, and M. Kao, "Failure Mechanism of 20-μm Pitch Microjoint Within a Chip Stacking Architecture," *IEEE/ECTC Proceedings*, Orlando, FL, June 2011, pp. 886–892.

[36] Lin, Y., C. Zhan, J. Juang, J. H. Lau, T. Chen, R. Lo, M. Kao, T. Tian, and K. N. Tu, "Electromigration in Ni/Sn Intermetallic Micro Bump Joint for 3D IC Chip Stacking," *IEEE/ECTC Proceedings*, Orlando, FL, June 2011, pp. 351–357.

[37] Selvanayagam, C., J. H. Lau, X. Zhang, S. Seah, K. Vaidyanathan, and T. Chai, "Nonlinear Thermal Stress/Strain Analysis of Copper Filled TSV (Through Silicon Via) and Their Flip-Chip Microbumps," *IEEE/ECTC Proceedings*, Orlando, FL, May 27–30, 2008, pp. 1073–1081. Also, *IEEE Transactions on Advanced Packaging*, Vol. 32, No. 4, November 2009, pp. 720–728.

[38] Zhang, X., T. Chai, J. H. Lau, C. Selvanayagam, K. Biswas, S. Liu, D. Pinjala, G. Tang, Y. Ong, S. Vempati, E. Wai, H. Li, B. Liao, N. Ranganathan, V. Kripesh, J. Sun, J. Doricko, and C. Vath, "Development of Through Silicon Via (TSV) Interposer Technology for Large Die (21 × 21 mm) Fine-Pitch Cu/Low-*k* FCBGA Package," *IEEE/ECTC Proceedings*, May 2009, pp. 305–312. Also, *IEEE Transactions on CPMT*, 2011, pp. 660–672.

[39] Khan, N., V. Rao, S. Lim, S. Ho, V. Lee, X. Zhang, R. Yang, E. Liao, Ranganathan, T. Chai, V. Kripesh, and J. H. Lau, "Development of 3D Silicon Module with TSV for System in Packaging," *IEEE/ECTC Proceedings*, May 2008, pp. 550–555.

Also, *IEEE Transactions on Components, Packaging and Manufacturing Technology,* Vol. 33, No. 1, March 2010, pp. 3–9.

[40] Yu, A., N. Khan, G. Archit, D. Pinjalal, K. Toh, V. Kripesh, S. Yoon, and J. H. Lau, "Development of silicon carriers with embedded thermal solutions for high power 3D package," *IEEE/ECTC Proceedings,* May 2008, pp. 24–28. Also, *IEEE Transactions on Components and Packaging Technology,* Vol. 32, No. 3, September 2009, pp. 566–571.

[41] Li, L., P. Su, J. Xue, M. Brillhart, J. H. Lau, P. Tzeng, C. Lee, C. Zhan, M. Dai, H. Chien, and S. Wu, "Addressing the Bandwidth Challenges in Next-Generation High-Performance Network Systems with 3D IC Integration," *IEEE/ECTC Proceedings* , 2012.

[42] Chien, H. C., J. H. Lau, Y. Chao, M. Dai, R. Tain, L. Li, P. Su, J. Xue, M. Brillhart, et al., "Thermal Evaluation and Analyses of 3D IC Integration SiP for Network System Applications," *IEEE/ECTC Proceedings,* 2012.

[43] Lau, J. H., *Ball Grid Array Technology,* McGraw-Hill, New York, 1995.

[44] Lau, J. H., Y. Chan, and R. Lee, "3D IC Integration with TSV Interposers for High-Performance Applications," *Chip Scale Review,* September–October, 2010, pp. 26–29.

[45] Lau, J. H., "TSV Interposers: The Most Cost-Effective Integrator for 3D IC Integration," *Chip Scale Review,* September–October, 2011, pp. 23–27.

[46] Wu, S., J. H. Lau, H. Chien, R. Tain, et al., "Ultra Low-Cost Through Silicon Holes (TSHs) Interposers for 3D IC Integration," *IEEE/ECTC Proceedings,* 2012.

[47] Lau, J. H., H. C. Chien, and R. Tain, "TSV Interposers with Embedded Microchannels for 3D IC and Multiple High-Power LEDs Integration SiP," ASME Paper no. InterPACK2011-52204.

[48] Lau, J. H., M. S. Zhang, and S. W. R. Lee, "Embedded 3D Hybrid IC Integration System-in-Package (SiP) for Opto-Electronic Interconnects in Organic Substrates," *ASME Transactions, Journal of Electronic Packaging,* Vol. 133, September 2011, pp. 1–7.

[49] Lau, J. H., *Solder Joint Reliability: Theory and Applications,* Van Nostrand Reinhold, New York, 1991.

1.12   General Readings

1.12.1   TSV and 3D Integration and Reliability

1. Moore, G., "Cramming More Components Onto Integrated Circuits," *Electronics,* Vol. 38, No. 8, April 19, 1965.
2. Lau, J. H., *Reliability of RoHS-Compliant 2D and 3D IC Interconnects,* McGraw-Hill, New York, 2011.
3. Lau, J. H., C. K. Lee, C. S. Premachandran, and A. Yu, *Advanced MEMS Packaging,* McGraw-Hill, New York, 2010.
4. Lau, J. H., "Overview and Outlook of TSV and 3D Integrations," *Journal of Microelectronics International,* Vol. 28, No. 2, 2011, pp. 8–22.
5. Lau, J. H., M. S. Zhang, and S. W. R. Lee, "Embedded 3D Hybrid IC Integration System-in-Package (SiP) for Opto-Electronic Interconnects in Organic Substrates," *ASME Transactions, Journal of Electronic Packaging,* Vol. 133, September 2011, pp. 031010: 1-7.
6. Lau, J. H., "Critical Issues of 3D IC Integrations," *IMAPS Transactions, Journal of Microelectronics and Electronic Packaging,* First Quarter, 2010, pp. 35–43.
7. Lau, J. H., "Design and Process of 3D MEMS Packaging," *IMAPS Transactions, Journal of Microelectronics and Electronic Packaging,* First Quarter, 2010, pp. 10–15.
8. Lau, J. H., R. Lee, M. Yuen, M., and P. Chan, "3D LED and IC Wafer Level Packaging," *Journal of Microelectronics International,* Vol. 27, No. 2, 2010, pp. 98–105.

9. Lau, J. H., "State-of-the-art and Trends in 3D Integration," *Chip Scale Review*, March–April 2010, pp. 22–28.
10. Lau, J. H., "TSV Manufacturing Yield and Hidden Costs for 3D IC Integration," *IEEE/ECTC Proceedings*, Las Vegas, NV, June 2010, pp. 1031–1041.
11. Lau, J. H., Y. S. Chan, and R. S. W. Lee, "3D IC Integration with TSV Interposers for High-Performance Applications," *Chip Scale Review*, September–October 2010, pp. 26–29.
12. Yu, A., J. H. Lau, S. Ho, A. Kumar, Y. Wai, D. Yu, M. Jong, V. Kripesh, D. Pinjala, and D. Kwong, "Study of 15-μm-Pitch Solder Microbumps for 3D IC Integration," *IEEE/ECTC Proceedings*, San Diego, CA, May 2009, pp. 6–10.
13. Yu, A., J. H. Lau, S. Ho, A. Kumar, H. Yin, J. Ching, V. Kripesh, D. Pinjala, S. Chen, C. Chan, C. Chao, C. Chiu, M. Huang, and C. Chen, "Three Dimensional Interconnects with High Aspect Ratio TSVs and Fine Pitch Solder Microbumps," *IEEE/ECTC Proceedings*, San Diego, CA, May 2009, pp. 350–354. Also, *IEEE Transactions in Advanced Packaging* (accepted).
14. Yu, A., A. Kumar, S. Ho, H. Yin, J. H. Lau, J. Ching, V. Kripesh, D. Pinjala, S. Chen, C. Chan, C. Chao, C. Chiu, M. Huang, and C. Chen, "Development of Fine Pitch Solder Microbumps for 3D Chip Stacking," *IEEE/EPTC Proceedings*, Singapore, December 2008, pp. 387–392. Also, *IEEE Transactions in Advanced Packaging* (accepted).
15. Yu, A., N. Khan, G. Archit, D. Pinjalal, K. Toh, V. Kripesh, S. Yoon, and J. H. Lau, "Development of Silicon Carriers with Embedded Thermal Solutions for High Power 3D Package," *IEEE Transactions on Components and Packaging Technology*, Vol. 32, No. 3, September 2009, pp. 566–571.
16. Tang, G., O. Navas, D. Pinjala, J. H. Lau, A. Yu, and V. Kripesh, "Integrated Liquid Cooling Systems for 3D Stacked TSV Modules," *IEEE Transactions on Components and Packaging Technologies*, Vol. 33, No. 1, 2010, pp. 184–195.
17. Chen, K., C. Premachandran, K. Choi, C. Ong, X. Ling, A. Khairyanto, B. Ratmin, P. Myo, and J. H. Lau, "C2W Bonding Method for MEMS Applications," *IEEE/ECTC Proceedings*, December 2008, pp. 1283–1287.
18. Premachandran, C. S., J. H. Lau, X. Ling, A. Khairyanto, K. Chen, and Myo Ei Pa Pa, "A Novel, Wafer-Level Stacking Method for Low-Chip Yield and Non-Uniform, Chip-Size Wafers for MEMS and 3D SIP Applications," *IEEE/ECTC Proceedings*, Orlando, FL, May 27–30, 2008, pp. 314–318.
19. Chai, T., X. Zhang, J. H. Lau, C. Selvanayagam, K. Biswas, S. Liu, D. Pinjala, G. Tang, Y. Ong, S. Vempati, E. Wai, H. Li, B. Liao, N. Ranganathan, V. Kripesh, J. Sun, J. Doricko, and C. Vath, "Development of Large Die Fine-Pitch Cu/Low-k FCBGA Package with Through Silicon Via (TSV) Interposer," *IEEE/ECTC Proceedings*, May 2009, pp. 305–312. Also, *IEEE Transactions on CPMT*, Vol. 1, No. 5, 2011, pp. 660-672.
20. Hoe, G., G. Tang, P. Damaruganath, C. Chong, J. H. Lau, X. Zhang, and K. Vaidyanathan, "Effect of TSV Interposer on the Thermal Performance of FCBGA Package," *IEEE/ECTC Proceedings*, Singapore, December 2009, pp. 778–786.
21. Choi, W. O., C. S. Premachandran, S. Ong, Ling, X., E. Liao, K. Ahmad, B. Ratmin, K. Chen, P. Thaw, and J. H. Lau, "Development of Novel Intermetallic Joints Using Thin Film Indium Based Solder by Low Temperature Bonding Technology for 3D IC Stacking," *IEEE/ECTC Proceedings*, San Diego, CA, May 2009, pp. 333–338.
22. Kumar, A., X. Zhang, Q. Zhang, M. Jong, G. Huang, V. Kripesh, C. Lee, J. H. Lau, D. Kwong, V. Sundaram, R. Tummula, and M. Georg, "Evaluation of Stresses in Thin Device Wafer Using Piezoresistive Stress Sensor," *IEEE Proceedingsof EPTC, December 2008*, , pp. 1270–1276. Also, *IEEE Transactions in Components and Packaging Technologies* Vol. 1, No. 6, June 2011, pp. 841-850.
23. Khan, N., L. Yu, P. Tan, S. Ho, N. Su, H. Wai, K. Vaidyanathan, D. Pinjala, J. H. Lau, T. Chuan, "3D Packaging with Through Silicon Via (TSV) for Electrical and Fluidic Interconnections," *IEEE/ECTC Proceedings*, San Diego, CA, May 2009, pp. 1153–1158.

24. Sekhar, V. N., S. Lu, A. Kumar, T. C. Chai, V. Lee, S. Wang, X. Zhang, C. S. Premchandran, V. Kripesh, and J. H. Lau, "Effect of Wafer Back Grinding on the Mechanical Behavior of Multilayered Low-k for 3D-Stack Packaging Applications," *IEEE/ECTC Proceedings*, Orlando, FL, May 27–30, 2008, pp. 1517–1524. Also, *IEEE Transactions on Components and Packaging Technologies* (in press).

25. Khan, N., V. Rao, S. Lim, S. Ho, V. Lee, X. Zhang, R. Yang, E. Liao, Ranganathan, T. Chai, V. Kripesh, and J. H. Lau, "Development of 3D Silicon Module with TSV for System in Packaging," *IEEE/ECTC Proceedings*, Orlando, FL, May 27–30, 2008, pp. 550–555.

26. Ho, S., S. Yoon, Q. Zhou, K. Pasad, V. Kripesh and J. H. Lau, "High RF Performance TSV for Silicon Carrier for High Frequency Application," *IEEE/ECTC Proceedings*, Orlando, FL, May 27–30, 2008, pp. 1956–1962.

27. Lim, S., V. Rao, H. Yin, W. Ching, V. Kripesh, C. Lee, J. H. Lau, J. Milla, and A. Fenner, "Process Development and Reliability of Microbumps," *IEEE/ECTC Proceedings*, December 2008, pp. 367–372. Also, *IEEE Transactions in Components and Packaging Technology*, Vol. 33, No. 4, December 2010, pp. 747-753.

28. Selvanayagam, C., J. H. Lau, X. Zhang, S. Seah, K. Vaidyanathan, and T. Chai, "Nonlinear Thermal Stress/Strain Analysis of Copper Filled TSV (Through Silicon Via) and Their Flip-Chip Microbumps," *IEEE/ECTC Proceedings*, Orlando, FL, May 27–30, 2008, pp. 1073–1081. Also, *IEEE Transactions on Advanced Packaging*, Vol. 32, No. 4, November 2009, pp. 720–728.

29. Zhang, X., A. Kumar, Q. X. Zhang, Y. Y. Ong, S. W. Ho, C. H. Khong, V. Kripesh, J. H. Lau, D.-L. Kwong, V. Sundaram, Rao R. Tummula, and G. Meyer-Berg, "Application of Piezoresistive Stress Sensors in Ultra Thin Device Handling and Characterization," *Journal of Sensors & Actuators: A. Physical*, Vol. 156, November 2009, pp. 2–7.

30. Lau, J. H., T. G. Yue, G. Y. Y. Hoe, X.W. Zhang, C. T. Chong, P. Damaruganath, and K. Vaidyanathan, "Effects of TSV (Through Silicon Via) Interposer/Chip on the Thermal Performances of 3D IC Packaging," ASME Paper no. IPACK2009–89380.

31. Lau, J. H., "State-of-the-Art and Trends in Through-Silicon Via (TSV) and 3D Integrations," ASME Paper no. IMECE2010-37783.

32. Lau, J. H., and G. Tang, "Thermal Management of 3D IC Integration with TSV (Through Silicon Via)," *IEEE/ECTC Proceedings*, San Diego, May 2009, pp. 635–640.

33. Carson, F., K. Ishibashi, S. Yoon, P. Marimuthu, and D. Shariff, "Development of Super Thin TSV PoP," *Proceedings of IEEE CPMT Symposium Japan*, August 2010, pp. 7–10.

34. Kohara, S., K. Sakuma, Y. Takahashi, T. Aoki, K. Sueoka, K. Matsumoto, P. Andry, C. Tsang, E. Sprogis, J. Knickerbocker, and Y. Orii, "Thermal Stress Analysis of 3D Die Stacks with Low-Volume Interconnections," *Proceedings of IEEE CPMT Symposium Japan*, August 2010, pp. 165–168.

35. Sekiguchi, M., H. Numata, N. Sato, T. Shirakawa, M. Matsuo, H. Yoshikawa, M. Yanagida, H. Nakayoshi, and K. Takahashi, "Novel Low Cost Integration of Through Chip Interconnection and Application to CMOS Image Sensor," *IEEE/ECTC Proceedings*, San Diego, CA, May 2006, pp. 1367–1374.

36. Takahashi, K., and M. Sekiguchi, "Through Silicon Via and 3D Wafer/Chip Stacking Technology," *IEEE Proceedings of Symposium on VLSI Circuits Digest of Technical Papers*, 2006, pp. 89–92.

37. Juergen, M., K. Wolf, A. Zoschke, R. Klumpp, M. Wieland, L. Klein, A. Nebrich, I. Heinig, W. Limansyah, O. Weber, O. Ehrmann, and H. Reichl, "3D Integration of Image Sensor SiP Using TSV Silicon Interposer," *IEEE/ECTC Proceedings*, December 2009, pp. 795–800.

38. Limansyah, I., M. J. Wolf, A. Klumpp, K.Zoschke, R. Wieland, M. Klein, H. Oppermann, L. Nebrich, A. Heinig, A. Pechlaner, H. Reichl, and W. Weber, "3D Image Sensor SiP with TSV Silicon Interposer," *IEEE/ECTC Proceedings*, May 2009, pp. 1430–1436.

39. Garrou, P., C. Bower, and P. Ramm, *3D Integration: Technology and Applications*, Wiley, Hoboken, NJ, 2009.
40. Ramm, P., M. Wolf, A. Klumpp, R. Wieland, B. Wunderle, B. Michel, and H. Reichl, "Through Silicon Via Technology: Processes and Reliability for Wafer-Level 3D System Integration," *IEEE/ECTC Proceedings*, Orlando, FL, May 2008, pp. 847–852.
41. Andry, P. S., C. K. Tsang, B. C. Webb, E. J. Sprogis, S. L. Wright, B. Bang, and D. G. Manzer, "Fabrication and Characterization of Robust Through-Silicon Vias for Silicon-Carrier Applications," *IBM Journal of Research and Development*, Vol. 52, No. 6, 2008, pp. 571–581.
42. Knickerbocker, J. U., P. S. Andry, B. Dang, R. R. Horton, C. S. Patel, R. J. Polastre, K. Sakuma, E. S. Sprogis, C. K. Tsang, B. C. Webb, and S. L. Wright, "3D Silicon Integration," *IEEE/ECTC Proceedings*, May 2008, pp. 538–543.
43. Kumagai, K., Y. Yoneda, H. Izumino, H. Shimojo, M. Sunohara, and T. Kurihara, "A Silicon Interposer BGA Package with Cu-Filled TSV and Multilayer Cu-Plating Interconnection," *IEEE/ECTC Proceedings*, Orlando, FL, May 2008, pp. 571–576.
44. Sunohara, M., T. Tokunaga, T. Kurihara, and M. Higashi, "Silicon Interposer with TSVs (Through Silicon Vias) and Fine Multilayer Wiring," *IEEE/ECTC Proceedings*, Orlando, FL, May 2008, pp. 847–852.
45. Lee, H. S., Y.-S. Choi, E. Song, K. Choi, T. Cho, and S. Kang, "Power Delivery Network Design for 3D SIP Integrated over Silicon Interposer Platform," *IEEE/ECTC Proceedings*, Reno, NV, May 2007, pp. 1193–1198.
46. Matsuo, M., N. Hayasaka, and K. Okumura, "Silicon Interposer Technology for High-Density Package," *IEEE/ECTC Proceedings*, May 2000, pp. 1455–1459.
47. Wong, E., J. Minz, and S. K. Lim, "Effective Thermal Via and Decoupling Capacitor Insertion for 3D System-on-Package," *IEEE/ECTC Proceedings*, San Diego, CA, May 2006, pp. 1795–1801.
48. Kang, U., H. Chung, S. Heo, D. Park, H. Lee, J. Kim, S. Ahn, S. Cha, J. Ahn, D. Kwon, J. Lee, H. Joo, W. Kim, D. Jang, N. Kim, J. Choi, T. Chung, J. Yoo, J. Choi, C. Kim, and Y. Jun, "8-Gb 3D DDR3 DRAM Using Through-Silicon-Via Technology," *IEEE Journal of Solid-State Circuits*, Vol. 45, No. 1, January 2010, pp. 111–119.
49. Lau, J. H., Y. Chan, and R. Lee, "Thermal-Enhanced and Cost-Effective 3D IC Integration with TSV (Through-Silicon Via) Interposers for High-Performance Applications," ASME Paper no. IMECE2010-40975.
50. Shi, X., P. Sun, Y. Tsui, P. Law, S. Yau, C. Leung, Y. Liu, C. Chung, S. Ma, M. Miao, and Y. Jin, "Development of CMOS-Process-Compatible Interconnect Technology for 3D Stacking of NAND Flash Memory Chips," *IEEE/ECTC Proceedings*, Las Vegas, NV, June 2010, pp. 74–78.
51. Kikuchi, K., C. Ueda, K. Takemura, O. Shimada, T. Gomyo, Y. Takeuchi, T. Ookubo, K. Baba, M. Aoyagi, T. Sudo, and K. Otsuka, "Low-Impedance Evaluation of Power Distribution Network for Decoupling Capacitor Embedded Interposers of 3D Integrated LSI System," *IEEE/ECTC Proceedings*, Las Vegas, NV, June 2010, pp. 1455–1460.
52. Sridharan, V., S. Min, V. Sundaram, V. Sukumaran, S. Hwang, H. Chan, F. Liu, C. Nopper, and R. Tummala, "Design and Fabrication of Bandpass Filters in Glass Interposer with Through-Package-Vias (TPV)," *IEEE/ECTC Proceedings*, Las Vegas, NV, June 2010, pp. 530–535.
53. Sukumaran, V., Q. Chen, F. Liu, N. Kumbhat, T. Bandyopadhyay, H. Chan, S. Min, C. Nopper, V. Sundaram, and R. Tummala, "Through-Package-Via Formation and Metallization of Glass Interposers," *IEEE/ECTC Proceedings*, Las Vegas, NV, June 2010, pp. 557–563.
54. Sakuma, K., K. Sueoka1, S. Kohara, K. Matsumoto, H. Noma, T. Aoki, Y. Oyama, H. Nishiwaki, P. S. Andry, C. K. Tsang, J. Knickerbocker, and Y. Orii, "IMC Bonding for 3D Interconnection," *IEEE/ECTC Proceedings*, Las Vegas, NV, June 2010, pp. 864–871.
55. Doany, F., B. Lee, C. Schow, C. Tsang, C. Baks, Y. Kwark, R. John, J. Knickerbocker, and J. Kash, "Terabit/s-Class 24-Channel Bidirectional Optical

Transceiver Module Based on TSV Si Carrier for Board-Level Interconnects," *IEEE/ECTC Proceedings*, Las Vegas, NV, June 2010, pp. 58–65.

56. Khan, N., D. Wee, O. Chiew, C. Sharmani, L. Lim, H. Li, and S. Vasarala, "Three Chips Stacking with Low Volume Solder Using Single Re-Flow Process," *IEEE/ECTC Proceedings*, Las Vegas, NV, June 2010, pp. 884–888.

57. Trigg, A., L. Yu, X. Zhang, C. Chong, C. Kuo, N. Khan, and D. Yu, "Design and Fabrication of a Reliability Test Chip for 3D-TSV," *IEEE/ECTC Proceedings*, Las Vegas, NV, June 2010, pp. 79–83.

58. Agarwal, R., W. Zhang, P. Limaye, R. Labie, B. Dimcic, A. Phommahaxay, and P. Soussan, "Cu/Sn Microbumps Interconnect for 3D TSV Chip Stacking," *IEEE/ECTC Proceedings*, Las Vegas, NV, June 2010, pp. 858–863.

59. Töpper, M., I. Ndip, R. Erxleben, L. Brusberg, N. Nissen, H. Schröder, H. Yamamoto, G. Todt, and H. Reichl, "3D Thin Film Interposer Based on TGV (Through Glass Vias): An Alternative to Si-Interposer," *IEEE/ECTC Proceedings*, Las Vegas, NV, June 2010, pp. 66–73.

60. Bouchoucha, M., P. Chausse, D. Henry, and N. Sillon, "Process Solutions and Polymer Materials for 3D-WLP Through Silicon Via Filling," *IEEE/ECTC Proceedings*, Las Vegas, NV, June 2010, pp. 1696–1698.

61. Liu, H., K. Wang, K. Aasmundtveit, and N. Hoivik, "Intermetallic Cu3Sn as Oxidation Barrier for Fluxless Cu-Sn Bonding," *IEEE/ECTC Proceedings*, Las Vegas, NV, June 2010, pp. 853–857.

62. Kang, I., G. Jung, B. Jeon, J. Yoo, and S. Jeong, "Wafer Level Embedded System in Package (WL-eSiP) for Mobile Applications," *IEEE/ECTC Proceedings*, Las Vegas, NV, June 2010, pp. 309–315.

63. Reed, J., M. Lueck, C. Gregory, A. Huffman, J. Lannon, Jr., and D. Temple, "High Density Interconnect at 10-μm Pitch with Mechanically Keyed Cu/Sn-Cu and Cu-Cu Bonding for 3D Integration," *IEEE/ECTC Proceedings*, Las Vegas, NV, June 2010, pp. 846–852.

64. Kraft, J., F. Schrank, J. Teva, J. Siegert, G. Koppitsch, C. Cassidy, E. Wachmann, F. Altmann, S. Brand, C. Schmidt, and M. Petzold, "3D Sensor Application with Open Through Silicon Via Technology," *IEEE ECTC Proceedings*, Orlando, FL, June 2011, pp. 560–566.

65. Wang, Y., and T. Suga, "Influence of Bonding Atmosphere on Low-Temperature Wafer Bonding," *IEEE/ECTC Proceedings*, Las Vegas, NV, June 2010, pp. 435–439.

66. Au, K., S. Kriangsak, X. Zhang, W. Zhu, and C. Toh, "3D Chip Stacking and Reliability Using TSV-Micro C4 Solder Interconnection," *IEEE/ECTC Proceedings*, Las Vegas, NV, June 2010, pp. 1376–1384.

67. Hsieh, M. C., S. Wu, C. Wu, J. H. Lau, R. Tain, and R. Lo, "Investigation of Energy Release Rate for Through Silicon Vias (TSVs) in 3D IC integration," *EuroSimE Proceedings*, April 2011, pp. 1/7–7/7.

68. Hsin, Y. C., C. Chen, J. H. Lau, P. Tzeng, S. Shen, Y. Hsu, S. Chen, C. Wn, J. Chen, T. Ku, and M. Kao, "Effects of Etch Rate on Scallop of Through-Silicon Vias (TSVs) in 200-mm and 300-mm Wafers," *IEEE ECTC Proceedings*, Orlando, FL, June 2011, pp. 1130–1135.

69. Chen, J. C., P. J. Tzeng, S. C. Chen, C. Y. Wu, J. H. Lau, C. C. Chen, C. H. Lin, Y. C. Hsin, T. K. Ku, and M. J. Kao, "Impact of Slurry in Cu CMP (Chemical Mechanical Polishing) on Cu Topography of Through Silicon Vias (TSVs), Re-distributed Layers, and Cu Exposure," *IEEE/ECTC Proceedings*, Orlando, FL, June 2011, pp. 1389–1394.

70. Chien, J., Y. Chao, J. H. Lau, M. Dai, R. Tain, M. Dai, P. Tzeng, C. Lin, Y. Hsin, S. Chen, J. Chen, C. Chen, C. Ho, R. Lo, T. Ku, and M. Kao, "A Thermal Performance Measurement Method for Blind Through Silicon Vias (TSVs) in a 300-mm Wafer," *IEEE ECTC Proceedings*, Orlando, FL, June 2011, pp. 1204–1210.

71. Tsai, W., H. Chang, C. Chien, J. H. Lau, H. Fu, C. W. Chiang, T. Y. Kuo, Y. H. Chen, R. Lo, and M. J. Kao, "How to Select Adhesive Materials for Temporary Bonding and De-Bonding of Thin-Wafer Handling in 3D IC Integration," *IEEE/ECTC Proceedings*, Orlando, FL, June 2011, pp. 989–998.

72. Lee, C. K., T. Chang, Y. Huang, H. Fu, J. H. Huang, Z. Hsiao, J. H. Lau, C. T. Ko, R. Cheng, K. Kao, Y. Lu, R. Lo, and M. J. Kao, "Characterization and Reliability Assessment of Solder Microbumps and Assembly for 3D IC Integration," *IEEE ECTC Proceedings*, Orlando, FL, June 2011, pp. 1468–1474.
73. Lin, Y., C. Zhan, J. Juang, J. H. Lau, T. Chen, R. Lo, M. Kao, T. Tian, and K. N. Tu, "Electromigration in Ni/Sn Intermetallic Micro Bump Joint for 3D IC Chip Stacking," *IEEE/ECTC Proceedings*, Orlando, FL, June 2011, pp. 351–357.
74. Zhan, C., J. Juang, Y. Lin, Y. Huang, K. Kao, T. Yang, S. Lu, J. H. Lau, T. Chen, R. Lo, and M. J. Kao, "Development of Fluxless Chip-on-Wafer Bonding Process for 3D chip Stacking with 30µm Pitch Lead-Free Solder Micro Bump Interconnection and Reliability Characterization," *IEEE/ECTC Proceedings*, Orlando, FL, June 2011, pp. 14–21.
75. Huang, S., T. Chang, R. Cheng, J. Chang, C. Fan, C. Zhan, J. H. Lau, T. Chen, R. Lo, and M. Kao, "Failure Mechanism of 20-µm Pitch Micro Joint Within a Chip Stacking Architecture," *IEEE/ECTC Proceedings*, Orlando, FL, June 2011, pp. 886–892.
76. Zhang, X., R. Rajoo, C. Selvanayagam, A. Kumar, V. Rao, N. Khan, V. Kripesh, J. H. Lau, D.-L. Kwong, V. Sundaram, and R. Tummula, "Application of Piezoresistive Stress Sensor in Wafer Bumping and Drop Impact Test of Embedded Ultra Thin Device," *IEEE/ECTC Proceedings*, Orlando, FL, June 2011, pp. 1276–1282.
77. Dorsey, P., "Xilinx Stacked Silicon Interconnect Technology Delivers Breakthrough FPGA Capacity, Bandwidth, and Power Efficiency," Xilinx white paper: Virtex-7 FPGAs, WP380, October 27, 2010, pp. 1–10.
78. Banijamali, B., S. Ramalingam, K. Nagarajan, and R. Chaware, "Advanced Reliability Study of TSV Interposers and Interconnects for the 28-nm Technology FPGA," *IEEE/ECTC Proceedings*, Orlando, FL, June 2011, pp. 285–290.
79. Banijamali, B., S. Ramalingam, N. Kim, and R. Wyland, "Ceramics vs. Low CTE Organic Packaging of TSV Silicon Interposers," *IEEE/ECTC Proceedings*, Orlando, FL, June 2011, pp. 573–576.
80. Kim, N., D. Wu, D. Kim, A. Rahman, and P. Wu, "Interposer Design Optimization for High Frequency Signal Transmission in Passive and Active Interposer Using Through Silicon Via (TSV)," *IEEE/ECTC Proceedings*, Orlando, FL, June 2011, pp. 1160–1167.
81. Khan, N., V. Rao, S. Lim, S. Ho, V. Lee, X. Zhang, R. Yang, E. Liao, Ranganathan, T. Chai, V. Kripesh, and J. H. Lau, "Development of 3D Silicon Module with TSV for System in Packaging," *IEEE Transactions on Components, Packaging and Manufacturing Technology*, Vol. 33, No. 1, March 2010, pp. 3–9.
82. Lau, J. H., H. C. Chien, and R. Tain, "TSV Interposers with Embedded Microchannels for 3D IC and Multiple High-Power LEDs Integration SiP," ASME paper no. InterPACK2011-52204, July 2011.
83. Lau, J. H., and X. Zhang, "Effects of TSV Interposer on the Reliability of 3D IC Integration SiP," ASME paper no. InterPACK2011-52205, July 2011.
84. Lau, J. H., "The Most Cost-Effective Integrator (TSV Interposer) for 3D IC Integration SiP," ASME paper no. InterPACK2011-52189, July 2011.
85. Tummula, R., and M. Swaminathan, *System-on-Package: Miniaturization of the Entire System*, McGraw-Hill, New York, 2008.
86. Lau, J. H., M. Dai, Y. Chao, W. Li, S. Wu, J. Hung, M. Hsieh, J. Chien, R. Tain, C. Tzeng, K. Lin, E. Hsin, C. Chen, M. Chen, C. Wu, J. Chen, J. Chien, C. Chiang, Z. Lin, L. Wu, H. Chang, W. Tsai, C. Lee, T. Chang, C. Ko, T. Chen, S. Sheu, S. Wu, Y. Chen, R. Lo, T. Ku, M. Kao, F. Hsieh, and D. Hu, "Feasibility Study of a 3D IC Integration System-in-Packaging (SiP)," *IEEE/ICEP Proceedings*, Nara, Japan, April 13, 2011, pp. 210–216.
87. Lau, J. H., "Evolution, Outlook, and Challenges of 3D IC/Si Integration," *IEEE/ICEP Proceedings* (Keynote), Nara, Japan, April 13, 2011, pp. 1–17.
88. Chai, T., X. Zhang, J. H. Lau, C. Selvanayagam, K. Biswas, S. Liu, D. Pinjala, G. Tang, Y. Ong, S. Vempati, E. Wai, H. Li, B. Liao, N. Ranganathan, V. Kripesh, J. Sun, J. Doricko, and C. Vath, "Development of Large Die Fine-Pitch Cu/Low-k FCBGA Package with Through Silicon Via (TSV) Interposer," *IEEE Transactions*

on *Components, Packaging and Manufacturing Technology*, Vol. 1, No. 5, 2011, pp. 660–672.

89. Kannan, S., S. Evana, A. Gupta, B. Kim, and L. Li, "3D Copper Based TSV for 60-GHz Applications," *IEEE/ECTC Proceedings*, Orlando, FL, June 2011, pp. 1168–1175.

90. Spiller, S., F. Molina, J. Wolf, J. Grafe, A. Schenke, D. Toennies, M. Hennemeyer, T. Tabuchi, and H. Auer, "Processing of Ultrathin 300-mm Wafers with Carrierless Technology," *IEEE/ECTC Proceedings*, Orlando, FL, June 2011, pp. 984–988.

91. Choi, K., K. Sung, H. Bae, J. Moon, and Y. Eom, "Bumping and Stacking Processes for 3D IC using Fluxfree Polymer," *IEEE/ECTC Proceedings*, Orlando, FL, June 2011, pp. 1746–1751.

92. Sukumaran, V., T. Bandyopadhyay, Q. Chen, N. Kumbhat, F. Liu, R. Pucha, Y. Sato, M. Watanabe, K. Kitaoka, M. Ono, Y. Suzuki, C. Karoui, C. Nopper, M. Swaminathan, V. Sundaram, and R. Tummala, "Design, Fabrication and Characterization of Low-Cost Glass Interposers with Fine-Pitch Through-Package-Vias," *IEEE/ECTC Proceedings*, Orlando, FL, June 2011, pp. 583–588.

93. Zhang, Y. C. King, Jr., J. Zaveri, Y. Kim, V. Sahu, Y. Joshi, and M. Bakir, "Coupled Electrical and Thermal 3D IC Centric Microfluidic Heat Sink Design and Technology," *IEEE/ECTC Proceedings*, Orlando, FL, June 2011, pp. 2037–2044.

94. Parekh, M., P. Thadesar, and M. Bakir, "Electrical, Optical and Fluidic Through-Silicon Vias for Silicon Interposer Applications," *IEEE/ECTC Proceedings*, Orlando, FL, June 2011, pp. 1992–1998.

95. Chen, Q., T. Bandyopadhyay, Y. Suzuki, F. Liu, V. Sundaram, R. Pucha, M. Swaminathan, and R. Tummala, "Design and Demonstration of Low Cost, Panel-Based Polycrystalline Silicon Interposer with Through-Package-Vias (TPVs)," *IEEE/ECTC Proceedings*, Orlando, FL, June 2011, pp. 855–860.

96. Liu, X., Q. Chen, V. Sundaram, M. Simmons-Matthews4, K. Wachtler, R. Tummala, and S. Sitaraman, "Thermo-Mechanical Behavior of Through Silicon Vias in a 3D Integrated Package with Inter-Chip Microbumps," *IEEE/ ECTC Proceedings*, Orlando, FL, June 2011, pp. 1190–1195.

97. Mitra1, J., M. Jung, S. Ryu, R. Huang, S. Lim, and D. Pan, "A Fast Simulation Framework for Full-chip Thermo-Mechanical Stress and Reliability Analysis of Through-Silicon-Via-Based 3D ICs," *IEEE/ECTC Proceedings*, Orlando, FL, June 2011, pp. 746–753.

98. Zhang, R., S. Lee, D. Xiao, and H. Chen, "LED Packaging Using Silicon Substrate with Cavities for Phosphor Printing and Copper-filled TSVs for 3D Interconnection," *IEEE/ECTC Proceedings*, Orlando, FL, June 2011, pp. 1616–1621.

99. Zhou, Z., C. Liu, X. Wang, X. Luo, and S. Liu, "Integrated Process for Silicon Wafer Thinning," *IEEE/ECTC Proceedings*, Orlando, FL, June 2011, pp. 1811–1814.

100. Wan, Z., X. Luo, and S. Liu, "Effect of Blind Hole Depth and Shape of Solder Joint on the Reliability of Through Silicon Via (TSV)," *IEEE/ECTC Proceedings*, Orlando, FL, June 2011, pp. 1657–1661.

101. Chen, Z., S. Zhou, Z. Lv, C. Liu, X. Chen, X. Jia, K. Zeng, B. Song , F. Zhu, M. Chen, X. Wang, H. Zhang, and S. Liu, "Expert Advisor for Integrated Virtual Manufacturing and Reliability for TSV/SiP Based Modules," *IEEE/ECTC Proceedings*, Orlando, FL, June 2011, pp. 1183–1189.

102. Lee, G., Y. Kim, S. Jeon, K. Byun, and D. Kwon, "Interfacial Reliability and Micropartial Stress Analysis Between TSV and CPB Through NIT and MSA," *IEEE/ECTC Proceedings*, Orlando, FL, June 2011, pp. 1436–1443.

103. Nimura, M., J. Mizuno, K. Sakuma, and S. Shoji, "Solder/Adhesive Bonding Using Simple Planarization Technique for 3D Integration," *IEEE/ECTC Proceedings*, Orlando, FL, June 2011, pp. 1147–1152.

104. Maria, J., B. Dang, S. Wright, C. Tsang, P. Andry, R. Polastre, Y. Liu, L. Wiggins and J. Knickerbocker, "3D Chip Stacking with 50-μm Pitch Lead-Free Micro-C4 Interconnections," *IEEE/ECTC Proceedings*, Orlando, FL, June 2011, pp. 268–273.

105. Sakuma, K., K. Toriyama, H. Noma, K. Sueoka, N. Unami, J. Mizuno, S. Shoji, and Y. Orii, "Fluxless Bonding for Fine-Pitch and Low-Volume Solder 3D Interconnections," *IEEE/ECTC Proceedings*, Orlando, FL, June 2011, pp. 7–13.

106. Andry, P., B. Dang, J. Knickerbocker, K. Tamura, and N. Taneichi, "Low-Profile 3D Silicon-on-Silicon Multi-chip Assembly," *IEEE/ECTC Proceedings*, Orlando, FL, June 2011, pp. 553–559.

107. Doany, F., C. Schow, B. Lee, R. Budd, C. Baks, R. Dangel, R. John, F. Libsch, and J. Kash, "Terabit/sec-Class Board-Level Optical Interconnects Through Polymer Waveguides Using 24-Channel Bidirectional Transceiver Modules," *IEEE/ECTC Proceedings*, Orlando, FL, June 2011, pp. 790–797.

108. Che, F., T. Chai, S. Lim, R. Rajoo, and X. Zhang, "Design and Reliability Analysis of Pyramidal Shape 3-Layer Stacked TSV Die Package," *IEEE/ECTC Proceedings*, Orlando, FL, June 2011, pp. 1428–1435.

109. Lee, J., D. Fernandez, M. Paing, Y. Yeo, and S. Gao, "Novel Chip Stacking Process for 3D Integration," *IEEE/ECTC Proceedings*, Orlando, FL, June 2011, pp. 1939–1943.

110. Au, K., J. Beleran, Y. Yang, Y. Zhang, S. Kriangsak, P. Wilson, Y. Drake, C. Toh, and C. Surasit, "Thru Silicon Via Stacking & Numerical Characterization for Multi-Die Interconnections using Full Array and Very Fine Pitch Micro C4 Bumps," *IEEE/ECTC Proceedings*, Orlando, FL, June 2011, pp. 296–303.

111. Lu, K., S. Ryu, Q. Zhao, K. Hummler, J. Im, R. Huang, and P. Ho, "Temperature-Dependent Thermal Stress Determination for Through-Silicon-Vias (TSVs) by Combining Bending Beam Technique with Finite Element Analysis," *IEEE/ECTC Proceedings*, Orlando, FL, June 2011, pp. 1475–1480.

112. Mitra, J., M. Jung, S. Ryu, R. Huang, S. Lim, and D. Pan, "A Fast Simulation Framework for Full-Chip Thermo-Mechanical Stress and Reliability Analysis of Through-Silicon-Via-Based 3D ICs," *IEEE/ECTC Proceedings*, Orlando, FL, June 2011, pp. 746–753.

113. Lin, Y., C. Hsieh, C. Yu, C. Tung, and D. Yu, "Study of the Thermo-Mechanical Behavior of Glass Interposer for Flip Chip Packaging Applications," *IEEE/ECTC Proceedings*, Orlando, FL, June 2011, pp. 634–638.

114. Lin, T., R. Wang, M. Chen, C. Chiu, S. Chen, T. Yeh, L. Lin, S. Hou, J. Lin, K. Chen, S. Jeng, and D. Yu, "Electromigration Study of Micro Bumps at Si/Si Interface in 3DIC Package for 28-nm Technology and Beyond," *IEEE/ECTC Proceedings*, Orlando, FL, June 2011, pp. 346–350.

115. Wei, C., C. Yu, C. Tung, R. Huang, C. Hsieh, C. Chiu1, H. Hsiao, Y. Chang, C. Lin, Y. Liang, C. Chen, T. Yeh, L. Lin, and D. Yu, "Comparison of the Electromigration Behaviors Between Micro-Bumps and C4 Solder Bumps," *IEEE/ECTC Proceedings*, Orlando, FL, June 2011, pp. 706–710.

116. Bouchoucha, M., P. Chausse, S. Moreau, L. Chapelon, N. Sillon, and O. Thomas, "Reliability Study of 3D-WLP Through Silicon Via with Innovative Polymer Filling Integration," *IEEE/ECTC Proceedings*, Orlando, FL, June 2011, pp. 567–572.

117. Shariff, D., P. Marimuthu, K. Hsiao, L. Asoy, C. Yee, A. Oo, K. Buchanan1, K. Crook, T. Wilby, and S. Burgess, "Integration of Fine-Pitched Through-Silicon Vias and Integrated Passive Devices," *IEEE/ECTC Proceedings*, Orlando, FL, June 2011, pp. 844–848.

118. Malta, D., C. Gregory, M. Lueck, D. Temple, M. Krause, F. Altmann, M. Petzold, M. Weatherspoon, and J. Miller, "Characterization of Thermo-Mechanical Stress and Reliability Issues for Cu-Filled TSVs," *IEEE/ECTC Proceedings*, Orlando, FL, June 2011, pp. 1815–1821.

119. Ramkumar, S., H. Venugopalan, and K. Khanna, "Novel Anisotropic Conductive Adhesive for 3D Stacking and Lead-Free PCB Packaging: A Review," *IEEE/ECTC Proceedings*, Orlando, FL, June 2011, pp. 246–254.

120. Oprins, H., V. Cherman, B. Vandevelde, C. Torregiani, M. Stucchi, G. Van der Plas, P. Marchal, and E. Beyne, "Characterization of the Thermal Impact of Cu-Cu Bonds Achieved Using TSVs on Hot Spot Dissipation in 3D Stacked ICs," *IEEE/ECTC Proceedings*, Orlando, FL, June 2011, pp. 861–868.

121. Tsukada, A., R. Sato, S. Sekine, R. Kimura, K. Kishi, Y. Sato, Y. Iwata, and H. Murata, "Study on TSV with New Filling Method and Alloy for Advanced 3D-SiP," *IEEE/ECTC Proceedings*, Orlando, FL, June 2011, pp. 861–868.

122. Yoon, S., K. Ishibashi, S. Dzafir, M. Prashant, P. Marimuthu, and F. Carson, "Development of Super Thin TSV PoP," *IEEE/ECTC Proceedings*, Orlando, FL, June 2011, pp. 274–278.

123. Meinshausen, L., K. Weide-Zaage, and M. Petzold, "Electro- and Thermomigration in Micro Bump Interconnects for 3D Integration," *IEEE/ECTC Proceedings*, Orlando, FL, June 2011, pp. 1444–1451.

124. Ko, Y., H. Fujii, Y. Sato, C. Lee, and S. Yoo, "Advanced Solder TSV Filling Technology Developed with Vacuum and Wave Soldering," *IEEE/ECTC Proceedings*, Orlando, FL, June 2011, pp. 2091–2095.

125. Jung, M., Y. Song, T. Yim, and J. Lee, "Evaluation of Additives and Current Mode on Copper Via Fill," *IEEE/ECTC Proceedings*, Orlando, FL, June 2011, pp. 1908–1912.

126. Choi, Y., J. Shin, and K. Paik, "A Study on the 3D-TSV Interconnection Using Wafer-Level Non-Conductive Adhesives (NCAs)," *IEEE/ECTC Proceedings*, Orlando, FL, June 2011, pp. 1126–1129.

127. Zoschke, K., J. Wolf, C. Lopper, I. Kuna, N. Jürgensen, V. Glaw, K. Samulewicz, J. Röder, M. Wilke, O. Wünsch, M. Klein, M. Suchodoletz, H. Oppermann, T. Braun, R. Wieland, and O. Ehrmann, "TSV Based Silicon Interposer Technology for Wafer Level Fabrication of 3D SiP Modules," *IEEE/ECTC Proceedings*, Orlando, FL, June 2011, pp. 836–843.

128. Brusberg, L., N. Schlepple, and H. Schröder, "Chip-to-Chip Communication by Optical Routing Inside a Thin Glass Substrate," *IEEE/ECTC Proceedings*, Orlando, FL, June 2011, pp. 805–812.

129. Schröder, H., L. Brusberg, N. Arndt-Staufenbiel, J. Hofmann, and S. Marx, "Glass Panel Processing for Electrical and Optical Packaging," *IEEE/ECTC Proceedings*, Orlando, FL, June 2011, pp. 625–633.

130. Redolfi, A., D. Velenis, S. Thangaraju, P. Nolmans, P. Jaenen, M. Kostermans, U. Baier, E. Van Besien, H. Dekkers, T. Witters, N. Jourdan, A. Van Ammel, K. Vandersmissen, S. Rodet, H. Philipsen, A. Radisic, N. Heylen, Y. Travaly, B. Swinnen, and E. Beyne, "Implementation of an Industry Compliant, SiN 50-μm Via-Middle TSV Technology on 300-mm Wafers," *IEEE/ECTC Proceedings*, Orlando, FL, June 2011, pp. 1384–1388.

131. Jourdain, A., T. Buisson, A. Phommahaxay, A. Redolfi, S. Thangaraju, Y. Travaly, E. Beyne, and B. Swinnen, "Integration of TSVs, Wafer Thinning and Backside Passivation on Full 300-mm CMOS Wafers for 3D Applications," *IEEE/ECTC Proceedings*, Orlando, FL, June 2011, pp. 1122–1125.

132. Pham, N., V. Cherman, B. Vandevelde, P. Limaye, N. Tutunjyan, R. Jansen, N. Van Hoovels, D. Tezcan, P. Soussan, E. Beyneand, and H. Tilmans, "Zero-Level Packaging for (RF-)MEMS Implementing TSVs and Metal Bonding," *IEEE/ECTC Proceedings*, Orlando, FL, June 2011, pp. 1588–1595.

133. Halder, S., A. Jourdain, M. Claes, I. Wolf, Y. Travaly, E. Beyne, B. Swinnen, V. Pepper, P. Guittet, G. Savage, and L. Markwort, "Metrology and Inspection for Process Control During Bonding and Thinning of Stacked Wafers for Manufacturing 3D SICs," *IEEE/ECTC Proceedings*, Orlando, FL, June 2011, pp. 999–1002.

134. Vos, J., A. Jourdain, M. Erismis, W. Zhang, K. De Munck, A. La Manna, D. Tezcan, and P. Soussan, "High Density 20-μm Pitch CuSn Microbump Process for High-End 3D Applications," *IEEE/ECTC Proceedings*, Orlando, FL, June 2011, pp. 27–31.

135. Zhang, W., B. Dimcic, P. Limaye, A. Manna, P. Soussan, and E. Beyne, "Ni/Cu/Sn Bumping Scheme for Fine-Pitch Micro-Bump Connections," *IEEE/ECTC Proceedings*, Orlando, FL, June 2011, pp. 109–113.

136. Che, F., H. Li, X. Zhang, S. Gao, and K. Teo, "Wafer Level Warpage Modeling Methodology and Characterization of TSV Wafers," *IEEE/ECTC Proceedings*, Orlando, FL, June 2011, pp. 1196–1203.

137. Kwon, W., J. Lee, V. Lee, J. Seetoh, Y. Yeo, Y. Khoo, N. Ranganathan, K. Teo, and S. Gao, "Novel Thinning/Backside Passivation for Substrate Coupling Depression of 3D IC," *IEEE/ECTC Proceedings*, Orlando, FL, June 2011, pp. 1395–1399.

138. Lee, J., V. Lee, J. Seetoh, S. Thew, Y. Yeo, H. Li, K. Teo, and S. Gao, "Advanced Wafer Thinning and Handling for Through Silicon Via Technology," *IEEE/ ECTC Proceedings*, Orlando, FL, June 2011, pp. 1852–1857.

139. Oprins, H., V. Cherman, B. Vandevelde, C. Torregiani, M. Stucchi, G. Van der Plas, P. Marchal, and E. Beyne, "Characterization of the Thermal Impact of Cu-Cu Bonds Achieved Using TSVs on Hot Spot Dissipation in 3D Stacked ICs," *IEEE/ECTC Proceedings*, Orlando, FL, June 2011, pp. 861–868.

140. Peng, L., H. Li, D. Lim, S. Gao, and C. Tan, "Thermal Reliability of Fine Pitch Cu-Cu Bonding with Self-Assembled Monolayer (SAM) Passivation for Wafer-on-Wafer 3D-Stacking," *IEEE/ECTC Proceedings*, Orlando, FL, June 2011, pp. 22–26.

141. Xie, L., W. Choi, C. Premachandran, C. Selvanayagam, K. Bai, Y. Zeng, S. Ong, E. Liao, A. Khairyanto, V. Sekhar, and S. Thew, "Design, Simulation and Process Optimization of AuInSn Low Temperature TLP Bonding for 3D IC Stacking," *IEEE/ECTC Proceedings*, Orlando, FL, June 2011, pp. 279–284.

142. Akasaka, Y., "Three-Dimensional IC Trends," *Proceedings of the IEEE*, Vol. 74, No. 12, December 1986, pp. 1703–1714.

143. Akasaka, Y., and Nishimura, T., "Concept and Basic Technologies for 3D IC Structure," *IEEE Proceedings of International Electron Devices Meetings*, Vol. 32, 1986, pp. 488–491.

144. Chen, K., S. Lee, P. Andry, C. Tsang, A. Topop, Y. Lin, Y., J. Lu, A. Young, M. Ieong, and W. Haensch, "Structure, Design and Process Control for Cu Bonded Interconnects in 3D Integrated Circuits," *IEEE Proceedings of International Electron Devices Meeting (IEDM 2006)*, San Francisco, CA, December 11–13, 2006, pp. 367–370.

145. Liu, F., R. Yu, A. Young, J. Doyle, X. Wang, L. Shi, K. Chen, X. Li, D. Dipaola, D. Brown, C. Ryan, J. Hagan, K. Wong, M. Lu, X. Gu, N. Klymko, E. Perfecto, A. Merryman, K. Kelly, S. Purushothaman, S. Koester, R. Wisnieff, and W. Haensch, "A 300-mm Wafer-Level Three-Dimensional Integration Scheme Using Tungsten Through-Silicon Via and Hybrid Cu-Adhesive Bonding," *IEEE Proceedings of IEDM*, December 2008, pp. 1–4.

146. Yu, R., F. Liu, R. Polastre, K. Chen, X. Liu, L. Shi, E. Perfecto, N. Klymko, M. Chace, T. Shaw, D. Dimilia, E. Kinser, A. Young, S. Purushothaman, S. Koester, and W. Haensch, "Reliability of a 300-mm-Compatible 3DI Technology Based on Hybrid Cu-Adhesive Wafer Bonding," *Proceedings of Symposium on VLSI Technology Digest of Technical Papers*, 2009, pp. 170–171.

147. Shigetou, A., T. Itoh, K. Sawada, and T. Suga, "Bumpless Interconnect of 6-µm Pitch Cu Electrodes at Room Temperature," *IEEE Proceedings of ECTC*, Lake Buena Vista, FL, May 27–30, 2008, pp. 1405–1409.

148. Tsukamoto, K., E. Higurashi, and T. Suga, "Evaluation of Surface Microroughness for Surface Activated Bonding," *Proceedings of IEEE CPMT Symposium Japan*, August 2010, pp. 147–150.

149. Kondou, R., C. Wang, and T. Suga, "Room-Temperature Si-Si and Si-SiN Wafer Bonding," *Proceedings of IEEE CPMT Symposium Japan*, August 2010, pp. 161–164.

150. Shigetou, A., T. Itoh, M. Matsuo, N. Hayasaka, K. Okumura, and T. Suga, "Bumpless Interconnect Through Ultrafine Cu Electrodes by Mans of Surface-Activated Bonding (SAB) Method," *IEEE Transaction on Advanced Packaging*, Vol. 29, No. 2, May 2006, p. 226.

151. Wang, C., and T. Suga, "A Novel Moire Fringe Assisted Method for Nanoprecision Alignment in Wafer Bonding," *IEEE/ECTC Proceedings*, San Diego, CA, May 25–29, 2009, pp. 872–878.

152. Wang, C., and T. Suga, "Moire Method for Nanoprecision Wafer-to-Wafer Alignment: Theory, Simulation and Application," *IEEE Proceedings of International Conference on Electronic Packaging Technology and High-Density Packaging*, August 2009, pp. 219–224.

153. Higurashi, E., D. Chino, T. Suga, and R. Sawada, "Au-Au Surface-Activated Bonding and Its Application to Optical Microsensors with 3D Structure," *IEEE Journal of Selected Topic in Quantum Electronics*, Vol. 15, No. 5, September–October 2009, pp. 1500–1505.
154. Burns, J., B. Aull, C. Keast, C. Chen, C. Chen, C. Keast, J. Knecht, V. Suntharalingam, K. Warner, P. Wyatt, and D. Yost, "A Wafer-Scale 3D Circuit Integration Technology," *IEEE Transactions on Electron Devices*, Vol. 53, No. 10, October 2006, pp. 2507–2516.
155. Chen, C., K. Warner, D. Yost, J. Knecht, V. Suntharalingam, C. Chen, J. Burns, and C. Keast, "Sealing Three-Dimensional SOI Integrated-Circuit Technology," *IEEE Proceedings of International SOI Conference*, 2007, pp. 87–88.
156. Chen, C., C. Chen, D. Yost, J. Knecht, P. Wyatt, J. Burns, K. Warner, P. Gouker, P. Healey, B. Wheeler, and C. Keast, "Three-Dimensional Integration of Silicon-on-Insulator RF Amplifier," *Electronics Letters*, Vol. 44, No. 12, June 2008, pp. 1–2.
157. Chen, C., C. Chen, D. Yost, J. Knecht, P. Wyatt, J. Burns, K. Warner, P. Gouker, P. Healey, B. Wheeler, and C. Keast, "Wafer-Scale 3D Integration of Silicon-on-Insulator RF Amplifiers," *IEEE Proceedings of Silicon Monolithic IC in RF Systems*, 2009, pp. 1–4.
158. Chen, C., C. Chen, P. Wyatt, P. Gouker, J. Burns, J. Knecht, D. Yost, P. Healey, and C. Keast, "Effects of Through-Box Vias on SOI MOSFETs," *IEEE Proceedings of VLSI Technology, Systems and Applications*, 2008, pp. 1–2.
159. Chen, C., C. Chen, J. Burns, D. Yost, K. Warner, J. Knecht, D. Shibles, and C. Keast, "Thermal Effects of Three-Dimensional Integrated Circuit Stacks," *IEEE Proceedings of International SOI Conference*, 2007, pp. 91–92.
160. Aull, B., J. Burns, C. Chen, B. Felton, H. Hanson, C. Keast, J. Knecht, A. Loomis, M. Renzi, A. Soares, V. Suntharalingam, K. Warner, D. Wolfson, D. Yost, and D. Young, "Laser Radar Imager Based on 3D Integration of Geiger-Mode Avalanche Photodiodes with Two SOI Timing Circuit Layers," *IEEE Proceedings of International Solid-State Circuits Conference*, 2006, p. 16.9.
161. Chatterjee, R., M. Fayolle, P. Leduc, S. Pozder, B. Jones, E. Acosta, B. Charlet, T. Enot, M. Heitzmann, M. Zussy, A. Roman, O. Louveau, S. Maitreqean, D. Louis, N. Kernevez, N. Sillon, G. Passemard, V. Pol, V. Mathew, S. Garcia, T. Sparks, and Z. Huang, "Three-Dimensional Chip Stacking Using a Wafer-to-Wafer Integration," *IEEE Proceedings of IITC*, 2007, pp. 81–83.
162. Ledus, P., F. Crecy, M. Fayolle, M. Fayolle, B. Charlet, T. Enot, M. Zussy, B. Jones, J. Barbe, N. Kernevez, N. Sillon, S. Maitreqean, D. Louis, and G. Passemard, "Challenges for 3D IC Integration: Bonding Quality and Thermal Management," *IEEE Proceedings of IITC*, 2007, pp. 210–212.
163. Poupon, G., N. Sillon, D. Henry, C. Gillot, A. Mathewson, L. Cioccio, B. Charlet, P. Leduc, M. Vinet, and P. Batude, "System on Wafer: A New Silicon Concept in Sip," *Proceedings of the IEEE*, Vol. 97, No. 1, January 2009, pp. 60–69.
164. Fujimoto, K., N. Maeda, H. Kitada, Y. Kim, A. Kawai, K. Arai, T. Nakamura, K. Suzuki, and T. Ohba, "Development of Multi-Stack Process on Wafer-on-Wafer (WoW)," *Proceedings of IEEE CPMT Symposium Japan*, August 2010, pp. 157–160.
165. Chen, Q., D. Zhang, Z. Wang, L. Liu, and J. Lu, "Chip-to-Wafer (C2W) 3D Integration with Well-Controlled Template Alignment and Wafer-Level Bonding," *IEEE/ECTC Proceedings*, Orlando, FL, June 2011, pp. 1–6.
166. Healy, M., and S. Lim, "Power Delivery System Architecture for Many-Tier 3D Systems," *IEEE/ECTC Proceedings*, Las Vegas, NV, June 2010, pp. 1682–1688.
167. Liu, F., X. Gu, K. A. Jenkins, E. A. Cartier, Y. Liu, P. Song, and S. J. Koester, "Electrical Characterization of 3D Through-Silicon-Vias," *IEEE/ECTC Proceedings*, Las Vegas, NV, June 2010, pp. 1100–1105.
168. Gu, X., B. Wu, M. Ritter, and L. Tsang, "Efficient Full-Wave Modeling of High Density TSVs for 3D Integration," *IEEE/ECTC Proceedings*, Las Vegas, NV, June 2010, pp. 663–666.
169. Okoro, C., R. Agarwal, P. Limaye, B. Vandevelde, D. Vandepitte, and E. Beyne1, "Insertion Bonding: A Novel Cu-Cu Bonding Approach for 3D Integration," *IEEE/ECTC Proceedings*, Las Vegas, NV, June 2010, pp. 1370–1375.

170. Huyghebaert, C., J. Olmen, O. Chukwudi, J. Coenen, A. Jourdain, M. Cauwenberghe, R. Agarwahl, R., A. Phommahaxay, M. Stucchi, and P. Soussan., "Enabling 10-μm Pitch Hybrid Cu-Cu IC Stacking with Through Silicon Vias," *IEEE/ECTC Proceedings*, Las Vegas, NV, June 2010, pp. 1083–1087.
171. Pak, J., J. Cho, J. Kim, J. Lee, H. Lee, K. Park, and J. Kim, "Slow Wave and Dielectric Quasi-TEM Modes of Metal-Insulator-Semiconductor (MIS) Structure Through Silicon Via (TSV) in Signal Propagation and Power Delivery in 3D Chip Package," *IEEE/ECTC Proceedings*, Las Vegas, NV, June 2010, pp. 667–672.
172. Cioccioa, L., P. Gueguena, E. Grouillera, L. Vandrouxa, V. Delayea, M. Rivoireb, J. Lugandb, and L. Claveliera, "Vertical Metal Interconnect Thanks to Tungsten Direct Bonding," *IEEE/ECTC Proceedings*, Las Vegas, NV, June 2010, 1359–1363.
173. Gueguena, P., L. Cioccioa, P. Morfoulib, M. Zussya, J. Dechampa, L. Ballya, and L. Claveliera, "Copper Direct Bonding: An Innovative 3D Interconnect," *IEEE/ECTC Proceedings*, Las Vegas, NV, June 2010, pp. 878–883.
174. Taibi, R., L. Ciocciob, C. Chappaz, L. Chapelon, P. Gueguenb, J. Dechampb, R. Fortunierc, and L. Clavelierb, "Full Characterization of Cu/Cu Direct Bonding for 3D Integration," *IEEE/ECTC Proceedings*, Las Vegas, NV, June 2010, pp. 219–225.
175. Lim, D., J. Wei, C. Ng, and C. Tan, "Low Temperature Bump-Less Cu-Cu Bonding Enhancement with Self-Assembled Monolayer (SAM) Passivation for 3D Integration," *IEEE/ECTC Proceedings*, Las Vegas, NV, June 2010, pp. 1364–1369.
176. Onkaraiah, S., and C. Tan, "Mitigating Heat Dissipation and Thermo-Mechanical Stress Challenges in 3D IC Using Thermal Through Silicon Via (TTSV)," *IEEE/ECTC Proceedings*, Las Vegas, NV, June 2010, pp. 411–416.
177. Campbell, D., "Process Characterization Vehicles for 3D Integration," *IEEE/ECTC Proceedings*, Las Vegas, NV, June 2010, pp. 1112–1116.
178. Bieck, F., S. Spiller, F. Molina, M. Töpper, C. Lopper, I. Kuna, T. Seng, and T. Tabuchi, "Carrierless Design for Handling and Processing of Ultrathin Wafers," *IEEE/ECTC Proceedings*, Las Vegas, NV, June 2010, pp. 316–322.
179. Itabashi, T., and M. Zussman, "High Temperature Resistant Bonding Solutions Enabling Thin Wafer Processing (Characterization of Polyimide Base Temporary Bonding Adhesive for Thinned Wafer Handling)," *IEEE/ECTC Proceedings*, Las Vegas, NV, June 2010, pp. 1877–1880.
180. Lee, G., H. Son, J. Hong, K. Byun, and D. Kwon, "Quantification of Micropartial Residual Stress for Mechanical Characterization of TSV Through Nanoinstrumented Indentation Testing," *IEEE/ECTC Proceedings*, Las Vegas, NV, June 2010, pp. 200–205.
181. Dang, B., P. Andry, C. Tsang, J. Maria, R. Polastre, R. Trzcinski, A. Prabhakar, and J. Knickerbocker, "CMOS Compatible Thin Wafer Processing Using Temporary Mechanical Wafer, Adhesive and Laser Release of Thin Chips/Wafers for 3D Integration," *IEEE/ECTC Proceedings*, Las Vegas, NV, June 2010, pp. 1393–1398.
182. Kang, U., H. Chung, S. Heo, D. Park, H. Lee, J. Kim, S. Ahn, S. Cha, J. Ahn, D. Kwon, J. Lee, H. Joo, W. Kim, D. Jang, N. Kim, J. Choi, T. Chung, J. Yoo, J. Choi, C. Kim, and Y. Jun, "8-Gb 3D DDR3 DRAM Using Through-Silicon-Via Technology," *IEEE Journal of Solid-State Circuits*, Vol. 45, No. 1, January 2010, pp. 111–119.
183. Farooq, M. G., T. L. Graves-Abe, W. F. Landers, C. Kothandaraman, B. A. Himmel, P. S. Andry, C. K. Tsang, E. Sprogis, R. P. Volant, K. S. Petrarca, K. R. Winstel, J. M. Safran, T. D. Sullivan, F. Chen, M. J. Shapiro, R. Hannon, R. Liptak, D. Berger, and S. S. Iyer, "3D Copper TSV Integration, Testing and Reliability," *Proceedings of IEEE IEDM*, Washington, DC, December 2011, pp. 7.1.1–7.1.4.
184. Dorsey, P., "Xilinx Stacked Silicon Interconnect Technology Delivers Breakthrough FPGA Capacity, Bandwidth, and Power Efficiency," Xilinx white paper: Virtex-7 FPGAs, WP380, October 27, 2010, pp. 1–10.

185. Banijamali, B., S. Ramalingam, K. Nagarajan, and R. Chaware, "Advanced Reliability Study of TSV Interposers and Interconnects for the 28-nm Technology FPGA," *IEEE/ECTC Proceedings*, Orlando, FL, June 2011, pp. 285–290.

186. Kim, N., D. Wu, D. Kim, A. Rahman, and P. Wu, "Interposer Design Optimization for High Frequency Signal Transmission in Passive and Active Interposer using Through Silicon Via (TSV)," *IEEE/ECTC Proceedings*, Orlando, FL, June 2011, pp. 1160–1167.

187. Lau, J. H., M. Dai, Y. Chao, W. Li, S. Wu, J. Hung, M. Hsieh, J. Chien, R. Tain, C. Tzeng, K. Lin, E. Hsin, C. Chen, M. Chen, C. Wu, J. Chen, J. Chien, C. Chiang, Z. Lin, L. Wu, H. Chang, W. Tsai, C. Lee, T. Chang, C. Ko, T. Chen, S. Sheu, S. Wu, Y. Chen, R. Lo, T. Ku, M. Kao, F. Hsieh, and D. Hu, "Feasibility Study of a 3D IC Integration System-in-Packaging (SiP)," *IEEE/ICEP Proceedings*, Nara, Japan, April 13, 2011, pp. 210–216.

188. Lau, J. H., C.-J. Zhan, P.-J. Tzeng, C.-K. Lee, M.-J. Dai, H.-C. Chien, Y.-L. Chao, W. Li, S.-T. Wu, J.-F. Hung, R.-M. Tain, C.-H. Lin, Y.-C. Hsin, C.-C. Chen, S.-C. Chen, C.-Y. Wu, J.-C. Chen, C.-H. Chien, C.-W. Chiang, H. Chang, W.-L. Tsai, R.-S. Cheng, S.-Y. Huang, Y.-M. Lin, T.-C. Chang, C.-D. Ko, T.-H. Chen, S.-S. Sheu, S.-H. Wu, Y.-H. Chen, W.-C. Lo, T.-K. Ku, M.-J. Kao, and D.-Q. Hu, "Feasibility Study of a 3D IC Integration System-in-Packaging (SiP) from a 300-mm Multi-Project Wafer (MPW)," *Proceedings of IMAPS International Conference*, Long Beach, CA, October 2011, pp. 446–454.

189. Zhan, C.-J., J. H. Lau, P.-J. Tzeng, C.-K. Lee, M.-J. Dai, H.-C. Chien, Y.-L. Chao, W. Li, S.-T. Wu, J.-F. Hung, R.-M. Tain, C.-H. Lin, Y.-C. Hsin, C.-C. Chen, S.-C. Chen, C.-Y. Wu, J.-C. Chen, C.-H. Chien, C.-W. Chiang, H. Chang, W.-L. Tsai, R.-S. Cheng, S.-Y. Huang, Y.-M. Lin, T.-C. Chang, C.-D. Ko, T.-H. Chen, S.-S. Sheu, S.-H. Wu, Y.-H. Chen, W.-C. Lo, T.-K. Ku, M.-J. Kao, and D.-Q. Hu, "Assembly Process and Reliability Assessment of TSV/RDL/IPD Interposer with Multi-Chip-Stacking for 3D IC Integration SiP," *IEEE/ECTC Proceedings* (in press).

190. Sheu, S., Z. Lin, J. Hung, J. H. Lau, P. Chen, S. Wu, K. Su, C. Lin, S. Lai, T. Ku, W. Lo, and M. Kao, "An Electrical Testing Method for Blind Through Silicon Vias (TSVs) for 3D IC Integration," *Proceedings of IMAPS International Conference*, Long Beach, CA, October 2011, pp. 208–214.

191. Wu, C., S. Chen, P. Tzeng, J. H. Lau, Y. Hsu, J. Chen, Y. Hsin, C. Chen, S. Shen, C. Lin, T. Ku, and M. Kao, "Oxide Liner, Barrier and Seed Layers, and Cu-Plating of Blind Through Silicon Vias (TSVs) on 300-mm Wafers for 3D IC Integration," *Proceedings of IMAPS International Conference*, Long Beach, CA, October 2011, pp. 1–7.

192. Chien, H., J. H. Lau, Y. Chao, R. Tain, M. Dai, S. Wu, W. Lo, and M. Kao, "Thermal Performance of 3D IC Integration with Through-Silicon Via (TSV)," *Proceedings of IMAPS International Conference*, Long Beach, CA, October 2011, pp. 25–32.

193. Chang, H. H., J. H. Lau, W. L. Tsai, C. H. Chien, P. J. Tzeng, C. J. Zhan, C. K. Lee, M. J. Dai, H. C. Fu, C. W. Chiang, T. Y. Kuo, Y. H. Chen, W. C. Lo, T. K. Ku, and M. J. Kao, "Thin Wafer Handling of 300-mm Wafer for 3D IC Integration," *Proceedings of IMAPS International Conference*, Long Beach, CA, October 2011, pp. 202–207.

194. Lau, J. H., P.-J. Tzeng, C.-K. Lee, C.-J. Zhan, M.-J. Dai, Li Li, C.-T. Ko, S.-W. Chen, H. Fu, Y. Lee, Z. Hsiao, J. Huang, W. Tsai, P. Chang, S. Chung, Y. Hsu, S.-C. Chen, Y.-H. Chen, T.-H. Chen, W.-C. Lo, T.-K. Ku, M.-J. Kao, J. Xue, and M. Brillhart, "Wafer Bumping and Characterizations of Fine-Pitch Lead-Free Solder Microbumps on 12-inch (300-mm) Wafer for 3D IC Integration," *Proceedings of IMAPS International Conference*, Long Beach, CA, October 2011, pp. 650–656.

195. Hsin, Y. C., C. Chen, J. H. Lau, P. Tzeng, S. Shen, Y. Hsu, S. Chen, C. Wn, J. Chen, T. Ku, and M. Kao, "Effects of Etch Rate on Scallop of Through-Silicon Vias (TSVs) in 200-mm and 300-mm Wafers," *IEEE/ECTC Proceedings*, Orlando, FL, June 2011, pp. 1130–1135.

196. Chen, J. C., J. H. Lau, P. J. Tzeng, S. C. Chen, C. Y. Wu, C. C. Chen, C. H. Lin, Y. C. Hsin, T. K. Ku, and M. J. Kao, "Impact of Slurry in Cu CMP (Chemical Mechanical Polishing) on Cu Topography of Through Silicon Vias (TSVs), Re-distributed Layers, and Cu Exposure," *IEEE/ECTC Proceedings*, Orlando, FL, June 2011, pp. 1389–1394. Also, *IEEE Transactions on CPMT* (in press).

197. Chien, J., Y. Chao, J. H. Lau, M. Dai, R. Tain, M. Dai, P. Tzeng, C. Lin, Y. Hsin, S. Chen, J. Chen, C. Chen, C. Ho, R. Lo, T. Ku, and M. Kao, "A Thermal Performance Measurement Method for Blind Through Silicon Vias (TSVs) in a 300-mm Wafer," *IEEE/ECTC Proceedings*, Orlando, FL, June 2011, pp. 1204–1210.

198. Tsai, W., H. H. Chang, C. H. Chien, J. H. Lau, H. C. Fu, C. W. Chiang, T. Y. Kuo, Y. H. Chen, R. Lo, and M. J. Kao, "How to Select Adhesive Materials for Temporary Bonding and De-Bonding of Thin-Wafer Handling in 3D IC Integration?" *IEEE/ECTC Proceedings*, Orlando, FL, June 2011, pp. 989–998.

199. Chang, H., J. Huang, C. Chiang, Z. Hsiao, H. Fu, C. Chien, Y. Chen, W. Lo, and K. Chiang, "Process Integration and Reliability Test for 3D Chip Stacking with Thin Wafer Handling Technology," *IEEE/ECTC Proceedings*, Orlando, FL, June 2011, pp. 304–311.

200. Lee, C. K., T. C. Chang, Y. Huang, H. Fu, J. H. Huang, Z. Hsiao, J. H. Lau, C. T. Ko, R. Cheng, K. Kao, Y. Lu, R. Lo, and M. J. Kao, "Characterization and Reliability Assessment of Solder Microbumps and Assembly for 3D IC Integration," *IEEE/ECTC Proceedings*, Orlando, FL, June 2011, pp. 1468–1474.

201. Zhan, C., J. Juang, Y. Lin, Y. Huang, K. Kao, T. Yang, S. Lu, J. H. Lau, T. Chen, R. Lo, and M. J. Kao, "Development of Fluxless Chip-on-Wafer Bonding Process for 3D Chip Stacking with 30-μm Pitch Lead-Free Solder Micro Bump Interconnection and Reliability Characterization," *IEEE/ECTC Proceedings*, Orlando, FL, June 2011, pp. 14–21.

202. Cheng, R., K. Kao, J. Chang, Y. Hung, T. Yang, Y. Huang, S. Chen, T. Chang, Q. Hunag, R. Guino, G. Hoang, J. Bai, and K. Becker, "Achievement of Low Temperature Chip Stacking by a Pre-Applied Underfill Material," *IEEE/ECTC Proceedings*, Orlando, FL, June 2011, pp. 1858–1863.

203. Huang, S., T. Chang, R. Cheng, J. Chang, C. Fan, C. Zhan, J. H. Lau, T. Chen, R. Lo, and M. Kao, "Failure Mechanism of 20-μm Pitch Micro Joint Within a Chip Stacking Architecture," *IEEE/ECTC Proceedings*, Orlando, FL, June 2011, pp. 886–892.

204. Lin, Y., C. Zhan, J. Juang, J. H. Lau, T. Chen, R. Lo, M. Kao, T. Tian, and K. N. Tu, "Electromigration in Ni/Sn Intermetallic Micro Bump Joint for 3D IC Chip Stacking," *IEEE/ECTC Proceedings*, Orlando, FL, June 2011, pp. 351–357.

205. Pang, X., T. T. Chua, H. Y. Li, E. B. Liao, W. S. Lee, and F. X. Che, "Characterization and Management of Wafer Stress for Various Pattern Densities in 3D Integration Technology," *IEEE/ECTC Proceedings*, Las Vegas, NV, June 2010, pp. 1866–1869.

206. Zoschke, K., M. Wegner, M. Wilke, N. Jürgensen, C. Lopper, I. Kuna, V. Glaw, J. Röderl, O. Wünsch1, M. J. Wolf, O. Ehrmann, and H. Reichl, "Evaluation of Thin Wafer Processing Using a Temporary Wafer Handling System as Key Technology for 3D System Integration," *IEEE/ECTC Proceedings*, Las Vegas, NV, June 2010, pp. 1385–1392.

207. Charbonnier, J., R. Hida, D. Henry, S. Cheramy, P. Chausse, M. Neyret, O. Hajji, G. Garnier, C. Brunet-Manquat, P. Haumesser, L. Vandroux, R. Anciant, N. Sillon, A. Farcy, M. Rousseau, J. Cuzzocrea, G. Druais, and E. Saugier, "Development and Characterisation of a 3D Technology Including TSV and Cu Pillars for High Frequency Applications," *IEEE/ECTC Proceedings*, Las Vegas, NV, June 2010, pp. 1077–1082.

208. Sun, Y., X. Li, J. Gandhi, S. Luo, and T. Jiang, "Adhesion Improvement for Polymer Dielectric to Electrolytic-Plated Copper," *IEEE/ECTC Proceedings*, Las Vegas, NV, June 2010, pp. 1106–1111.

209. Kawano, M., N. Takahashi, M. Komuro, and S. Matsui, "Low-Cost TSV Process Using Electroless Ni Plating for 3D Stacked DRAM," *IEEE/ECTC Proceedings*, Las Vegas, NV, June 2010, pp. 1094–1099.

210. Malta, D., C. Gregory, D. Temple, T. Knutson, C. Wang, T. Richardson, Y. Zhang, and R. Rhoades, "Integrated Process for Defect-Free Copper Plating and Chemical-Mechanical Polishing of Through-Silicon Vias for 3D Interconnects," *IEEE/ECTC Proceedings*, Las Vegas, NV, June 2010, pp. 1769–1775.

211. Campbell, D., "Yield Modeling of 3D Integrated Wafer Scale Assemblies," *IEEE/ECTC Proceedings*, Las Vegas, NV, June 2010, pp. 1935–1938.

212. Archard, D., K. Giles, A. Price, S. Burgess, and K. Buchanan, "Low Temperature PECVD of Dielectric Films for TSV Applications," *IEEE/ECTC Proceedings*, Las Vegas, NV, June 2010, pp. 764–768.

213. Shigetou, A., and T. Suga, "Modified Diffusion Bonding for Both Cu and SiO$_2$ at 150°C in Ambient Air," *IEEE/ECTC Proceedings*, Las Vegas, NV, June 2010, pp. 872–877.

214. Amagai, M., and Y. Suzuki, "TSV Stress Testing and Modeling," *IEEE/ECTC Proceedings*, Las Vegas, NV, June 2010, pp. 1273–1280.

215. Gupta, A., S. Kannan, B. Kim, F. Mohammed, and B. Ahn, "Development of Novel Carbon Nanotube TSV Technology," *IEEE/ECTC Proceedings*, Las Vegas, NV, June 2010, pp. 1699–1702.

216. Kannan, S., A. Gupta, B. Kim, F. Mohammed, and B. Ahn, "Analysis of Carbon Nanotube Based Through Silicon Vias," *IEEE/ECTC Proceedings*, Las Vegas, NV, June 2010, pp. 51–57.

217. Lu, K., S. Ryu, Q. Zhao, X. Zhang, J. Im, R. Huang, and P. Ho, "Thermal Stress Induced Delamination of Through Silicon Vias in 3D Interconnects," *IEEE/ECTC Proceedings*, Las Vegas, NV, June 2010, pp. 40–45.

218. Miyazaki, C., H. Shimamoto, T. Uematsu, Y. Abe, K. Kitaichi, T. Morifuji, and S. Yasunaga, "Development of High Accuracy Wafer Thinning and Pickup Technology for Thin Wafer (Die)," *Proceedings of IEEE CPMT Symposium Japan*, August 2010, pp. 139–142.

219. Cho, J., K. Yoon, J. Pak, J. Kim, J. Lee, H. Lee, K. Park, and J. Kim, "Guard Ring Effect for Through Silicon Via (TSV) Noise Coupling Reduction," *Proceedings of IEEE CPMT Symposium Japan*, August 2010, pp. 151–154.

220. Nonake, T., K. Fujimaru, A. Shimada, N. Asahi, Y. Tatsuta, H. Niwa, and Y. Tachibana, "Wafer and/or Chip Bonding Adhesives for 3D Package," *Proceedings of IEEE CPMT Symposium Japan*, August 2010, pp. 169–172.

221. Lau, J. H., "Heart and Soul of 3D IC Integration," posted at 3D InCites Conference, June 29, 2010; http://www.semineedle.com/posting/34277.

222. Lau, J. H., "Who Invented the TSV and When?" posted at 3D InCites Conference, April 24, 2010; http://www.semineedle.com/posting/31171.

223. Gat, A., L. Gerzberg, J. Gibbons, T. Mages, J. Peng, and J. Hong, "CW Laser of Polyerystalline Silicon: Crystalline Structure and Electrical Properties," *Applied Physics Letters*, Vol. 33, No. 8, October 1978, pp. 775–778.

224. Lau, J. H., *Flip Chip Technology*, McGraw-Hill, New York, 1996.

225. Lau, J. H., *Low Cost Flip Chip Technologies*, McGraw-Hill, New York, 2000.

226. Hu, G., H. Kalyanam, S. Krishnamoorthy, and L. Polka, "Package Technology to Address the Memory Bandwidth Challenge for Tera-Scale Computing," *INTEL Technology Journal*, Vol. 11, 2007, pp. 197–206.

227. Ong, Y., S. Ho, K. Vaidyanathan, V. Sekhar, M. Jong, S. Long, V. Lee, C. Leong, V. Rao, J. Ong, X. Ong, X. Zhang, Y. Seung, J. H. Lau, Y. Lim, D. Yeo, K. Chan, Z. Yanfeng, J. Tan, and D. Sohn. "Design, Assembly and Reliability of Large Die and Fine-Pitch Cu/Low-*k* Flip Chip Package," *Journal of Microelectronics Reliability*, Vol. 50, 2010, pp. 986–994.

228. Lau, J. H., and R. S. W. Lee, *Microvias for Low-Cost, High-Density Interconnects*, McGraw-Hill, New York, 2001.

229. Lau, J. H., *Ball-Grid Array Technology*, McGraw-Hill, New York, 1995.

230. Lau, J. H., T. Tseng, and D. Cheng, "Heat Spreader with a Placement Recess and Bottom Saw-Teeth for Connection to Ground Planes on a This Two-Side Single-Core BGA Substrate," U.S. Patent No. 6,057,601; granted May 2, 2000; filed November 27, 1998.

231. Lau, J. H., and K. L. Chen, "Thermal and Mechanical Evaluations of a Cost-Effective Plastic Ball Grid Array Package," *ASME Transactions, Journal of Electronic Packaging*, Vol. 119, September 1997, pp. 208–212.
232. Lau, J. H., and T. Chen, "Cooling Assessment and Distribution of Heat Dissipation of a Cavity Down Plastic Ball Grid Array Package: NuBGA," *IMAPS Transactions, International Journal of Microelectronics & Electronic Packaging*, Vol. 21, No. 1, 1998, pp. 20–28.
233. Lau, J. H., and T. Chou, "Electrical Design of a Cost-Effective Thermal Enhanced Plastic Ball Grid Array Package: NuBGA," *IEEE Transactions on CPMT, Part B*, Vol. 21, No. 1, February 1998, pp. 35–42.
234. Lau, J. H., T. Chen, and R. Lee, "Effect of Heat Spreader Sizes on the Thermal Performance of Large Cavity-Down Plastic Ball Grid Array Packages," *ASME Transactions, Journal of Electronic Packaging*, Vol. 121, No. 4, 1999, pp. 242–248.
235. Lau, J. H., "Design, Manufacturing, and Testing of a Novel Plastic Ball Grid Array Package," *Journal of Electronics Manufacturing*, Vol. 9, No. 4, December 1999, pp. 283–291.
236. Lau, J. H., and T. Chen, "Low-Cost Thermal and Electrical Enhanced Plastic Ball Grid Array Package: NuBGA," *Microelectronics International*, 1999.
237. Lau, J. H., "Solder Joint Reliability of Flip Chip and Plastic Ball Grid Array Assemblies Under Thermal, Mechanical, and Vibration Conditions," *IEEE Transaction on CPMT, Part B*, Vol. 19, No. 4, November 1996, pp. 728–735.
238. Lau, J. H., with R. Lee, "Design for Plastic Ball Grid Array Solder Joint Reliability," *Journal of the Institute of Interconnection Technology*, Vol. 23, No. 2, January 1997, pp. 11–13.
239. Lau, J. H., with W. Jung and Y. Pao, "Nonlinear Analysis of Full-Matrix and Perimeter Plastic Ball Grid Array Solder Joints," *ASME Transactions, Journal of Electronic Packaging*, Vol. 119, September 1997, pp. 163–170.
240. Lau, J. H., with R. Lee, "Solder Joint Reliability of Cavity-Down Plastic Ball Grid Array Assemblies," *Journal of Soldering and Surface Mount Technology*, Vol. 10, No. 1, February 1998, pp. 26–31.
241. Lau, J. H., *Solder Joint Reliability: Theory and Applications*, Van Nostrand Reinhold, New York, 1991.

1.12.2 3D MEMS and IC Integration

1. Madou, M. J., *Fundamentals of Microfabrication: The Science of Miniaturization*, CRC Press, Boca Raton, FL, 2002.
2. Nguyen, C., "MEMS Technology for Timing and Frequency Control," *IEEE Trans. Ultrason. Ferroelect. Freq. Contr.* 54:251–270, 2007.
3. Gad-el-Hak, M. *The Mems Handbook*, CRC Press, Boca Raton, FL, 2002.
4. Hacker, J. B., R. E. Mihailovich, M. Kim, and J. F. DeNatale, "A Ka-Band 3-Bit rf MEMS True-Time-Delay Network," *IEEE Trans. Microwave Theory Tech.* 51:305–308, 2003.
5. Menz, W., J. Mohr, and O. Paul, *Microsystem Technology*, Wiley-VCH, Hoboken, NJ, 2001.
6. Goldsmith, C. L., Z. Yao, S. Eshelman, and D. Denniston, "Performance of Low-Loss RF MEMS Capacitive Switches," *IEEE Microwave Wireless Compon. Lett.* 8:269–271, 1998.
7. Senturia, S. D., *Microsystem Design*, Springer, New York, 2000.
8. Rebeiz, G. M., *RF MEMS: Theory, Design and Technology*, Wiley, Hoboken, NJ, 2003.
9. Anagnostou, D. E., C. G. Christodoulou, G. Tzeremes, T. S. Liao, and P. K. L. Yu, "Fractal Antennas with RF-MEMS Switches for Multiple Frequency Applications," *Proceedings of the IEEE APS/URSI International Symposium*, Vol. 2, San Antonio, TX, June 2002, pp. 22–25.
10. Yano, M., F. Yamagishi, and T. Tsuda, "Optical MEMS for Photonic Switching Compact and Stable Optical Crossconnect Switches for Simple, Fast, and Flexible Wavelength Applications in Recent Photonic Networks," *J. Selected Topics Quantum Elect.* 11:383–394, 2005.

11. Anagnostou, D. E., G. Zheng, M. Chryssomallis, J. C. Lyke, G. E. Ponchak, J. Papapolymerou, and C. G. Christodoulou, "Design, Fabrication and Measurements of a Self-Similar Re-Configurable Antenna with RF-MEMS Switches," *IEEE Trans. Antennas Propagat.* 54:422–432, 2006.
12. Liu, A. Q., and X. M. Zhang, "A Review of MEMS External-Cavity Tunable Lasers," *J. Micromech. Microeng.* 17:R1–R13, 2007.
13. Huff, G. H., and J. T. Bernhard, "Integration of Packaged rf MEMS Switches with Radiation Pattern Reconfigurable Square Spiral Microstrip Antennas," *IEEE Trans. Antennas Propagat.* 54:464–469, 2006.
14. Kingsley, N., D. E. Anagnostou, M. Tentzeris, and J. Papapolymerou, "RF MEMS Sequentially Reconfigurable Sierpinski Antenna on a Flexible Organic Substrate with Novel DC-Biasing Technique," *IEEE/ASME J. Microelectromech. Syst.* 16:1185–1192, 2007.
15. Van Caekenberghe, K., and K. Sarabandi, "A 2-Bit Ka-Band RF MEMS Frequency Tunable Slot Antenna," *IEEE Antennas Wireless Propagat. Lett.* 7:179–182, 2008.
16. Nguyen, C., "MEMS Technology for Timing and Frequency Control," *IEEE Trans. Ultrason. Ferroelect. Freq. Contr.* 54:251–270, 2007.
17. Young, R. M., J. D. Adam, C. R. Vale, T. T. Braggins, S. V. Krishnaswamy, C. E. Milton, D. W. Bever, L. G. Chorosinski, Li-Shu Chen, D. E. Crockett, C. B. Freidhoff, S. H. Talisa, E. Capelle, R. Tranchini, J. R. Fende, J. M. Lorthioir, and A. R. Tories, "Low-Loss Bandpass RF Filter Using MEMS Capacitance Switches to Achieve a One-Octave Tuning Range and Independently Variable Bandwidth," *IEEE MTT-S Int. Microwave Symp. Digest* 3:1781–1784, 2003.
18. Tan, G. L., R. E. Mihailovich, J. B. Hacker, J. F. DeNatale, and G. M. Rebeiz, "Low-Loss 2- and 4-bit TTD MEMS Phase Shifters Based on SP4T Switches," *IEEE Trans. Microwave Theory Tech.* 51:297–304, 2003.
19. Hacker, J. B., R. E. Mihailovich, M. Kim, and J. F. DeNatale, "A Ka-Band 3-Bit RF MEMS True-Time-Delay Network," *IEEE Trans. Microwave Theory Tech.* 51:305–308, 2003.
20. Ford, J. E., K. W. Goossen, J. A. Walker, D. T. Neilson, D. M. Tennant, S. Y. Park, and J. W. Sulhoff, "Interference-Based Micromechanical Spectral Equalizers," *IEEE J. Selected Topics Quantum Elect.* 10:579–587, 2004.
21. Nordquist, C. D., C. W. Dyck, G. M. Kraus, I. C. Reines, L. Goldsmith, D. Cowan, T. A. Plut, F. Austin, IV, P. S. Finnegan, M. H. Ballance, and T. Sullivan, "A DC to 10 GHz 6-Bit RF MEMS Time Delay Circuit," *IEEE Microwave Wireless Compon. Lett.* 16:305–307, 2006.
22. Perruisseau-Carrier, J., R. Fritschi, P. Crespo-Valero, and A. K. Skrivervik, "Modeling of Periodic Distributed MEMS Application to the Design of Variable True-Time-Delay Lines," *IEEE Trans. Microwave Theory Tech.* 54:383–392, 2006.
23. Kim, C.-H., N. Park, and Y.-K. Kim, "MEMS Reflective Type Variable Optical Attenuator Using Off-Axis Misalignment," *Proc. IEEE/LEOS Int. Conf. Opt. MEMS*, Lugano, Switzerland, 2002, pp. 55–56.
24. Lakshminarayanan, B., and T. M. Weller, "Design and Modeling of 4-Bit Slow wave MEMS Phase Shifters," *IEEE Trans. Microwave Theory Tech.* 54:120–127, 2006.
25. Lakshminarayanan, B., and T. M. Weller, "Optimization and Implementation of Impedance-Matched True-Time-Delay Phase Shifters on Quartz Substrate," *IEEE Trans. Microwave Theory Tech.* 55:335–342, 2007.
26. Van Caekenberghe, K., and T. Vaha-Heikkila, "An Analog RF MEMS Slotline True-Time-Delay Phase Shifter," *IEEE Trans. Microwave Theory Tech.* 56:2151–2159, 2008.
27. Maciel, J. J., J. F. Slocum, J. K. Smith, and J. Turtle, "MEMS Electronically Steerable Antennas for Fire Control Radars," *IEEE Aerosp. Electron. Syst. Mag.*, November 2007, pp. 17–20.
28. Yeow, T.-W., K. L. E. Law, and A. Goldenberg, "MEMS Optical Switches," *IEEE Commun. Mag.* 39:158–163, 2001.

29. Herrick, K. J., G. Jerinic, R. P. Molfino, S. M. Lardizabal, and B. Pillans, "S-Ku Band Intelligent Amplifier Microsystem," *Proc. SPIE* 6232, May 2006.
30. Neukermans, A., and R. Ramaswami, "MEMS Technology for Optical Networking," *IEEE Commun. Mag.* 39:62–69, 2001.
31. Pranonsatit, S., A. S. Holmes, I. D. Robertson, and S. Lucyszyn, "Single-Pole Eight-Throw RF MEMS Rotary Switch," *IEEE/ASME J. Microelectromech. Syst.* 15:1735–1744, 2006.
32. Lin, L. Y., and E. L. Goldstein, "Opportunities and Challenges for MEMS in Lightwave Communications," *IEEE J. Selected Topics Quantum Elect.* 8:163–172, 2002.
33. Vaha-Heikkila, T., K. Van Caekenberghe, J. Varis, J. Tuovinen, and G. M. Rebeiz, "RF MEMS Impedance Tuners for 6–24-GHz Applications," *Wiley Int. J. RF Microwave Computer-Aided Eng.* 17: 265–278, 2007.
34. Syms, R. A., and D. F. Moore, "Optical MEMS for Telecoms," *Materials Today* 5:26–35, 2002.
35. Schoebel, J., T. Buck, M. Reimann, M. Ulm, M. Schneider, A. Jourdain, G. Carchon, and H. Tilmans, "Design Considerations and Technology Assessment of Phased Array Antenna Systems with RF MEMS for Automotive Radar Applications," *IEEE Trans. Microwave Theory Tech.* 53:1968–1975, 2005.
36. Wu, M. C., O. Solgaard, and J. E. Ford, "Optical MEMS for Lightwave Communication," *J. Lightwave Technol.* 24:4433–4454, 2006.
37. Mailloux, R. J. *Phased Array Antenna Handbook*, Artech House, New York, 2005.
38. Hoffmann. M., and E. Voges, "Bulk Silicon Micromachining for MEMS in Optical Communication Systems," *J. Micromech. Microeng.* 12:349–360, 2002.
39. Jung, C., M. Lee, G. P. Li, and F. D. Flaviis, "Reconfigurable Scan-Beam Singlearm Spiral Antenna Integrated with RF MEMS Switches," *IEEE Trans. Antennas Propagat.* 54:455–463, 2006.
40. Chang-Hasnain, C. J. "Tunable VCSEL," *J. Selected Topics Quantum Elect.* 6: 978–987, 2000.
41. Lijie, J. Z., and D. Uttamchandani, "Integrated Self-Assembling and Holding Technique Applied to a 3D MEMS Variable Optical Attenuator," *IEEE J. Microelectromech. Syst.* 13:83–90, 2004.
42. Wikipedia, "Microelectromechanical Systems," available at http:// en.wikipedia .org/wiki/MEMS; accessed April 1, 2008.
43. Yole Development, "World MEMS Markets: The 2006–2012 MEMS Market Database," 2008.
44. Lau, J. H., C. P. Wong, N. C. Lee, and R. Lee, *Electronics Manufacturing with Lead-Free, Halogen-Free, and Adhesive Materials*, McGraw-Hill, New York, 2003.
45. Lau, J. H., and R. Lee, *Microvias for Low-Cost, High-Density Interconnects*, McGraw-Hill, New York, 2001.
46. Lau, J. H., *Low-Cost Flip-Chip Technologies for DCA, WLCSP, and PBGA Assemblies*, McGraw-Hill, New York, 2000.
47. Lau, J. H., and R. Lee, *Chip Scale Package Design: Materials, Process, Reliability, and Applications*, McGraw-Hill, New York, 1999.
48. Lau, J. H., C. P. Wong, J. Prince, and W. Nakayama, *Electronic Packaging: Design, Materials, Process, and Reliability*, McGraw-Hill, New York, 1998.
49. Lau, J. H., and Y. Pao, *Solder Joint Reliability of BGA, CSP, Flip Chip, and Fine Pitch SMT Assemblies*, McGraw-Hill, New York, 1997.
50. Lau, J. H., *Flip Chip Technologies*, McGraw-Hill, New York, 1996.
51. Lau, J. H., *Ball Grid Array Technology*, McGraw-Hill, New York, 1995.
52. Lau, J. H., *Chip on Board Technologies for Multichip Modules*, Van Nostrand Reinhold, New York, 1994.
53. Lau, J. H., *Handbook of Fine Pitch Surface Mount Technology*, Van Nostrand Reinhold, New York, 1994.
54. Lau, J. H., *Thermal Stress and Strain in Microelectronics Packaging*, Van Nostrand Reinhold, New York, 1993.
55. Lau, J. H., *Handbook of Tape Automated Bonding*, Van Nostrand Reinhold, New York, 1992.
56. Lau, J. H., *Solder Joint Reliability: Theory and Applications*, Van Nostrand Reinhold, New York, 1991.

57. Mohamed Gad-el-Hak. *The MEMS Handbook*, CRC Press, Boca Raton, FL, 2002.
58. Van Caekenberghe, K., and K. Sarabandi, "A 2-Bit Ka-Band RF MEMS Frequency Tunable Slot Antenna," *IEEE Antennas and Wireless Propagat. Lett.* 7:179–182, 2008.
59. Premachandran, C. S., M. Chew, W. Choi, A. Khairyanto, K. Chen, J. Singh, S. Wang, Y. Xu, N. Chen, C. Sheppard, M. Olivo, and J. H. Lau, "Influence of Optical Probe Packaging on a 3D MEMS Scanning Micro Mirror for Optical Coherence Tomography (OCT) Applications," *IEEE Proceedings of Electronic Components and Technology Conference*, Orlando, FL, May 27–30, 2008, pp. 829–833.
60. Premachandran, C. S., J. H. Lau, X. Ling, A. Khairyanto, K. Chen, and Myo Ei Pa Pa, "A Novel, Wafer-Level Stacking Method for Low-Chip Yield and Non-Uniform, Chip-Size Wafers for MEMS and 3D SiP Applications," *IEEE Proceedings of Electronic Components and Technology Conference*, Orlando, FL, May 27–30, 2008, pp. 314–318.
61. Lau, J. H., "3D MEMS Packaging," *IMAPS Proceedings*, San Jose, CA, November 2009.
62. Chen, K., C. Premachandran, K. Choi, C. Ong, X. Ling, A. Ratmin, M. Pa, and J. H. Lau, "C2W Low Temperature Bonding Method for MEMS Applictions," *IEEE Proceedings of Electronics Packaging Technology Conference*, Singapore, December 2008, pp. 1–7.
63. Klumpp, A., "Vertical System Integration by Using Inter-Chip Vias and Solid-Liquid Interdiffusion Bonding," *Jpn. J. Appl. Phys.* 43:L829–L830, 2004.
64. Chen, K., "Microstructure Examination of Copper Wafer Bonding," *J. Electron. Mater.* 30:331–335, 2001.
65. Morrow, P. R., "Three-Dimensional Wafer Stacking via Cu-Cu Bonding Integrated with 65-nm Strained-Si/Low-k CMOS Technology," *IEEE Electron. Device Lett.* 27:335–337, 2006.
66. Shimbo, M., K. Furukawa, K. Fukuda, and K. Tanzawa, "Silicon-to-Silicon Direct Bonding Method," *J. Appl. Phys.* 60:2987–2989, 1986.
67. Made, R., C. L. Gan, L. Yan, A. Yu, S. U. Yoon, J. H. Lau, and C. Lee, "Study of Low Temperature Thermocompression Bonding in Ag-In Solder for Packaging Applications," *J. Electron. Mater.* 38:365–371, 2009.
68. Yan, L.-L., C.-K. Lee, D.-Q. Yu, A.-B. Yu, W.-K. Choi, J. H. Lau, and S.-U. Yoon, "A Hermetic Seal Using Composite Thin Solder In/Sn as Intermediate Layer and Its Interdiffusion Reaction with Cu," *J. Electron. Mater.* 38:200–207, 2009.
69. Yan, L.-L., V. Lee, D. Yu, W. K. Choi, A. Yu, S.-U. Yoon, and J. H. Lau, "A Hermetic Chip to Chip Bonding at Low Temperature with Cu/In/Sn/Cu Joint," *IEEE/ECTC Proceedings*, Orlando, FL, May 2008, pp. 1844–1848.
70. Yu, A., C. Lee, L. Yan, R. Made, C. Gan, Q. Zhang, S. Yoon, and J. H. Lau, "Development of Wafer Level Packaged Scanning Micromirrors," *Proc. Photon. West* 6887:1–9, 2008.
71. Lee, C., A. Yu, L. Yan, H. Wang, J. Han, Q. Zhang, and J. H. Lau, "Characterization of Intermediate In/Ag Layers of Low Temperature Fluxless Solder Based Wafer Bonding for MEMS Packaging," *J. Sensors Actuators* (in press).
72. Yu, D.-Q., C. Lee, L. L. Yan, W. K. Choi, A. Yu, and J. H. Lau, "The Role of Ni Buffer Layer on High-Yield, Low-Temperature Hermetic Wafer Bonding Using In/Sn/Cu Metallization," *Appl. Phys. Lett.* 94 (3):034105 – 034105-3, 2009.
73. Yu, D. Q., L. L. Yan, C. Lee, W. K. Choi, S. U. Yoon, and J. H. Lau, "Study on High Yield Wafer to Wafer Bonding Using In/Sn and Cu Metallization," *Proceedings of the Eurosensors Conference*, Dresden, Germany, 2008, pp. 1242–1245.
74. Yu, D., C. Lee, and J. H. Lau, "The Role of Ni Buffer Layer Between InSn Solder and Cu Metallization for Hermetic Wafer Bonding," *Proceedings of the International Conference on Electronics Materials and Packaging*, Taipei, Taiwan, October 22–24, 2008, pp. 335–338.
75. Yu, D., L. Yan, C. Lee, W. Choi, M. Thew, C. Foo, and J. H. Lau, "Wafer Level Hermetic Bonding Using Sn/In and Cu/Ti/Au Metallization," *IEEE/ECTC Proceeding*, Singapore, December 2008, pp. 1–6.

76. Choi, W., D. Yu, C. Lee, L. Yan, A. Yu, S. Yoon, J. H. Lau, M. Cho, Y. Jo, and H. Lee, "Development of Low Temperature Bonding Using In-Based Solders," *IEEE/ECTC Proceedings*, Orlando, FL, May 2008, pp. 1294–1299.

77. Choi, W., C. Premachandran, C. Ong, X. Liang, E. Liao, A. Khairyanto, K. Chen, K., P. Thaw, and J. H. Lau, "Development of Novel Intermetallic Joints Using Thin Film Indium Based Solder by Low Temperature Bonding Technology for 3D IC Stacking," *IEEE/ECTC Proceedings*, San Diego, CA, May 2009, pp. 333–338.

78. Chen, K., C. Premachandran, K. Choi, C. Ong, X. Ling, A. Khairyanto, B. Ratmin, P. Myo, and J. H. Lau, "C2W Bonding Method for MEMS Applications, *IEEE/ECTC Proceedings*, Singapore, December 2008, pp. 1283–1287.

79. Premachandran, C. S., J. H. Lau, X. Ling, A. Khairyanto, K. Chen, and Myo Ei Pa Pa. "A Novel, Wafer-Level Stacking Method for Low-Chip Yield and Non-Uniform, Chip-Size Wafers for MEMS and 3D SIP Applications," *IEEE Proceedings of Electronic Components and Technology Conference*, Orlando, FL, May 27–30, 2008, pp. 314–318.

80. Simic, V., and Z. Marinkovic, "Room Temperature Interactions in Ag–Metals Thin Film Couples." *Thin Solid Films* 61:149–160, 1979.

81. Lin, J.-C., "Solid-Liquid Interdiffusion Bonding Between In-Coated Silver Thick Films," *Thin Solid Films* 61:212–221, 1979.

82. Roy, R., and S. K. Sen, "The Kinetics of Formation of Intermetallics in Ag/In Thin Film Couples," *Thin Solid Films* 61:303–318, 1979.

83. Chuang, R. W., and C. C. Lee, "Silver-Indium Joints Produced at Low Temperature for High Temperature Devices," *IEEE Trans. Components Packag. Technol.* A25:453–458, 2002.

84. Chuang, R. W., and C. C. Lee, "High-Temperature Non-Eutectic Indiumtin Joints Fabricated by a Fluxless Process," *Thin Solid Films* 414:175–179, 2002.

85. Lee, C. C., and R. W. Chuang, "Fluxless Non-eutectic Joints Fabricated using gold-tin multilayer composite," *IEEE Trans. Components Packag. Technol.* A26:416–422, 2003.

86. Humpston, G., and D. Jacobson, *Principles of Soldering and Brazing*, ASM International, Materials Park, MD, 1993, pp 128–132.

87. Chuang, T., H. Lin, and C. Tsao, "Intermetallic Compounds Formed During Diffusion Soldering of Au/Cu/Al$_2$O$_3$ and Cu/Ti/Si with Sn/In Interlayer," *J. Electron. Mater.* 35:1566–1570, 2006.

88. Lee, C., and S. Choe, "Fluxless In-Sn Bonding Process at 140°C," *Mater. Sci. Eng.* A333:45–50, 2002.

89. Lee, C. "Wafer Bonding by Low-Temperature Soldering," *Sensors & Actuators* 85:330–334, 2000.

90. Tazzoli, A., M. Rinaldi, and G. Piazza, "Ovenized High Frequency Oscillators Based on Aluminum Nitride Contour-Mode MEMS Resonators," *IEEE/IEDM Proceedings*, December 2011, pp. 20.2.1–20.2.4.

91. Hwang, E., A. Driscoll, and S. Bhave, "Platform for JFET-Based Sensing of RF MEMS Resonators in CMOS Technology," *IEEE/IEDM Proceedings*, December 2011, pp. 20.4.1–20.4.4.

92. Liu, H., C. Lee1, T. Kobayashi, C. Tay, and C. Quan, "A MEMS-Based Wideband Piezoelectric Energy Harvester System Using Mechanical Stoppers," *IEEE/ IEDM Proceedings*, December 2011, pp. 29.6.1–29.6.4.

93. Vigna, B., "Tri-Axial MEMS Gyroscopes and Six Degree-Of-Freedom Motion Sensors," *IEEE/IEDM Proceedings*, December 2011, pp. 29.1.1–29.1.4.

94. Elfrink, R., S. Matova, C. de Nooijer, M. Jambunathan, M. Goedbloed, J. van de Molengraft, V. Pop, R.J.M. Vullers, M. Renaud, and R. van Schaijk, "Shock Induced Energy Harvesting with a MEMS Harvester for Automotive Applications," *IEEE/IEDM Proceedings*, December 2011, pp. 29.5.1–29.5.4.

95. Lin, Y., T. Riekkinen, W. Li, E. Alon, and C. Nguyen, "A Metal Micromechanical Resonant Switch for On-Chip Power Applications," *IEEE/IEDM Proceedings*, December 2011, pp. 20.6.1–20.6.4.

96. Sdeghi, M., R. Peterson, and K. Najafi, "Micro-Hydraulic Structure for High Performance Bio-Mimetic Air Flow Sensor Arrays," *IEEE/IEDM Proceedings*, December 2011, pp. 29.4.1–29.4.4.

97. Bhugra, H., Y. Wang, W. Pan, D. Lei, and S. Lee, "High Performance pMEMS Oscillators: The Next Generation Frequency," *IEEE/IEDM Proceedings*, December 2011, pp. 20.1.1–20.1.4.
98. Vullers, R., R. Schaijk, M. Goedbloed, R. Elfrink, Z. Wang, and C. Hoof, "Process Challenges of MEMS Harvesters and Their Effect on Harvester Performance," *IEEE/IEDM Proceedings*, December 2011, pp. 10.2.1–10.2.4.
99. Lau, J. H., *Reliability of RoHS-Compliant 2D and 3D IC Interconnects*, McGraw-Hill, New York, 2011.
100. Lai, J. H., C. Lee, C. Premachandran, and A. Yu, *Advanced MEMS Packaging*, McGraw-Hill, New York, 2010.

1.12.3 Semiconductor IC Packaging

1. Lau, J. H., *Through-Silicon Vias for 3D Integration*, McGraw-Hill, New York, 2013.
2. Lau, J. H., *Reliability of RoHS-Compliant 2D and 3D IC Interconnects*, McGraw-Hill, New York, 2011.
3. Lau, J. H., C. K. Lee, C. S. Premachandran, and A. Yu, *Advanced MEMS Packaging*, McGraw-Hill, New York, 2010.
4. Lau, J. H., C. P. Wong, N. C. Lee, and S. W. Lee, *Electronics Manufacturing with Lead-Free, Halogen-Free and Conductive-Adhesive Materials*, McGraw-Hill, New York, 2003.
5. Lau, J. H, and S. W. Ricky Lee, *Microvias for Low Cost, High Density Interconnects*, McGraw-Hill, New York, 2001.
6. Lau, J. H., *Low-Cost Flip-Chip Technologies*, McGraw-Hill, New York, 2000.
7. Lau, J. H., and S. W. R. Lee, *Chip-Scale Package*, McGraw-Hill, New York, 1999.
8. Lau, J. H., C. P. Wong, J. Prince, and W. Nakayama, *Electronic Packaging: Design, Materials, Process, and Reliability*, McGraw-Hill, New York, 1998.
9. Lau, J. H. and Y.-H. Pao, *Solder Joint Reliability of BGA, CSP, Flip Chip and Fine Pitch SMT Assemblies*, McGraw-Hill, New York, 1997.
10. Lau, J. H. (ed.), *Flip Chip Technologies*, McGraw-Hill, New York, 1996.
11. Lau, J. H. (ed.), *Ball Grid Array Technology*, McGraw-Hill, New York, 1995.
12. Lau, J. H. (ed.), *Chip-on-Board: Technologies for Multichip Modules*, Van Nostrand Reinhold, New York, 1994.
13. Lau, J. H. (ed.), *Handbook of Fine Pitch Surface Mount Technology*, Van Nostrand Reinhold, New York, 1994.
14. Frear, D., H. Morgan, S. Burchett, and J. H. Lau (eds.), *The Mechanics of Solder Alloy Interconnects*, Van Nostrand Reinhold, New York, 1994.
15. Lau, J. H. (ed.), *Thermal Stress and Strain in Microelectronics Packaging*, Van Nostrand Reinhold, New York, 1993.
16. Lau, J. H. (ed.), *Handbook of Tape Automated Bonding*, Van Nostrand Reinhold, New York, 1992.
17. Lau, J. H. (ed.), *Solder Joint Reliability: Theory and Applications*, Van Nostrand Reinhold, New York, 1991.
18. Tummala, R. R., and Madhavan Swaminathan, *System on Package: Miniaturization of the Entire System*, McGraw-Hill, New York, 2008.
19. Tummala, R. R., *Fundamentals of Microsystems Packaging*, McGraw-Hill, New York, 2001.
20. Tummala, R. R., and E. J. Rymaszewski (eds.), *Microelectronics Packaging Handbook*, Van Nostrand Reinhold, New York, 1989.
21. Tummala, R. E., E. J. Rymasewski, and A. G. Klopfenstein (eds.), *Microelectronics Packaging Handbook: Semiconductor Packaging*, Part II, 2nd ed., Chapman & Hall, New York, 1997.
22. Tummala, R. E., E. J. Rymasewski, and A. G. Klopfenstein (eds.), *Microelectronics Packaging Handbook: Subsystem Packaging*, Part III, 2nd ed., Chapman & Hall, New York, 1997.
23. Wong, C. P., *Polymers for Electronic and Photonic Application*, Academic Press, New York, 1993.

24. Lu, D., and C. P. Womg (eds.), *Materials for Advanced Packaging*, Springer, New York, 2008.
25. Suhir, E., Y. C. Lee, and C. P. Wong (eds.), *Micro- and Opto-Electronic Materials and Structures: Physics, Mechanics, Design, Reliability, Packaging*, Springer, New York, 2007.
26. Garrou, P., C. Bower, and P. Ramm (eds.), *Handbook of 3D Integration*, Wiley, New York, 2008.
27. Wu, B., A. Kumar, and S. Ramaswami (eds.), *3D IC Stacking Technology*, McGraw-Hill, New York, 2011.
28. Xie, Y., J. Cong, and S. Sapatnekar (eds.), *Three-Dimensional Integrated Circuit Design*, Springer, New York, 2010.
29. Flick, E. W., *Adhesives, Sealants and Coatings for the Electronics Industry*, 2nd ed., Noyes Publications, Park Ridge, NJ, 1992.
30. Bar-Cohen, A., and A. D. Kraus (eds.), *Advances in Thermal Modeling of Electronic Components and Systems*, Vol. 1, Hemisphere Publishing Corporation, New York, 1988.
31. Bar-Cohen, A., and A. D. Kraus (eds.), *Advances in Thermal Modeling of Electronic Components and Systems*, Vol. 2, ASME Press, New York, 1990.
32. Bar-Cohen, A., and A. D. Kraus (eds.), *Advances in Thermal Modeling of Electronic Components and Systems*, Vol. 3, ASME Press, New York, 1993.
33. Bar-Cohen, A., and A. D. Kraus (eds.), *Advances in Thermal Modeling of Electronic Components and Systems*, Vol. 4, ASME Press, New York, 1998.
34. Shockley, W., *Electrons and Holes in Semiconductors: With Applications to Transistor Electronics*, Van Nostrand Co., New York, 1950.
35. Terano, T., K. Asai, and M. Sugeno, *Applied Fuzzy Systems*, Academic Press Professional, Cambridge, MA, 1989.
36. Hwang, J. S., *Ball Grid Array and Fine Pitch Peripheral Interconnections: A Handbook of Technology and Applications for Microelectronics/Electronics Manufacturing*, Electrochemical Publications, Ayr, Scotland, 1995.
37. Rahn, A., *The Basics of Soldering*, Wiley, New York, 1993.
38. Moore, T. M., and R. G. McKenna (eds.), *Characterization of Integrated Circuit Packaging Materials*, Butterworth-Heinemann, Boston, MA, 1993.
39. Bakoglu, H. B., *Circuits, Interconnections, and Packaging for VSLI*, Addison-Wesley, New York, 1990.
40. Solberg, V., *Design Guidelines for Surface Mount and Fine Pitch Technology*, 2nd ed., McGraw-Hill, New York, 1995.
41. Parker, C. B. (ed.), *Dictionary of Scientific and Technical Terms*, 5th ed., McGraw-Hill, New York, 1994.
42. Kavanagh, P., *Downsizing for Client/Server Applications*, Academic Press Professional, Cambridge, MA, 1995.
43. ASM International Handbook Committee, *Electronic Materials Handbook:* Vol. I, *Packaging*, ASM, Materials Park, OH, 1989.
44. Landers, T. L., W. D. Brown, E. W. Fant, E. M. Malstrom, and N. M. Schmitt, *Electronics Manufacturing Processes*, Prentiss-Hall, Englewood Cliffs, NJ, 1994.
45. Baker, D., D. C. Koehler, W. O. Fleckenstein, C. E. Roden, and R. Sabia, *Physical Design of Electronic Systems:* Vol. III: Integrated Device and Connection Technology, Prentice-Hall, Englewood Cliffs, NJ, 1971.
46. Keyser, C. A., *Materials: Science in Engineering*, 2nd ed., Merrill Publishing Co., Columbus, OH, 1968.
47. Harper, C. A., and R. M. Sampson, *Electronic Materials and Processes Handbook*, 2nd ed., McGraw-Hill, New York, 1994.
48. Harper, C. A. (ed.), *Electronic Packaging and Interconnection Handbook*, McGraw-Hill, New York, 1991.
49. Klir, G. J., and T. A Folger, *Fuzzy Sets, Uncertainty, and Information*, Prentice-Hall, Englewood Cliffs, NJ, 1988.
50. Morris, J. E. (ed.), *Electronics Packaging Forum*, Vol. 1, Van Nostrand Reinhold, New York, 1990.
51. Dieter, G. E., *Engineering Design: A Materials and Processing Approach*, McGraw-Hill, New York, 1991.

52. Morris, J. E. (ed.), *Electronics Packaging Forum*, Vol. 2, Van Nostrand Reinhold, New York, 1991.
53. Ruoff, A. L., *Introduction to Materials Science*, Prentice-Hall, Englewood Cliffs, NJ, 1972.
54. Morris, J. E. (ed.), *Electronics Packaging Forum: Multichip Module Technology Issues*, IEEC Press, New York, 1994.
55. Tittel, E., and M. Robbins, *E-mail Essentials*, Academic Press Professional, Cambridge, MA, 1994.
56. *Engineered Materials Handbook*: Vol. 1, *Composites*, ASM International, New York, 1987.
57. *Engineered Materials Handbook:* Vol. 2, *Engineering Plastics*, ASM International, New York, 1988.
58. *Engineered Materials Handbook:* Vol. 3, *Adhesives and Sealants*, ASM International, New York, 1990.
59. *Engineered Materials Handbook:* Vol. 4, *Ceramics and Glasses*, ASM International, New York, 1991.
60. Marcoux, P. P., *Fine Pitch Surface Mount Technology: Quality, Design, and Manufacturing Techniques*, Van Nostrand Reinhold, New York, 1992.
61. Fjelstad, J., *Flexible Circuit Technology*, Silicon Valley Publishers Group, Campbell, CA 1994.
62. Stearns, T. H., *Flexible Printed Circuitry*, McGraw-Hill, New York, 1996.
63. McNeill, F. M., and E. Thro, *Fuzzy Logic: A Practical Approach*, Academic Press Professional, Cambridge, MA, 1994.
64. Cox, E., *The Fuzzy Systems Handbook: A Practitioner's Guide to Building, Using, and Maintaining Fuzzy Systems*, Academic Press Professional, Cambridge, MA, 1994.
65. *General Requirements for Implementation of Statistical Process Control*, ANSI/IPC-PC-90, IPC, Lincolnwood, IL, October 1990.
66. *Guidelines for Multichip Module Technology Utilization*, IPC-MC-790, IPC, Lincolnwood, IL, July 1992.
67. Pecht, M. (ed.), *Handbook of Electronic Package Design*, Marcel Dekker, New York, 1991.
68. Matisoff, B. S., *Handbook of Electronics Packaging Design and Engineering*, 2nd ed., Van Nostrand Reinhold, New York, 1990.
69. Harper, C. A. (ed.), *Handbook of Plastics, Elastomers, and Composites*, 2nd ed., McGraw-Hill, New York, 2002.
70. Johnson, H. W., and M. Graham, *High-Speed Digital Design: A Handbook of Black Magic*, Prentice-Hall, Englewood Cliffs, NJ, 1993.
71. Licari, J. J., and L. R. Enlow, *Hybrid Microcircuit Technology Handbook: Materials, Processes, Design, Testing and Production*, Noyes Publications, Park Ridge, NJ, 1988.
72. Sergent, J. E., and C. A. Harper (eds.), *Hybrid Microelectronics Handbook*, 2nd ed., McGraw-Hill, Inc., New York, 1995.
73. Pecht, M., *Integrated Circuit, Hybrid, and Multichip Module Package Design Guidelines: A Focus on Reliability*, Wiley, New York, 1994.
74. O'Mara, W. C., *Liquid Crystal Flat Panel Display: Manufacturing Science and Technology*, Van Nostrand Reinhold, New York, 1993.
75. Kirschman, R. K. (ed.), *Low-Temperature Electronics*, IEEE Press, New York, 1986.
76. *Mastering and Implementing BGA Technology*, AEIC, New York, 1995.
77. Yost, F. G., F. M. Hosking, and D. R. Frear, *The Mechanics of Solder Alloy Wetting and Spreading*, Van Nostrand Reinhold, New York, 1993.
78. Grovenor, C. R. M., *Microelectronic Materials*, Institute of Physics Publishing, Washington, DC, 1994.
79. Tewksbury, S. K. (ed.), *Microelectronic System Interconnections: Performance and Modeling*, IEEE Press, New York, 1994.
80. Landzberg, A. H. (ed.), *Microelectronics Manufacturing Diagnostics Handbook*, Van Nostrand Reinhold, New York, 1993.
81. Motorola Semiconductor Products, Inc., *Microprocessor Applications Manual*, McGraw-Hill, New York, 1975.

82. Lowenheim, F. (ed.), *Modern Electroplating*, 3rd ed., Wiley, New York, 1974.
83. Doane, D. A., and Franzon, P. D., *Multichip Module Technologies and Alternatives: The Basics*, Van Nostrand Reinhold, New York, 1993.
84. Johnson, R. W., R. K. F. Teng, and J. W. Balde (eds.), *Multichip Modules: Systems Advantages, Major Constructions, and Materials Technologies*. IEEE Press, New York, 1991.
85. Rao, G. K., *Multilevel Interconnect Technology*, McGraw-Hill, New York, 1993.
86. Tittel, E., and M. Robbins, *Network Design Essentials: Everything You Need to Know*, Academic Press Professional, Cambridge, MA, 1994.
87. Simon, A. R., *Network Re-Engineering: Foundations of Enterprise Computing*, Academic Press Professional, Cambridge, MA, 1994.
88. Beranek, L. L. (ed.), *Noise and Vibration Control*, rev. ed., McGraw-Hill, New York, 1971.
89. Montgomery, S. L., *Object-Oriented Information Engineering: Analysis, Design, and Implementation*, Academic Press Professional, Cambridge, MA, 1994.
90. Simon, A. R., and T. Wheeler, *Open Systems Handbook*, 2nd ed., Academic Press Professional, Cambridge, MA., 1995.
91. Dally, J. W., *Packaging of Electronic Systems: A Mechanical Engineering Approach*, McGraw-Hill, New York, 1990.
92. Neugebauer, C. A., A. F. Yerman, R. O. Carlson, J. F. Burgess, H. F. Webster, and H. H. Glascock, *The Packaging of Power Semiconductor Devices*, Electrocomponent Science Monographs, Vol. 7, Gordon and Breach Science Publishers, New York, 1986.
93. Hannemann, R. J., A. D. Kraus, and M. Pecht (eds.), *Physical Architecture of VLSI Systems*, Wiley, New York, 1994.
94. Pecht, M., *Placement and Routing of Electronic Modules*, Marcel Dekker, New York, 1993.
95. Pecht, M. G., L. T. Nguyen, and E. B. Hakim (eds.), *Plastic-Encapsulated Microelectronics: Materials, Processes, Quality, Reliability, and Applications*, Wiley, New York, 1995.
96. Manzione, L. T., *Plastic Packaging of Microelectronic Devices*, Van Nostrand Reinhold, New York, 1990.
97. Seraphim, D. P., R. Lasky, and C.-Y. Li, *Principles of Electronic Packaging*, McGraw-Hill, New York, 1989.
98. Coombs, C. F., Jr., *Printed Circuits Handbook*, 4th ed., McGraw-Hill, New York, 1996.
99. Stevens, R. T., *Quick Reference to Computer Graphics Terms*, Academic Press Professional, Cambridge, MA, 1993.
100. *Reliability Assessment of Wafer Scale Integration Using Finite Element Analysis*, Rome Laboratory, Griffiss AFB, New York, October 1991.
101. Harman, George G., *Reliability and Yield Problems of Wire Bonding in Microelectronics*, International Society for Hybrid Microelectronics, Washington, DC, 1991.
102. *Rome Laboratory Reliability Engineer's Toolkit: An Application Oriented Guide for the Practicing Reliability Engineer*, Systems Reliability Division, AFMC, Griffiss AFB, New York, 1993.
103. Frear, D. R., W. B. Jones, and K. R. Kinsman (ed.), *Solder Mechanics: A State of the Art Assessment*, Minerals, Metals and Materials Society, Washington, DC, 1990.
104. Manko, H. H., *Soldering Handbook for Printed Circuits and Surface Mounting*, 2nd ed., Van Nostrand Reinhold, New York, 1994.
105. Judd, M., and K. Brindley, *Soldering in Electronics Assembly, BH Newness*, East Kilbride, Scotland, 1992.
106. Pecht, M. (ed.), *Soldering Processes and Equipment*, Wiley, New York, 1993.
107. Manko, H. H., *Solders and Soldering: Materials, Design, Production, and Analysis for Reliable Bonding*, McGraw-Hill, New York, 1992.
108. Christian, J., and G.-A. Nazri, *Solid State Batteries: Materials Design and Optimization*, Kluwer Academic Publishers, Boston, MA 1994.

109. Suhir, E., *Structural Analysis in Microelectronic and Fiber-Optic Systems*: Vol. I, Van Nostrand Reinhold, New York, 1991.

110. Engel, P. A., *Structural Analysis of Printed Circuit Board Systems*, Mechanical Engineering Series, Springer-Verlag, New York, 1993.

111. Caswell, G. (organizer), *Surface Mount Technology*, International Society for Hybrid Microelectronics, Silver Spring, MD, 1984,.

112. Classon, F., *Surface Mount Technology for Concurrent Engineering and Manufacturing*, McGraw-Hill, New York, 1993.

113. Vardaman, J., *Surface Mount Technology: Recent Japanese Developments*, IEEE Press, New York, 1993.

114. Furkay, S. S., R. F. Kilburn, G. Monti, Jr. (eds.), *Thermal Management Concepts in Microelectronic Packaging: from Component to System*, International Society for Hybrid Microelectrinics, Silver Spring, MD, 1984.

115. Messner, G., I. Turlik, J. W. Balde, and P. E. Garrou, *Thin Film: Multichip Modules*, International Society of Hybrid Microelectronics, Reston, VA, 1992.

116. Perry, D. L., *VHDL*, 2nd ed., McGraw-Hill, New York, 1994.

117. Pick, J., *VHDL Techniques, Experiments, and Caveats*, McGraw-Hill, New York, 1996.

118. Steinberg, D. S., *Vibration Analysis for Electronic Equipment*, 2nd ed., Wiley, New York, 1988.

119. Bar-Cohen, A., and A. D. Kraus (eds.), *Advances in Thermal Modeling of Electronic Components and Systems*, Vol. 3, ASME Press, New York, 1993.

120. Leidheiser, H., Jr., *Corrosion Control by Coatings*, Science Press, Princeton, NJ, 1979.

121. Koch, W. E., *Engineering Applications of Lasers and Holography*, Plenum Press, New York, 1975.

122. Hinch, S. W., *Handbook of Surface Mount Technology*, Longman Scientific & Technical, Essex, England, 1988.

123. Mandelkern, L., *An Introduction to Macromolecules*, Springer-Verlag, New York, 1972.

124. Brisky, M., *Mastering SMT Manufacturing*, SMT Plus, Inc., New York, 1992.

125. Chung, D. D. L., *Materials for Electronic Packaging*, Butterworth-Heinemann, Boston, MA 1995.

126. Lea, C., *A Scientific Guide to Surface Mount Technology*, Electrochemical Publications, Ayr, Scotland, 1988.

127. Prasad, Ray P., *Surface Mount Technology: Principles and Practice*, Van Nostrand Reinhold, New York, 1989.

128. Prasad, R. P., *Surface Mount Technology: Principles and Practice*, 2nd ed., Chapman & Hall, New York, 1997.

129. Lall, P., M. G. Pecht, and E. B. Hakim, *Influence of Temperature on Microelectronics and System Reliability*, CRC Press, Boca Raton, FL, 1997.

130. Harper, C. A., *Electronic Packaging and Interconnection Handbook*, 2nd ed., McGraw-Hill, New York, 1997.

131. Woodgate, R. W., *The Handbook of Machine Soldering: SMT and TH*, 3rd ed., Wiley, New York, 1996.

132. Hwang, J. S., *Modern Solder Technology for Competitive Eelctronics Manufacturing*, McGraw-Hill, New York, 1996.

133. Licari, J. J., *Multichip Module Design, Fabrication, & Testing*, McGraw-Hill, New York, 1995.

134. Alvino, W. M., *Plastics for Electronics: Materials, Properties, and Design Applications*, McGraw-Hill, New York, 1995.

135. Pecht, M., A. Dasgupta, J. W. Evans, and J. Y. Evans (eds.), *Quality Conformance and Qualification of Microelectronic Packages and Interconnects*, Wiley, New York, 1994.

136. Harman, G., *Wire Bonding in Microelectronics: Materials, Processes, Reliability, and Yield*, McGraw-Hill, New York, 1997.

137. Intel Corp., *Packaging*, Intel Publications, Mt. Prospect, IL, 1997.

138. Fjelstad, J., *An Engineer's Guide to Flexible Circuit Technology*, Electrochemical Publications, 1997.
139. Barr, A., and W. J. Barr (eds.), *Smart Cards: Seizing Strategic Business Opportunities*, Irwin Professional Publishing, Chicago, 1997.
140. Elshabini-Riad, A., and F. D. Barlow, III, *Thin Film Technology Handbook*, McGraw-Hill, New York, 1997.
141. Lee, Y. C., and W. T. Chen (eds.), *Manufacturing Challenges in Electronic Packaging*, Chapman & Hall, London, 1998.
142. Konsowski, S. G., and A. R. Helland, *Electronic Packaging of High Speed Circuitry*, McGraw-Hill, New York, 1997.
143. DiGiacomo, G., *Reliability of Electronic Packages and Semiconductor Devices*, McGraw-Hill, New York, 1996.
144. Jawitz, M. W., *Printed Circuit Board Materials Handbook*, McGraw-Hill, New York, 1997.
145. Mahidhara, R. K., D. R. Frear, S. M. L. Sastry, K. L. Murty, P. K. Liaw, and W. L. Winterbottom (eds.), *Design and Reliability of Solders and Solder Interconnections*, TMS Minerals Metals Materials, Warrendale, PA, 1997.
146. Srihari, K., C. R. Emerson, S. Krishnan, Y. Hwang, C. H. Wu, and C. W. Yeh, *A "Design for Manufacturing" Environment for Surface Mount PCB Assembly: A Concise User's Guide*, Department of Mechanical and Industrial Engineering., Watson School, SUNY Binghamton, NY, 2012.
147. Ginsberg , G. L. (ed.), *Connectors and Interconnections Handbook*, Vol. 1: *Basic Technology*, Electronic Connector Study Group,. Fort Washington, PA, 1977, Lib. Congress No. 77-088086.
148. Ginsberg, G. L. (ed.), *Connectors and Interconnections Handbook*, Vol. 4: *Materials*, Electronic Connector Study Group, Fort Washington, PA, 1983.
149. Bardes, B. P., *Metals Handbook*, 9th ed., Vol. 1: *Properties and Selection: Irons and Steels*, American Society of Metals, New York, 1978.
150. Bardes, B. P. (ed.), *Metals Handbook*, 9th ed., Vol. 2: *Properties and Selection: Nonferrous Alloys and Pure Metals*, American Society of Metals, New York, 1979.
151. Bardes, B. P. (ed.), *Metals Handbook*, 9th ed., Vol. 3: *Properties and Selection: Stainless Steels, Tool Materials and Special-Purpose Metals*, American Society of Metals, New York, 1980.
152. Dallas, D. B. (ed.), *Tool and Manufacturing Engineers Handbook*, 3rd ed., McGraw-Hill, New York, 1976.
153. Gabriel, B. L., *SEM: A User's Manual for Materials Science*, American Society for Metals, New York, 1985.
154. Oechsner, H. (ed.), *Topics in Current Physics: Thin Film and Depth Profile Analysis*, Springer-Verlag, New York, 1984.
155. Sterling, D. J., Jr., *Technician's Guide to Fiber Optics*, Delmar Publishers, New York, 1987.
156. Senior, J. M., *Optical Fiber Communications: Principles and Practice*, Prentice-Hall, Englewood Cliffs, NJ, 1985.
157. Pecht, M. G., et al., *Electronic Packaging Materials and their Properties*, CRC Press, Boca Raton, FL, 1999.
158. McKeown, S. A., *Mechanical Analysis of Electronic Packaging Systems*, Marcel Dekker, New York, 1999.
159. Liu, J., *Conductive Adhesives for Electronics Packaging*, Electrochemical Publications, Isle of Man, UK, 1999.
160. Woishnis, W. A. (ed.), *Engineering Plastics and Composites*, 2nd ed., ASM International, Materials Park, OH, 1993.
161. Ross, R. J., C. Boit, and D. Staab (eds.), *Microelectronic Failure Analysis: Desk Reference*, 4th ed., ASM International, Materials Park, OH, 1999.
162. Blackwell, G. R. (ed.), *The Electronic Packaging Handbook*, CRC Press, Boca Raton, FL, 1999.
163. Brown, W. (ed.), *Advanced Electronic Packaging: With Emphasis on Multichip Modules*, IEEE Press, Piscataway, NJ, 1999.
164. Avallone, E. A., and T. Baumeister, III (eds.), *Marks' Standard Handbook for Mechanical Engineers*, 10th ed., McGraw-Hill, New York, 1996.

165. Shewhart, W. A., *Statistical Method from the Viewpoint of Quality Control*, Dover Publications, New York, 1986.
166. Mickelson, A. R., N. R. Basavanhally, Y.-C. Lee (eds.), *Optoelectronic Packaging*, Wiley, New York, 1997.
167. Wagner, L. C. (ed.), *Failure Analysis of Integrated Circuits: Tools and Techniques*, Kluwer Academic Publishers, Boston, MA 1999.
168. Harper, C. A. (ed.), *Electronic Packaging and Interconnection Handbook*, 3rd ed., McGraw-Hill, New York, 2000.
169. McCluskey, F. P., R. Grzybowski, and T. Podlesak (eds.), *High Temperature Electronics*, CRC Press, Boca Raton, FL, 1997.
170. Lyshevski, S. E., *Nano- and Microelectromechanical Systems: Fundamentals of Nano- and Microengineering*, CRC Press, Boca Raton, FL, 2001.
171. Remsburg, R., *Thermal Design of Electronic Equipment*, CRC Press, Boca Raton, FL, 2001.
172. Harris, M. C., and G. E. Moore (eds.), *Prospering in a Global Economy: Linking Trade and Technology Policies*, National Academy Press, Washington, DC, 1992.
173. Whitaker, J. C., *Microelectronics*, CRC Press, Boca Raton, FL, 2000.
174. Madou, M., *Fundamentals of Microfabrication*, CRC Press, Boca Raton, FL, 1997.
175. Gibson, D. D., and R. W. Smilor (eds.), *Technology Transfer in Consortia and Strategic Alliances*, Rowman & Littlefield, Lanham, MD, 1992.
176. Brett, A., D. V. Gibson, and R. W. Smilor (eds.), *University Spin-Off Companies*, Rowman & Littlefield, Lanham, MD, 1991.
177. Pecht, M., *Integrated Circuit, Hybrid, and Multichip Module Package Design Guidelines: A Focus on Reliability*, Wiley, New York, 1994.
178. Chigrinov, V. G., *Liquid Crystal Devices: Physics and Applications*, Artech House, Boston, MA 1999.
179. Mohamed Gad-el-Hak (ed.), *The MEMS Handbook*, CRC Press, Boca Raton, FL, 2002.
180. Pecht, M. (ed.), *Product Reliability, Maintainability, and Supportability Handbook*, CRC Press, Boca Raton, FL, 1995.
181. Azar, K. (ed.), *Thermal Measurements in Electronic Cooling*, CRC Press, Boca Raton, FL, 1997.

CHAPTER 2

Through-Silicon Via Technology

2.1 Introduction

Through-silicon via (TSV) technology is the heart and most important key enabling technology of three-dimensional (3D) Si integration and 3D integrated circuit (IC) integration [1,2]. It provides the opportunity for the shortest chip-to-chip interconnects and the smallest pad size and pitch of interconnects. Compared with other interconnection technologies, such as wire bonding, the advantages of TSV include [1–47]

1. Better electrical performance
2. Lower power consumption
3. Wider data width and thus bandwidth
4. Higher density
5. Smaller form factor
6. Lighter weight
7. Lower cost (hopefully)

There are six key steps in making a TSV, and they will be the focus of this chapter:

1. Via formation by either deep reactive-ion etch (DRIE) or laser drilling
2. Dielectric layer deposition by either thermal oxidation for passive interposers or plasma-enhanced chemical vapor deposition (PECVD)
3. Barrier and seed layer deposition by physical vapor deposition (PVD)
4. Cu plating to fill the vias or W (tungsten) sputtering (CVD) for very tiny vias

5. Chemical and mechanical polishing (CMP) of Cu plating residues (overburden)

6. TSV revealing

In general, the order of cost is PVD > PECVD > CMP > Cu plating > DRIE. The inventor of TSV will be briefly mentioned first.

2.2   Who Invented TSV and When

TSV was invented more than 50 years ago [3] by the 1956 Nobel Laureate in Physics, William Shockley. (Yes, the same Shockley who coinvented the transistor, which is generally considered the greatest invention in semiconductor industry.) He filed the patent "Semiconductive wafer and method of making the same" on October 23, 1958, and was granted the U.S. patent (3,044,909) on July 17, 1962. One of the key claims is shown in Fig. 2.1, which gets the semiconductor world so excited today. Basically, the "deep pits" (which are called TSVs today) on the wafer allow the signals from its top side to its bottom side and vice versa. The term *through-silicon via* was coined by Sergey Savastiouk, "Industry Insights: Moore's Law – the Z Dimension," in *Solid State Technology*, January 2000.

2.3   High-Volume Products with TSV Technology

For high-volume commercial products with TSV technology, even Hewlett-Packard's (HP's) coplanar GaAs radiofrequency (RF) monolithic microwave-integrated circuit (MMIC) has been using via-hole grounding technology since 1976 (Fig. 2.2), but it was not for 3D integration. Generally, the industry considers Toshiba's complementary metal-oxide semiconductor (CMOS) image sensor with through-chip vias (TCVs) [4] to be the first high-volume 3D integration product (2008) (Fig. 2.3). Speaking straight, however, just like HP's MMIC, it is not a 3D integration

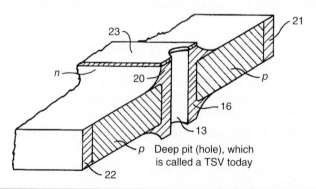

FIGURE 2.1   TSV invented by William Shockley, U.S. Patent No. 3,044,909, filed on October 23, 1958.

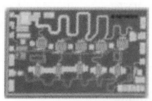

Coplanar GaAs RF MMIC
(monolithic microwave IC)

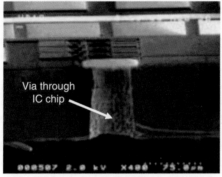

Via-hole grounding technology

FIGURE 2.2   HP's high-volume product (MMIC) with via through an IC chip
(1976).

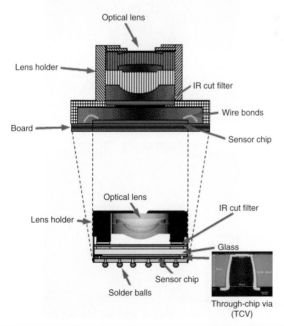

FIGURE 2.3   Toshiba's CMOS image sensor (converted from wire bond to flip
chip) with TCV (2008).

product either. There are some 3D IC microelectromechanical system (MEMS) integration products [1,2], but the volume is very small.

2.4   Via Forming

2.4.1   DRIE versus Laser Drilling

There are at least two methods to form the TSV. One is by laser drilling and the other by deep reactive-ion etch (DRIE). Laser drilling is a single-point operation, and thus it is cost-effective for small number of vias per chip. DRIE operates on a whole wafer and thus is more suitable for high-density (via) application. Depending on the kind of laser, the TSV hole size and pitch are limited [for a CO_2 laser, via hole size = 65 µm (top) and 25 µm (bottom), pitch = 90 µm; for ultraviolet (UV) laser, via hole size = 50 µm (top) and 25 µm (bottom), pitch = 125 µm; and for Excimer laser, via hole size = 18 µm (top) and 12 µm (bottom), pitch = 35 µm].

On the other hand, with DRIE, the TSV hole size and pitch can go down to as small as 1 and 5 µm, respectively. Most potential applications of TSV are in the range of 5 µm ≤ TSV hole size ≤ 20 µm and 20 µm ≤ pitch. Also, it should be pointed out that the surface condition of the laser-drilled vias is very rough, as shown in Fig. 2.4. Some chemical polishing

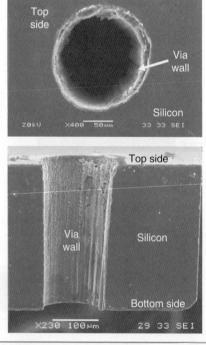

FIGURE 2.4   TSV fabricated by laser, top view (*above*) and cross-sectional view (*below*). The roughness is obvious.

is needed to obtain a smooth wall, which increases the manufacturing steps and cost. Only DRIE will be discussed in this book.

In most of the potential TSV 3D integration applications, the thickness of the passive/active interposers/chips ranges from 50 to 200 μm, and that of the stacking memory chips ranges from 20 to 50 μm. Thus all TSVs fabricated are blind vias [1,2], and the aspect ratios (thickness/diameter) of most TSVs are at least 2 and could be as large as 50 or more.

Based on the literatures, more than 95 percent of vias have been formed by DRIE Bosch technology, which is a highly anisotropic etching process. The silicon vias can be formed in an inductive-coupled plasma (ICP)–based DRIE system such as that from Applied Materials and SPTS (Fig. 2.5). Although the ICP system is designed mainly for performing deep reactive-ion etching of silicon by a specially designed switched etch and passivation process, also known as the *Bosch process*, it also can be used in reactive-ion etch mode, referred as the *non-Bosch process* [14–16].

Scientists at Bosch have developed a revised process to further improve process control in DRIE [14–16]. Instead of relying on the balance between plasma etching and polymer passivation, the Bosch process uses alternate etching and passivation steps, as shown in Figs. 2.6 and 2.7. The passivation is achieved by deposition of a polymer using CF_x^+, which is decomposed from C_4F_8 feedgas. The subsequent step is a reactive-ion etching process using such species as F^+ and SF_x^+, here as the polymer on the bottom of the trench is removed owing to the directional etching effect of ion bombardment. The normal cycle time for each deposition or etch step is a few seconds.

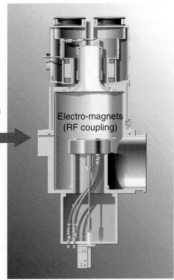

Top power (13 MHz):
_High gas dissociation
_Wafer-less cleans

Bias power (13 MHz):
_Ion assistance

Electro-magnets
(RF coupling)

FIGURE 2.5 The SPTS DRIE chamber.

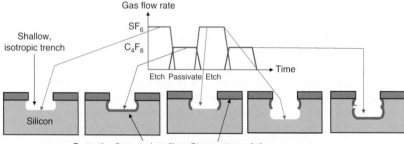

1. Usually, SF_6 will be used in etch cycle whereas C_4F_8 in passivation cycle.
2. If etch cycle time is too long and passivation cycle time is too short, then the via will not be straight and sidewall will be rough (large scallops).
3. On the other hand, if the passivation cycle time is too long, then it slows down the etch rate and throughputs.
4. Thus, the etch rate should be balanced and optimized.

FIGURE 2.6   TSV DRIE. SF_6 is used for the etch silicon and passivation cycle, whereas C_4F_8 is used for the passivation cycle.

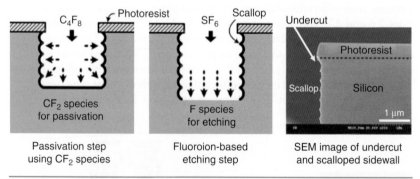

FIGURE 2.7   The process cycles between an etch step using SF_6 gas and a polymer-deposition step using C_4F_8. The polymer protects the sidewalls from etching by the reactive fluorine radicals.

The resulting etch rate is 1 to 15 μm/min typically. An aspect ratio of the etched trench of 30:1 can be achieved. The general issue of DRIE is that the etch rate is a function of the density of the opening area and trench width. Investigation of an etched dummy wafer is necessary for etch-rate calibration when users have a new lithography mask layout [14–16].

The SPTS system consists of ICP electronics, a load-lock and carousel wafer-loading unit, and a process chamber, as shown in Fig. 2.5. The plasma is generated by a coil assembly that is inductively coupled at 13 MHz via the matching unit in a ceramic plasma chamber. This provides high-density plasma capable of achieving high etch rates with little substrate damage. Independent biasing of the platen is made available by a separate 13-MHz RF biasing circuit on the bottom electrode that comes with automatic power control and impedance matching. Process gas is introduced to the chamber through the upper electrode assembly. Wafers are clamped by electrostatic chuck (ESC) on the lower electrode, which is powered at 13 MHz. The platen temperature is kept at 10°C by using recirculating deionizer (DI) water through a chiller system.

2.4.2 Tapered Via by DRIE

The high-aspect-ratio tapered silicon vias are formed by three independently controlled process steps: (1) the straight via-formation step by the Bosch etch process, (2) the via-tapering step by a controlled isotropic etch process, and (3) the corner-rounding step by a global isotropic etch process.

The first etching step is designed mainly to achieve high etch rates with the Bosch etch process. Typically, a high-etch-rate Bosch process tends to give a vertical to slightly reentrant profile, as shown in Fig. 2.8 (*left*). The gases used in the Bosch process are mainly sulfur hexafluoride (SF$_6$) plus oxygen (O$_2$) in the etch phase and C$_4$F$_8$ in the passivation phase. The recipes are shown in Table 2.1 [14–16]. It should be noted that as a consequence of the cyclic etch/passivation process, the sidewalls become scalloped or rough. In this step, the via

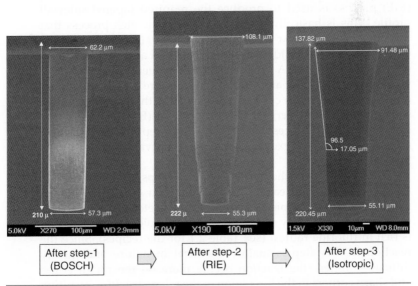

FIGURE 2.8 Step 1 shows results achieved after 50-μm diameter vias are etched to a depth in the range of 200 to 220 μm. Step 2 shows the evolution of a tapered profile after etch by a controlled isotropic etch process. Step 3 shows the final tapered via profile after the top corner is rounder by a global isotropic etch process.

Step-1- Straight (BOSCH) etch process	Etch cycle: APC: 77% (26 mt); 130-sccm SF_6; 13-sccm O_2; 600-W coil power; 20-W platen power; 6 sec
	Passivation cycle: APC: 77% (17 mt); 85-sccm C_4F_8; 600-W coil; 5 sec
	Platen temperature: 10°C Total process time: 60 min Etch rate: 3–3.5 µm/min on 15–20% exposed area
Step-2- Via tapering (RIE) process	APC: 78% (30 mt); 84-sccm SF_6; 67-sccm O_2; 59-sccm Ar; 600-W coil power; 30-W platen power Platen temperature: 10°C Process time: 60 min Etch rate: 3.5–4.0 µm/min on 15–20% exposed area
Step-3- Via corner rounding (*isotropic etch*) process	APC: 65% (12–13 mtorr); 180-sccm SF_6; 18-sccm O_2; 600-W coil power; 30-W platen power Platen temperature: 10°C Process time: 10 min Etch rate: 1.5–2.0 µm/min on blanket silicon wafer

TABLE 2.1   Summary of the Three-Step Via Tapering Process

is etched to approximately 50 to 60 percent of the final depth. This step defines the dimension of the via at the bottom and the depth.

A non-Bosch etch process consisting of a reactive-ion etching (RIE) process is used to produce the required tapered-sidewall profile. This is basically a controlled isotropic etch process that uses SF_6 plus O_2 plus Ar (argon) etch chemistry (see Table 2.1). The oxygen used helps in sidewall passivation and also controls excessive lateral etch rate. A proper balance between SF_6 and O_2 provides the desired taper angle to the via structure. As a consequence of this process, a sharp curvature is formed at the top of the via, as shown in Fig. 2.8 (*center*). At the end of step 1 (Bosch process) and step 2 (non-Bosch process), the required via depth and taper angle are almost achieved, except for the sharp curvature at the top of the via.

After completing the earlier two etch steps, the etch mask is fully stripped and cleaned. The wafer then is subjected to a maskless global isotropic etch process. In this etch step, the etched via patterns are mainly subjected to an isotropic etch plasma that is rich in fluorine radical (see Table 2.1 for the recipes). Since the reaction in this step is mainly chemical in nature and mostly diffusion-limited, the reaction occurs more on the rough edges and sharp corners in the top region of the microstructure, resulting in well-rounded, smooth sidewalls inside the vias, as shown in Fig. 2.8 (*right*). Today, most TSVs published are straight vias. In these cases, only step 1 is needed.

2.4.3   Straight Via by DRIE

As mentioned earlier, most TSVs are straight vias formed by DRIE, and the etch rate is the most important factor for obtaining high-quality TSVs with smooth sidewalls, that is, minimal sidewall scallop and undercut, as shown in Fig. 2.6 [38–43]. Usually, SF_6 will be used in the etch silicon and passivation cycle, whereas C_4F_8 will be used in the passivation cycle, as shown in Fig. 2.7. If etch cycle time is too long and passivation cycle time is too short, the via will not be straight, and the sidewall will be rough (large scallops). On the other hand, if the passivation cycle time is too long, then it slows down the etch rate and throughput. Thus the etch rate should be balanced and optimized.

A design of experiment (DoE) is planned and executed to determine the optimal etch rate of TSVs in 200- and 300-mm wafers. Various etch rates ranging from 1.7 to 18 μm/min and TSV diameters (1, 10, 20, 30, and 50 μm) will be considered. The backside helium (BHE) in/out pressures (2666 Pa/2666 Pa) are the same for all cases. Characterization of the sidewall scallop will be performed by cross sections and scanning electron microscopy (SEM). Furthermore, with the same etch recipe, mask, and nine (5, 10, 15, 20, 25, 30, 40, 55, and 65 μm) TSV diameters, the etch results, including the etch rate, TSV depth, and sidewall scallop, of 200- and 300-mm wafers are provided and compared. Finally, a set of useful process guidelines and recipes for optimal TSV etching are presented [10].

DoE for Straight TSV Etching on 300-mm Wafer

Table 2.2 shows the DoE study for TSV etching on 300-mm wafers. It can be seen that (1) 11 different recipes are considered, (2) the etch rate ranges from 1.7 to 18 μm/min, (3) the TSV diameter ranges from 1 to 50 μm, and (4) the BHE in/out pressures (2666 Pa/2666 Pa) are the same for all cases. For the 1-μm-diameter TSVs, the aspect ratio (AR) is 20 (depth = 20 μm); for the 10-μm-diameter TSVs, the AR is 7.5 (depth = 75 μm); for the 20-μm-diameter TSVs, the AR is 4.2 (depth = 85 μm); for the 30-μm-diameter TSVs, the AR is 3 (depth = 90 μm); and for the 50-μm-diameter TSVs, the AR is 2 (depth = 100 μm).

DoE Results for Straight TSV Etching on 300-mm Wafer

The DoE results are shown in Fig. 2.9, where the scallops on the TSV sidewall for different etch rates and TSV diameters are given. The scallops are determined by cross sections and SEM measurements. It can be seen that (1) for all the cases, in general, the higher the etch rates, the larger the scallops, (2) for 1-μm-diameter TSVs, the effect of etch rate on the scallop is very small, and the scallop ranges from 57 to 83 nm (for etch rates ranging from 1.7 to 2.13 μm/min), as shown in Fig. 2.10, (3) for 10-μm-diameter TSVs, the effect of etch rate on the scallop is visible, and the scallop ranges from 107 to 278 nm (for etch rates ranging from 3.5 to 5.8 μm/min), as shown in Fig. 2.11, (4) for 20-μm-diameter TSVs, the effect of etch rate on the scallop is sizable,

Recipe	Dep Time (sec)	Etch1 Time (sec)	Etch2 Time (sec)	BHE (Pa) in/out	CD Size (um)	Max Scallop (nm)	Etch Rate (um/min)
1-1st	1.8	1.3	1.0	2666/2666	1	83	2.1
1-2nd	1.6	1.3	1.0	2666/2666	1	68	18
1-3rd	1.6	1.3	1.0	2666/2666	1	57	1.7
10-1st 20-1st	1.8	1.3	2.0	2666/2666	10 20	278 225	5.8 8.8
10-2nd 20-2nd	1.6	1.3	1.0	2666/2666	10 20	107 93	3.5 4.2
30-1st	1.8	1.3	1.8	2666/2666	30	258	9.55
30-2nd	1.6	1.3	1.3	2666/2666	30	179	9.1
30-3rd	1.8	1.3	1.3	2666/2666	30	146	8.2
30-4th	1.1	1.3	0.6	2666/2666	30	136	5.8
30-5th 50-2nd	1.5	1.3	0.6	2666/2666	30 50	97 99	4.6 5.2
50-1st	1.8	1.3	2.0	2666/2666	50	235	11.0

TABLE 2.2   DoE of TSV Etching on 300-mm Wafers

and the scallop ranges from 93 to 225 nm (for etch rates ranging from 4.2 to 8.8 μm/min), as shown in Fig. 2.12; (5) for 30-μm-diameter TSVs, the effect of etch rate on the scallop is considerably larger, and the scallop ranges from 97 to 258 nm (for etch rates ranging from 4.6 to 9.5 μm/min), as shown in Fig. 2.13, and (6) for 50-μm-diameter TSVs, the effect of etch rate on the scallop is large, and the scallop ranges from 99 to 235 nm (for etch rates ranging from 5.2 to 11 μm/min), as shown in Fig. 2.14.

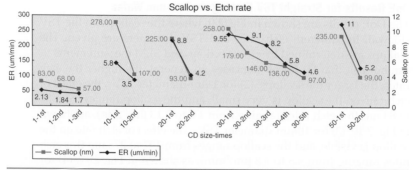

FIGURE 2.9   Etch rate (ER) versus TSV diameter (CD) versus scallop.

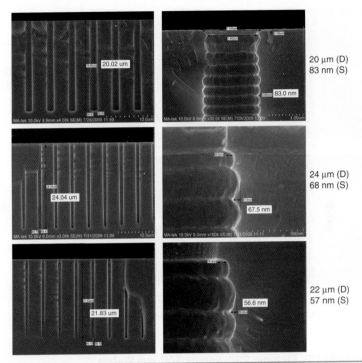

FIGURE 2.10 Etching results for 1-μm-diameter TSVs on a 300-mm wafer: (*above*) 20-μm depth with 83-nm scallops; (*center*) 24-μm depth with 68-nm scallops; (*below*) 22-μm depth with 57-nm scallops.

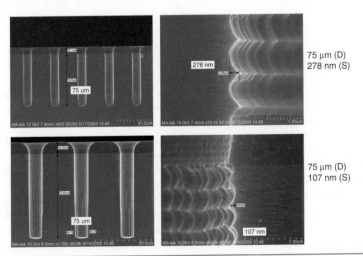

FIGURE 2.11 Etching results for 10-μm-diameter TSVs on 300-mm wafer: (*above*) 75-μm depth with 278-nm scallops; (*below*) 75-μm depth with 107-nm scallops.

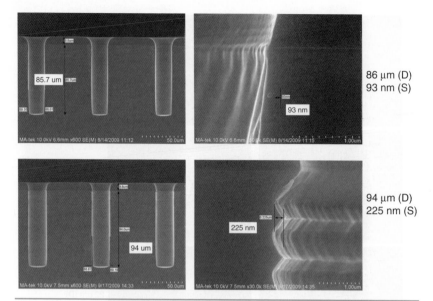

85.7 um

93 nm

86 µm (D)
93 nm (S)

94 um

225 nm

94 µm (D)
225 nm (S)

FIGURE 2.12   Etching results for 20-µm-diameter TSVs on 300-mm wafer:
(*above*) 86-µm depth with 93-nm scallops; (*below*) 94-µm depth with 225-nm scallops.

Straight TSV Etching: 200- versus 300-mm Wafers

A single 300-mm (SPTS-ASE-V300) chamber is used for the etching experiments on 200- and 300-mm wafers. The photoresist mask used is the same for both cases. The 200- and 300-mm wafers are etched by the same recipe as shown in Table 2.3. It can be seen that the passivation (C_4F_8) deposition time is 1.8 seconds, the etch 1 time (SF_6 to remove the passivation) is 1.3 seconds, and the etch 2 time (SF_6 to vertically etch off the silicon) is 2 seconds. A mask with nine different diameters, namely, 5, 10, 15, 20, 25, 30, 40, 55, and 65 µm, is used for the etching.

The etch depth versus TSV diameter for the 200- and 300-mm wafers is shown in Fig. 2.15. It can be seen that (1) for both wafers and all etch cycles, in general, the larger the TSV diameter, the deeper is the TSV depth [for example, the depth corresponding to the 5-µm-diameter TSVs (with etch cycles = X) for the 200-mm wafer is 51 µm and 49 µm for the 300-mm wafer, but for the 20-µm-diameter TSVs, it is 83 µm for the 200-mm wafer and 81 µm for the 300-mm wafer], (2) for both wafers, the larger the etch cycles (longer etching), the deeper is the depth [for example, the depth corresponding to the 30-µm-diameter TSVs (with etch cycles = X) for the 200-mm wafer is 97 µm and 94 µm for the 300-mm wafer, but for etch cycles = 2.25X, it is 212 µm for the 200-mm wafer and 215 µm for the 300-mm wafer], and (3) for all the diameters and etch cycles, the depth between the 200- and 300-mm wafers is very similar.

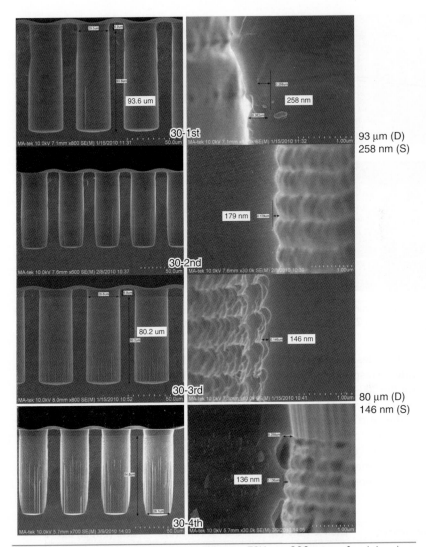

30-1st   93 μm (D)
258 nm (S)

30-2nd

30-3rd   80 μm (D)
146 nm (S)

30-4th

FIGURE 2.13   Etching results for 30-μm-diameter TSVs on 300-mm wafer: (*above*) 93-μm depth with 258-nm scallops; (*below*) 80-μm depth with 146-nm scallops.

The etch rate versus TSV diameter for the 200- and 300-mm wafers is shown in Fig. 2.16. It can be seen that (1) for both wafers and all etch cycles, in general, the higher the TSV etch rate, the larger is the TSV diameter [e.g., the etch rate corresponding to the 5-μm-diameter TSVs (with etch cycles = X) for the 200-mm wafer is 5 μm/min and 4.8 μm/min for the 300-mm wafer, but for the 40-μm-diameter TSVs it is 10 μm/min for the 200-mm wafer and 9.7 μm/min for the 300-mm wafer], (2) for both wafers, the larger the

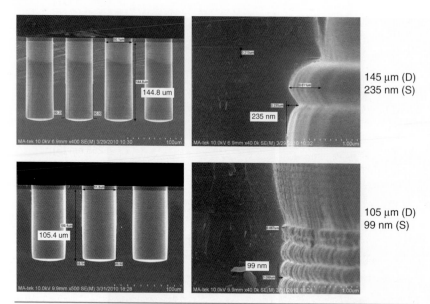

145 µm (D)
235 nm (S)

105 µm (D)
99 nm (S)

FIGURE 2.14 Etching results for 50-µm-diameter TSVs on 300-mm wafer: (*above*) 93-µm depth with 258-nm scallops; (*below*) 80-µm depth with 146-nm scallops.

etch cycles (longer etching), the lower is the etch rate [e.g., the etch rate corresponding to the 30-µm-diameter TSVs (with etch cycles = X) for the 200-mm wafer is 9.5 µm/min and 9.2 µm/min for the 300-mm wafer, but for etch cycles = 2.25X it is 6.9 µm/min for the 200-mm wafer and 7 µm/min for the 300-mm wafer], and (3) for all the diameters and etch cycles, the etch rate between the 200- and 300-mm wafers is very similar.

The maximum scallop versus TSV diameter for the 200- and 300-mm wafers is shown in Fig. 2.17. It can be seen that (1) for both wafers

Dep.Time (s)	Etch 1 Time (s)	Etch 2 Time (s)	BHE (Pa) in/out
1.8	1.3	2.0	2666/2666

- A single 300-mm (SPTS-ASE-V300) chamber is used for the etching experiments on 200- and 300-mm wafers.
- The photoresist mask used is the same for both cases.
- Same recipe for 200- and 300-mm wafers.
- Only etch cycle (time) changes in the different experiments. (For example, the CD = 5-µm and depth = 80-µm profile uses up more cycle time than the CD = 5-µm and depth = 50-µm profile.)

TABLE 2.3 Same recipes for 200- and 300-mm Wafers. Only Etch Cycle (time) Changes in the Different Experiments. (For example, the CD = 5-µm and depth = 80-µm profile uses up more cycle times than CD = 5-µm and depth = 50-µm profile.)

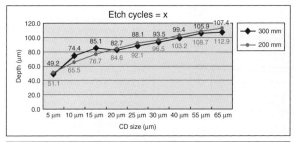

For both wafers and all etch cycles, the larger the TSV diameter the deeper the TSV depth, e.g., the depth corresponding to the 5-μm-diameter (with etch cycles = X) for the 200-mm wafer is 51 μm and 49 μm for the 300-mm wafer, but to the 20-μm-diameter is 83 μm for the 200-mm wafer and is 81 μm for the 300-μm wafer.

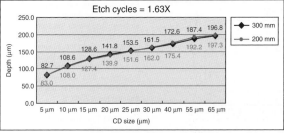

For both wafers and all etch cycles, the larger the TSV diameter the deeper the TSV depth, e.g., the depth corresponding to the 5-μm-diameter (with etch cycles = X) for the 200-mm wafer is 51 μm and 49 μm for the 300-mm wafer, but to the 20-μm-diameter is 83 μm for the 200-mm wafer and is 81 μm for the 300-μm wafer.

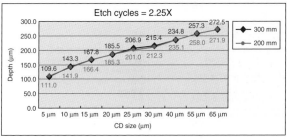

For all the diameters and etch cycles, the depth between the 200-and 300-mm wafers is very similar.

FIGURE 2.15   TSV depth versus diameter with different etch cycles for 200- and 300-mm wafers.

and all etch cycles, in general, the larger the TSV diameter, the larger are the maximum scallop, even though there are ups and downs, (2) for both wafers, the larger the etch cycles (longer etching), the smaller are the scallops, and (3) the maximum scallop is quite different between the 200- and 300-mm wafers; the difference is smaller at lower etch cycles but is larger at higher etch cycles. Typical cross sections of the 200- and 300-mm wafers are shown in Fig. 2.18 with 5-μm-diameter TSVs and various etch cycles: X, 1.63X, and 2.25X.

Straight TSV Etching Process Guidelines

Table 2.4 shows the key items that need to be considered for a high-quality (less scallop and undercut) TSV etching. It can be seen that (1) the key elements of TSV etching are etch rate, power, etch cycle, wet strip, and scallop; (2) a balance of etch rate, power, and etch cycle (time) is necessary; and (3) a good starting recipe is deposition

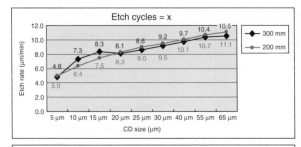

For both wafers and all etch cycles, the higher the TSV etch rate the larger the TSV diameter, e.g., the etch rate corresponding to the 5-μm-diameter (with etch cycles = X) for the 200-mm wafer is 5 μm/min and 4.8 μm/min for the 300-mm wafer, but to the 40-μm-diameter is 10 μm/min for the 200-mm wafer and is 9.7 μm/min for the 300-μm wafer.

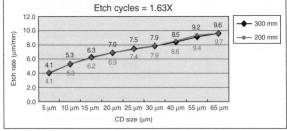

For both wafers, the larger the etch cycles (longer etching) the smaller the etch rate, e.g., the etch rate corresponding to the 30-μm-diameter (with etch cycles = X) for the 200-mm wafer is 9.5 μm/min and 9.2 μm/min for the 300-mm wafer, but to etch cycles = 225X is 6.9 μm/min for the 200-mm wafer and 7 μm/min for the 300-mm wafer.

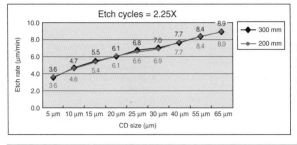

For all the diameters and etch cycles, the etch rate between the 200-and 300-mm wafers is very similar.

FIGURE 2.16   Etch rate versus diameter with different etch cycles for 200- and 300-mm wafers.

(to form a thin layer of passivation) time = 1.8 s, etch 1 (to remove passivation) time = 1.3 s, and etch 2 (to vertically etch off silicon) time = 2 s.

Summary and Recommendations

The effects of etch rate on TSV sidewall performance (scallop) have been investigated. A DoE has been executed to determine the optimal etch rate of TSVs in 300-mm wafers. Also, with the same etch recipe, mask, and TSV diameters, the etch results such as etch rate, TSV depth, and scallop of 200- and 300-mm wafers have been determined and compared. Some important results and recommendations are summarized in the following [10]:

1. DoE of TSV etching for the 300-mm wafers:
   - For all the cases, the higher the etch rate, the larger the scallops.

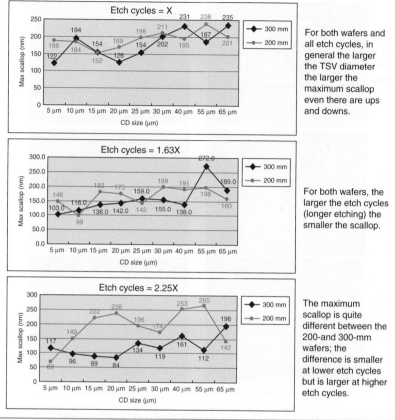

For both wafers and all etch cycles, in general the larger the TSV diameter the larger the maximum scallop even there are ups and downs.

For both wafers, the larger the etch cycles (longer etching) the smaller the scallop.

The maximum scallop is quite different between the 200-and 300-mm wafers; the difference is smaller at lower etch cycles but is larger at higher etch cycles.

FIGURE 2.17   Maximum scallop versus diameter with different etch cycles for 200- and 300-mm wafers.

- For 1-μm-diameter TSVs, the effect of etch rate on the scallop is very small, and the scallop ranges from 57 to 83 nm (for etch rates ranging from 1.7 to 2.13 μm/min).

- For 10-μm-diameter TSVs, the effect of etch rate on the scallop is visible, and the scallop ranges from 107 to 278 nm (for etch rates ranging from 3.5 to 5.8 μm/min).

- For 20-μm-diameter TSVs, the effect of etch rate on the scallop is sizable, and the scallop ranges from 93 to 225 nm (for etch rates ranging from 4.2 to 8.8 μm/min).

- For 30-μm-diameter TSVs, the effect of etch rate on the scallop is considerably large, and the scallop ranges from 97 to 258 nm (for etch rates ranging from 4.6 to 9.5 μm/min).

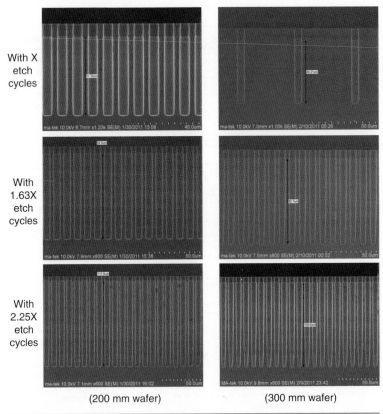

With X
etch
cycles

With
1.63X
etch
cycles

With
2.25X
etch
cycles

(200 mm wafer)   (300 mm wafer)

FIGURE 2.18   Etched TSVs with 5-µm diameter and X etch cycles (*above*); 1.63X etch cycles (*center*), and 2.25X etch cycles (*below*). The 200-mm wafer is on the left, whereas the 300-mm wafer is on the right.

- For 50-µm-diameter TSVs, the effect of etch rate on the scallop is large, and the scallop ranges from 99 to 235 nm (for etch rates ranging from 5.2 to 11 µm/min).

2. Comparison of the etch rate, depth, and scallop between the 200- and 300-mm wafers:

   - For both wafers and all etch cycles, the larger the TSV diameter, the deeper the TSV depth.
   - For both wafers, the larger the etch cycle, the deeper the depth.
   - For all the diameters and etch cycles, the TSV depth between the 200- and 300-mm wafers is very similar.
   - For both wafers and all etch cycles, the higher the TSV etch rate, the larger the TSV diameter.
   - For both wafers, the larger the etch cycle, the smaller the etch rate.

Items to Be Considered	Conditions
(1) Scallop issue	Passivation time (C_4F_8 step) increases, then scallop decreases Etch time (SF_6 step) increases, then scallop increases
(2) Etch rate issue	Passivation time (C_4F_8 step) increases, then etch rate decreases + striation & residue Etch time (SF_6 step) increases, then etch rate increases
(3) Power issue	Power increases, then etch rate increases and scallop increases + Sidewall passivation not enough and lead to striation Power decreases, then etch rate decreases and scallop decreases
(4) Etch cycle issue	The larger the etch cycles, the larger the TSV diameter; the larger the TSV depth, the smaller the etch rate
(5) Wet strip issue	Wet strip after removing photo resist is suggested. X-section + SEM inspection

TABLE 2.4 Key Elements That Need to Be Considered for High-Quality TSV Etching

- For all diameters and etch cycles, the etch rate between the 200- and 300-mm wafers is very similar.

- For both wafers and all etch cycles, in general, the larger the TSV diameter, the larger is the maximum scallop.

- For both wafers, the larger the etch cycle, the smaller the scallop.

- The maximum scallop is quite different between the 200- and 300-mm wafers; the difference is smaller at lower etch cycles but is larger at higher etch cycles.

2.5 Dielectric Isolation Layer (Oxide Liner) Deposition

Oxide liner deposition can be performed by either thermal oxidation or plasma-enhanced chemical vapor deposition (PECVD) methods. In general, the SiO_2 fabricated by the thermal oxidation method is very uniform (but with very high temperatures >1000°C), and that by the PECVD method is with steps (but with very low temperatures <250°C). Since silicon is an electrical conductive material, in order to prevent current leakage and cross-talk from TSV to TSV, well-isolated SiO_2 (≥0.1 μm) for TSVs is necessary.

2.5.1 Tapered Oxide Liner by Thermal Oxidation

For passive interposers (which do not have any devices such as transistors), thermal oxidation can be used to deposit the isolation layer. The 1-µm SiO_2 can be grown in a furnace such as the one (Ellipsiz) shown in Fig. 2.19. Hydrogen and oxygen are mixed in the quartz bulb, where they react to create a flame and steam (this is why it is also called a wet process) at 1050°C. The oxidation time of Si is 3½ hours. Conformal SiO_2 is shown in Fig. 2.20. The top, middle, and bottom of the TSV are covered by uniform SiO_2. Uniformity of SiO_2 is within ±5 percent. The sidewall roughness decreases from 200 to 250 nm to less than 100 nm because of silicon consumption in wet oxidation [14–16]. Unfortunately, for device wafers, the thermal oxidation method cannot be used because the allowable temperature for most device chips is less than 400 to 450°C.

FIGURE 2.19   Ellipsiz furnace system.

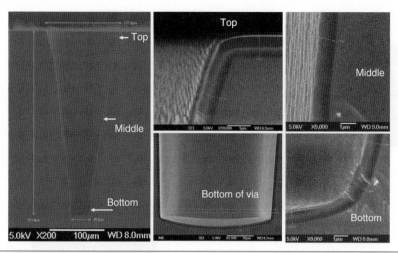

FIGURE 2.20   Dielectric isolation layer (SiO_2) deposited by wet thermal oxidation (1080°C).

2.5.2 Tapered Oxide Liner by PECVD

The SiO_2 also can be deposited by the plasma-enhanced chemical vapor deposition (PECVD) dry process with equipment provided by such companies as Applied Materials and SPTS (Fig. 2.21). It can be seen in Fig. 2.22 that the deposited oxide thickness is 1.9 to 2 µm on

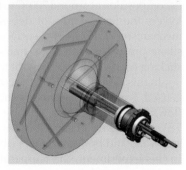

FIGURE 2.21   The SPTS plasma-enhanced chemical vapor deposition (PECVD). There is an air-cooled platen for low-temperature deposition of thick via liner films.

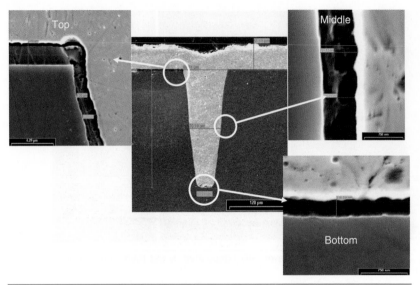

FIGURE 2.22   Dielectric isolation layer (SiO$_2$) fabricated by plasma-enhanced chemical vapor deposition (PECVD).

the top side of the TSV, 1.3 to 1.4 μm on the top sidewall, 0.7 to 0.8 μm on the middle sidewall, and 0.35 to 0.45 μm on the via bottom. Thus the SiO$_2$ thickness deposited by the dry process is not as uniform as that done by the wet process. In this case, the process temperature is less than 250°C [16], which is good for most wafers.

2.5.3   DoE for Straight Oxide Liner by PECVD

Regardless of some developed materials for forming SiO$_2$, the focus in this section is on tetraethyl orthosilicate [Si(OC$_2$H$_5$)$_4$; TEOS] research because it is a mature precursor for the semiconductor interlayer. Some important characteristics of TEOS are good conformity of coating, relatively inert material, liquid at room temperature, thermally decomposition at around 700°C to form SiO$_2$, and plasma enhancement lowers the temperature of deposition to below 500°C. The deposit mechanism of TEOS is a surface reaction, and compared with the silane precursor with mass transfer mechanism, TEOS has a lower sticking coefficient that allows it to travel to the TSV bottom easier [5].

In this experiment [23], the conventional 300-mm AMAT PECVD is used. Some key parameters such as RF power, TEOS flow, oxygen flow, and pressure are split by DoE on 10- and 30-μm-diameter TSVs. The main observed areas are step coverage of TSV bottom sidewalls and bottom corners (the thinnest place of oxide liner) [6]. In this experiment, the temperature (180°C) is fixed because most adhesive materials for temporary bonding in 3D IC integration cannot withstand high temperatures. The guideline for the PE-TEOS oxide process in low temperature is extracted here.

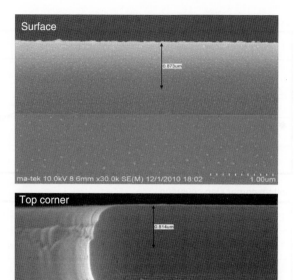

FIGURE 2.23   Surface oxide thickness at the iso-field is 8730 Å (*above*). Surface oxide thickness near at the dense via shrinks to 8140 Å (*below*).

Step coverage is defined by via oxide thickness divided by the surface thickness. However, the surface thickness at the iso-field is different from that of the dense via owing to the loading effect at the silicon surface, as shown in Fig. 2.23. It can be seen that the surface oxide thickness at the iso-field (8730 Å) has been shrunk to 8140 Å near the dense via. In this study, the surface thickness at the iso-field is chosen as the denominator of step coverage [23].

2.5.4   DoE Results for Straight Oxide Liner by PECVD

The DoE results of (10-μm-diameter and 30-μm-thick) TSVs are shown in Fig. 2.24. It can be seen that (1) a distinct way to improve step coverage is by raising RF power and lowering TEOS flow, which means a lower deposit rate, (2) the oxygen flow seems to be a minor factor in step coverage, and (3) the pressure is divided in two directions; thus further experiments on the pressure effect are executed in the following.

The higher-aspect-ratio (10-μm-diameter and 60-μm-thick) TSVs are used, and recipes with higher RF power, lower TEOS flow, and medium oxygen flow are chosen. Recipes with high and low pressures are set, respectively, as recipe 1 and recipe 2, and the results are shown in Table 2.5. It can be seen that (1) for high-pressure recipe 1, the step coverage

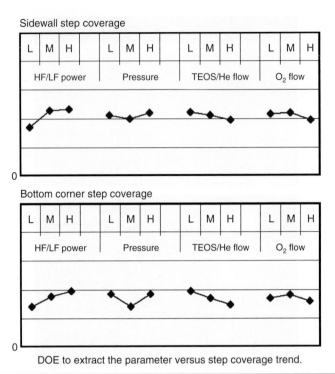

FIGURE 2.24   DoE result at sidewall step coverage (*above*) and bottom corner step coverage (*below*) on (10-μm-diameter and 30-μm-thick) TSV.

	Recipe 1	Recipe 2	Recipe 3
HF/LF power	H	H	H
Pressure	H	L	L
TEOS/He flow	L	L	L
O_2 flow	M	M	M
Temperature	180°C	180°C	250°C
Step coverage			
Top corner	41.5%	40%	37.1%
Top sidewall	22.6%	20.2%	34.1%
Middle sidewall	11.7%	13.7%	15%
Bottom sidewall	8.6%	9.7%	11.7%
Bottom corner	7.9%	8.3%	10.3%
Bottom	8.3%	11.7%	16.9%

TABLE 2.5   DoE Results for (10-μm-diameter and 60-μm-depth) Blind TSVs

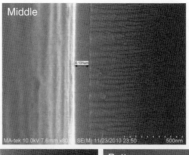

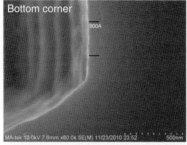

FIGURE 2.25   TSV bottom sidewall (*above*), TSV bottom corner (*below, left*), and TSV bottom (*below, right*). (Recipe 3 on 10-μm-diameter and 60-μm-thick TSV.)

of the bottom sidewall, bottom corner, and bottom is 8.6, 7.9, and 8.3 percent, respectively, and (2) for low-pressure recipe 2, the step coverage of bottom sidewall, bottom corner, and bottom is improved to 9.7, 8.3, and 11.7 percent, respectively. The reason that low pressure contributes to higher step coverage is the longer mean free path for TEOS delivery.

Moreover, when the temperature is raised to 250°C (based on the thermal budget of temporary bonding adhesives) and set as recipe 3, an obvious improvement in step coverage is observed, especially in bottom region, of 16.9 percent. Figure 2.25 shows SEM images of recipe 3, and it is clear that temperature is one of the keys of step coverage performance. Thus higher temperature, higher RF power, lower pressure, and lower TEOS flow lead to better TSV step coverage.

2.5.5   Summary and Recommendations

Some important results and recommendations are summarized in the following:

- The important parameters for making the TSV oxide liner are RF power, TEOS flow, pressure, temperature, and oxygen flow.

- It has been found that a combination of higher RF power, lower TEOS flow, higher temperature, and lower pressure leads to better TSV step coverage.

- The oxygen flow seems to be a minor factor.
- The bottom sidewall and bottom corner (the thinnest place of oxide liner) of TSVs are the most critical areas in step coverage.

2.6 Barrier (Adhesion) Layer and Seed (Metal) Layer Deposition

The most common adhesion materials used for the barrier layer of TSVs are titanium (Ti) and tantalum (Ta). Ti is a chemical element with atomic number 22 in the periodic table. It is a low-density, strong, lustrous, corrosion-resistant transition metal with a silver color. The atomic weight, ionization energy, melting point, boiling point, and density of Ti are, respectively, 47.88, 658 kJ/mol, 1660°C, 3287°C, and 4.5 g/cm³.

On the other hand, Ta is a chemical element with atomic number 73 in the periodic table. It is a rare, hard, blue-gray, lustrous transition metal that is highly corrosion-resistant. The atomic weight, ionization energy, melting point, boiling point, and density of Ta are, respectively, 180.9479, 761 kJ/mol, 2996°C, 5425°C, and 16.654 g/cm³. The most common seed layer of TSVs is copper (Cu). The deposition of the barrier layer and seed layer is by physical vapor deposition (PVD), and the equipment is provided by such companies as Tango, Applied Materials, and SPTP (Fig. 2.26).

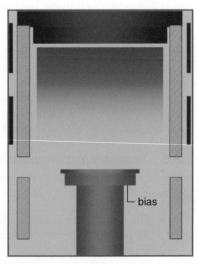

FIGURE 2.26   The SPTS physical vapor deposition (PVD).

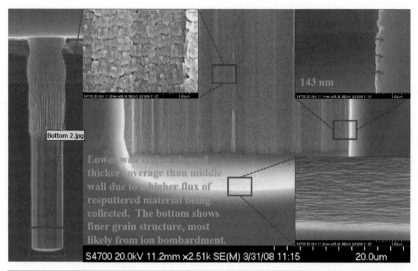

FIGURE 2.27   Ti barrier/adhesion layer and Cu seed layer fabricated by Tango AXCECA physical vapor deposition (PVD).

2.6.1   Tapered TSV with Ti Barrier Layer and Cu Seed Layer

Figure 2.27 shows the Ti (3000-Å) barrier and Cu (2-μm) seed layer on the walls of the via shown in Fig. 2.22. They are deposited by physical vapor deposition (PVD) with Tango System's AXCECA chamber, and the key process parameters are shown in Table 2.6. It can be seen that very good coverage at the via bottom and via sidewall has been achieved [15].

The Ti barrier and Cu seed layer also can be deposited by a wet process. Depositions of the insulation and barrier layers inside

Process	Ti	Cu
Thickness	3 kÅ	2 μm
DC power	8 kW	8 kW
Coil power	600 W	600 W
Bias power	1000 W	1000 W
Ar flow/Ch. Pressure	10/1.5 mT	20/2 mT
Bias voltage	100 V	150 V
Target voltage	550 V	630 V

TABLE 2.6   Important Process Parameters for Fabricating the Ti Barrier and Cu Seed Layer

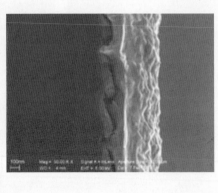

Figure 2.28   Ti barrier/adhesion layer and Cu seed layer deposition by Alchimer's electrografting process.

high-aspect-ratio TSVs have been achieved by Alchimer's electrografting (eG), which initiates chemical bond formation by means of a small electric current, followed by electroless chemical propagation. Figure 2.28 shows the results of wet deposition, and it can be seen that good coverage on via sidewalls and via bottom has been achieved. (Ti = 100 nm and Cu = 1 µm).

2.6.2   Straight TSV with Ta Barrier Layer and Cu Seed Layer

To overcome the difficulty of deep-via deposition, PVD systems have been developed from original planar diode sputtering to long-throw, collimate, and RF sputtering. The later ionized-metal plasma (IMP) PVD systems are a progressive enhancement. In the AMAT PVD system shown schematically in Fig. 2.29, the further advanced revolution self-ionized plasma (SIP) is used. High direct-current (DC) target power of several tens of kilowatts enhances the sputtering yield and causes the target metal to self-ionize without the using of RF coil of the IMP system. Better anisotropic deposition is contributed by the extremely low process pressure and ultrahigh ionized energy of plasma produced by high DC target power. In this barrier and seed layer step-coverage experiment, the parameters DC target power and substrate power, which influence perpendicularity the most, are examined [23].

2.6.3   Straight TSV with Ta Barrier Layer Experiments and Results

For the Ta barrier experiment, 10-µm-diameter and 60-µm-thick TSVs are used as observed vias. We consider three levels of DC target

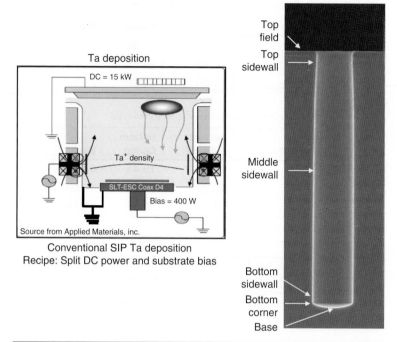

Source from Applied Materials, inc.

FIGURE 2.29   Step coverage of Ta barrier on (10-μm-diameter and 60-μm-thick) TSV with different DC power settings and substrate biases.

power and three levels of substrate power. The mask layout with SEM bar design ensures accurate cutting on the via center of the cross section SEM images. Table 2.7 shows the experiment results. It can be seen that (1) an obvious increase in step coverage is observed with lower dc power, (2) in general, with increasing ionization power, the plasma will have higher energy to attend the deep via, (3) however, owing to the ultrahigh ionization rate of the SIP system, raising the DC power further will reduce the plasma mean free path and step coverage, (4) the higher substrate bias makes the ionized metal easier to attend the via bottom, and (5) substrate bias is essential for the SIP system; otherwise, the highly ionized metal will lose straightness and collide randomly, which are reflected in the invisible step coverage of zero substrate bias.

With fine-tuning of PVD power parameters, Ta step coverage of 10-μm-diameter and 60-μm-thick TSVs by low DC target power and high substrate bias can be as high as 30 percent at the bottom corner, which is typically the minimal region for PVD of Ta. Figure 2.30 shows the cross sections and SEM images.

Sub. Bias (W)	Bottom Sidewall	Bottom Corner	Base
Zero	0%	0%	13.5%
L	20.2%	12.7%	17.7%
H	25.6%	21.9%	23.5%

Higher substrate bias shows better avg. step coverage.

DC Power (W)	Bottom Sidewall	Bottom Corner	Base
L	21.5%	15.4%	23.2%
M	15.1%	10.5%	15.8%
H	9.2%	8.6%	15.7%

Lower dc power shows better avg. step coverage.

TABLE 2.7   (Top) Step Coverage of Ta Barrier on (10-μm diameter and 60-μm depth) Blind TSVs with Different Substrate Power. (Bottom) Step Coverage of Ta Barrier on (10-μm diameter and 60-μm depth) Blind TSVs with Different DC Power

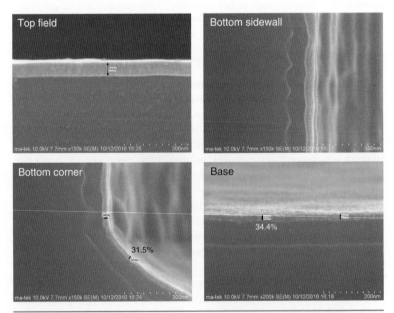

FIGURE 2.30   SEM images of Ta barrier step coverage on (10-μm diameter and 60-μm thick) TSV with fine-tuning parameters (DC power = low and substrate bias = high power). With these conditions, up to 30 percent Ta step coverage is seen.

2.6.4   Straight TSV with Cu Seed Layer Experiments and Results

For the Cu seed layer experiment, the 10-μm-diameter and 30-μm-thick TSVs are used as observed vias, as shown in Fig. 2.31. Figures 2.32 and 2.33 show the results, and it can be seen that (1) for the substrate bias split experiment, just as in the Ta barrier case, higher substrate bias leads to better step coverage, which also can be attributed to the higher attraction of ionized metals to substrate bias; (2) even with a further increase in substrate bias, the deposition rate will be sacrificed a little, but the step coverage and uniformity (which are more important) are improved; and (3) as for the DC power split experiment, although high DC power results in thicker bottom deposition, the critical areas of step coverage for deep vias are on the sidewall and the corner, which can be improved by lower DC power. The Cu seed layer recipe with the best parameters of low power and high substrate bias is used for the Cu plating experiment.

2.6.5   Summary and Recommendations

Some important results and recommendations are summarized in the following:

- The important parameters for making the TSV barrier layer (Ta) and seed layer (Cu) are the DC power and the substrate power (which is a must).

- It has been found that for SIP PVD, lower dc power leads to better TSV step coverage and that higher substrate bias makes the ionized metal easier to attend the via bottom. Thus a combination of lower DC power and higher substrate power lead to better TSV step coverage.

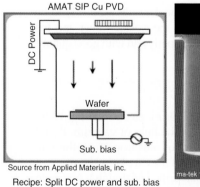
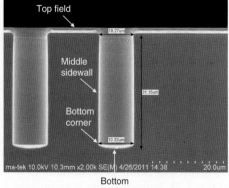

FIGURE 2.31   Step coverage of Cu seed on (10-μm-diameter and 30-μm-thick) TSV with different split DC power levels and substrate bias.

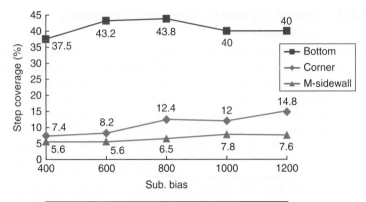

Sub. Bias (w)	Step coverage (combine with Ta)		
High DC power	Middle sidewall	Bottom corner	Bottom
L	5.6%	7.4%	37.5%
M-L	5.6%	8.2%	43.2%
M	6.5%	12.4%	43.8%
H-M	7.8%	12%	40%
H	7.6%	14.8%	40%

FIGURE 2.32   The step coverage of Cu seed on (10-μm-diameter and 30-μm-thick) TSV with a different substrate power setting (high DC power).

- The critical areas for barrier and seed layers are the same as those for the oxide liner.
- Based on leakage-current measurement results (<50 pA) to be shown in Sec. 2.7.3, it is found that the isolation liner and the barrier layer have been fabricated properly and with good quality.

2.7   TSV Filling by Cu Plating

TSVs can be filled with Cu, Ti, Al, or solder by electroplating, W by sputtering, or polymers by printing with the help of vacuum. Based on the literatures, most TSVs are filled by electroplating the Cu, which will be discussed in this book. Most of the Cu plating equipment is provided by Semitool (a subsidiary of Applied Materials), as shown in Fig. 2.34. Full Cu filling is significant for the 3D IC TSV process. Seams or voids generated during Cu plating may cause some potential electrical, thermal, and mechanical reliability problems.

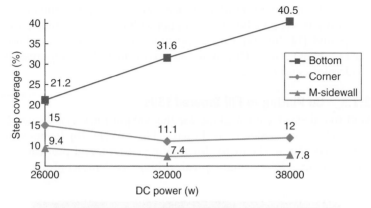

DC power	Step coverage (combine with Ta)		
(w) H-M Sub. bias	Middle sidewall	Bottom corner	Bottom
L	9.4%	15%	21.2%
M	7.4%	11.1%	31.6%
H	7.8%	12%	40.5%

FIGURE 2.33   The step coverage of Cu seed on (10-µm-diameter and 30-µm-thick) TSV with a different DC target power setting (H-M substrate bias).

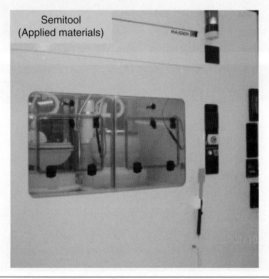

FIGURE 2.34   Cu electroplating by Semitool (a subsidiary of Applied Materials).

The Cu electroplating solution for deep and high-aspect-ratio via filling applications can be either copper sulfate or a cyanide-based compound [14–16]. Typical composition of an electrolyte includes $CuSO_4$, H_2SO_4, and Cl^-, with additives including suppressors, accelerators, and levelers.

2.7.1   Cu Plating to Fill Tapered TSVs

Void-free wafer-level Cu filling for the 300-µm-thick vias shown in Fig. 2.8 has been achieved, as shown in Fig. 2.35 [14–16]. It should be noted that plating chemistry, low plating current, and pretreatment (i.e., wetting + DI rinsing + preabsorbing accelerator) are the key

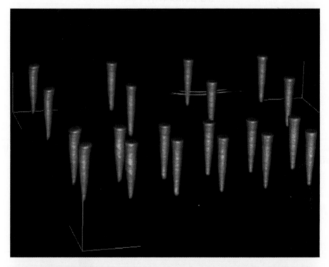

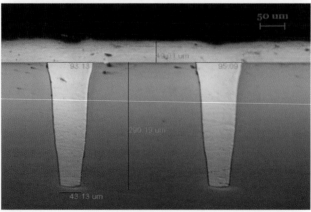

FIGURE **2.35**   TSV filled with electroplated Cu. (*Above*) 3D X-ray image with the computed tomographic (CT) reconstruction of the electroplated Cu. (*Below*) TSV with voidless filled Cu (after CMP).

contributors to void-free Cu filling for deep and high-aspect-ratio tapered vias. The electroplating process uses pulsed reverse plating, which employs a two-component additive system consisting of brightener and leveler [14–16].

2.7.2 Cu Plating to Fill Straight TSVs

During the Cu plating process, TSV filling as monitored by the bottom-up ratio is extraordinarily important. Typically, the *bottom-up ratio* is defined as the ratio of bottom plating thickness and top-sidewall plating thickness during the half-full status of TSV filling. Some TSV Cu plating experimental results are shown in Figs. 2.36 through 2.38. It can be seen in Fig. 2.36 (*above*) that the bottom-up ratio is 14.3 [40/(1.6 − 0.2) × 2], which is much higher than the aspect ratio (60/10) of this TSV. This means that it is possible to finish full Cu TSV filling and still have enough margins for recipe fine-tuning. However, based on the same recipe but with increased total plating time, the TSVs are completely filled with Cu without any voids or seams, as shown in Fig. 2.36 (*below*).

Figure 2.37 (*above*) shows the experimental result of a bottom-up ratio of 4.6 [35/(4 − 0.2) × 2], which is lower than the aspect ratio 6.7

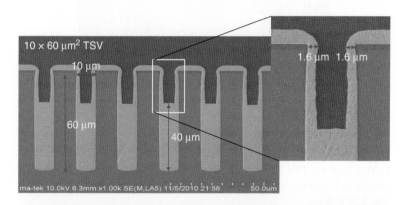

FIGURE 2.36 Bottom-up ratio of 14.3 for this TSV filling (*above*). Full and void-free TSV filling is achieved based on the recipe with a bottom-up ratio of 14.3 and an aspect ratio of 6 (*below*).

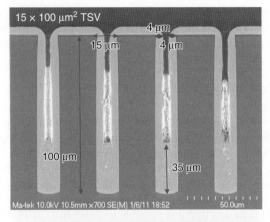

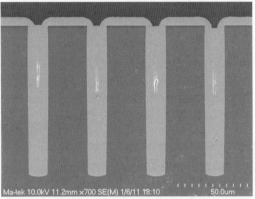

FIGURE **2.37** Bottom-up ratio of 4.6 for this TSV filling (*above*). Small seams are observed in this TSV filling based on the recipe with a bottom-up ratio of 4.6 and an aspect ratio of 6.7 (*below*).

(100/15) of the TSV. When the plating time increases as a proportion, it results in Cu filling with seams generated in the via center, as shown in Fig. 2.37 (*below*).

It should be pointed out that under the premise of the same chemistry bath conditions, the most significant process parameters to affect Cu plating performance are current density and waveform. With respect to current density, a larger current density accelerates the duration of plating. However, owing to current crowding at the via top and mass transport limitations to the via bottom, the excessive use of high current density will result in the formation of TSV voids [7]. In this situation, to reduce the overall plating time, we adjust the time distribution of current density and moderately drop the current density of the main bottom-up step equivalently with total current. After this adjustment, the plating time of the 10-μm-diameter and 60-μm-thick TSVs is

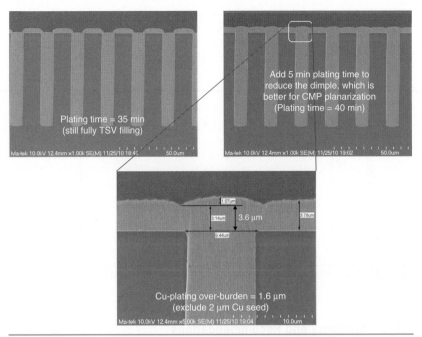

FIGURE 2.38 (*Above, left*) After the current adjustment, the plating time for (10-μm-diameter and 60-μm-thick) voidless TSVs is 35 minutes. (*Above, right*) Add 5 minutes more of plating time to reduce the dimple (plating time = 40 min). (*Below*) Magnified view.

reduced from 60 to 35 minutes without affecting the filling performance, as shown in Fig. 2.38 (*above, left*).

Regarding to the CMP planarization performance, five more minutes of Cu plating time is added to achieve a planar plating surface, as shown in Fig. 2.38 (*above, right*). Additionally, because of this current density adjustment, the top-field Cu plating overburden is reduced from 2.6 to 1.6 μm, which is also beneficial to the CMP process.

After Cu TSV plating, sufficient annealing temperature and annealing time are needed. Otherwise, owing to bulk Cu grain growth, the Cu-filled TSVs may protrude above the via, which may cause problems for the following process.

2.7.3   Leakage-Current Test of Blind Straight TSVs

The leakage performance of 10-μm-diameter and 60-μm-thick TSVs fabricated by using the current optimized oxide liner, barrier layer, seed layer, and Cu plating recipes is measured. Figure 2.39 shows the measurement results. It can be seen that the leakage current between two TSVs with a pitch of 150 μm shows good isolation and barrier properties with a current level of less than 50 pA [23].

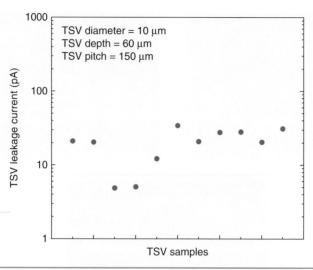

FIGURE 2.39   Leakage-current measurement of the (10-µm-diameter and 60-µm-thick) blind TSV.

2.7.4   Summary and Recommendations

Some important results and recommendations are summarized as follows:

- Under the same chemical bath conditions, the important parameters for making the Cu-filled TSV by electroplating are current density and waveform.

- It has been found that for void-free Cu plating, the time distribution of current density and the current density of the main bottom-up step in the total equivalent current amount must be adjusted.

- To achieve a planar surface that is beneficial to the following CMP process, a few more minutes of Cu plating time are recommended. Also, to prevent the Cu-filled TSVs from protruding above the vias, sufficient annealing temperature and time are recommended right after the Cu plating.

2.8   Chemical-Mechanical Polishing of Cu Plating Residues

2.8.1   CMP for Tapered TSVs

A thick layer (30 to 50 µm) of Cu overburden (Fig. 2.40) is plated onto the wafer surface because of the long plating time required to fill the voidless 300-µm-thick vias shown in Figs. 2.8 and 2.22 [14–16].

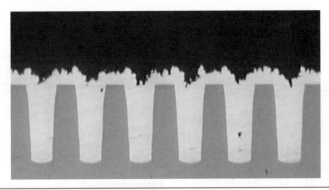

FIGURE 2.40   Cu plating residue owing to long plating time for voidless TSVs.

The wafer bows because of the thick Cu on the surface. Conventional chemical etching takes a long time to remove the thick Cu, and nonuniform etching has been observed. Okamoto GNX 200 is used for the higher-stress Cu chemical-mechanical polishing (CMP). A soft polishing pad and strong-removal-rate slurry from Rohms and Hass are applied for thicker copper removal. The load of polishing is 320 to 350 g/cm^2. The speed of pad and chuck are 90 rpm. The speed of slurry feeding is 170 to 200 mL/min.

2.8.2   CMP for Straight TSVs

In this section, the optimization of Cu CMP performance (dishing) for removing thick Cu plating overburden owing to Cu plating for deep TSVs in a 300-mm wafer is investigated. In order to obtain a minimum Cu dishing in the TSV region, a proper selection of Cu slurries is proposed for the current two-step Cu polishing process. The bulk of Cu is removed with a slurry with a high Cu removal rate in the first step, and the Cu surface is planarized with a slurry with a high Cu passivation capability in the second step. The Cu dishing can be improved up to 97 percent for 10-μm-diameter TSVs on a 300-mm wafer. The dishing/erosion of the metal/oxide can be reduced with respect to a correspondingly optimized Cu plating overburden for TSVs and redistribution layers (RDLs).

Problem Definition

Recently, low-resistivity Cu by the damascene technique was used as the TSV conducting material [24]. However, owing to long Cu plating times for deep TSVs, the Cu plating residues (overburden) on top of the wafer must be removed by CMP, as shown in Fig. 2.41. A thick Cu overburden layer of 30 to 50 μm resulting from the filling process of deep vias has been presented in Refs. 25 and 26. Two kinds of CMP tools can be used to remove the thick Cu layer and minimize metal dishing. One is the grinder/polisher using a slurry with a high etch

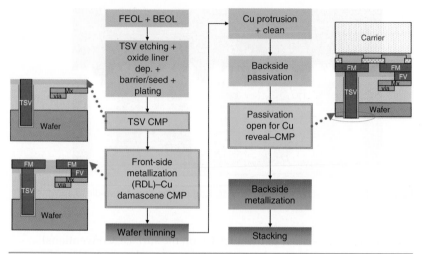

FIGURE 2.41   Possible applications of CMP in a TSV process.

rate, and the other is the front-end Cu CMP tool [25]. With the use high-etch-rate slurry, process control is important. Previous studies [27] show that with various slurries and polishing times, good dishing performance (<0.4 μm dishing) can be obtained.

To achieve a high Cu removal rate and minimum metal dishing, the slurry is the key-enabling material for CMP. Slurries with both high Cu attacking and high Cu passivation abilities are investigated in this study [11]. Concerning the long Cu plating time to fill up the deep TSV structures, the metal overburden affect CMP performance of both TSVs and RDL patterns. The process concept of the front-side metal CMP for TSVs (Fig. 2.41) is similar to the traditional Cu CMP in the CMOS front-end-of-line (FEOL) process. On the other hand, for the CMP process of a thinned wafer that is attached to a carrier (support wafer), as shown in Fig. 2.41, a backside CMP process also should be considered. (See Sec. 2.9.1.) The backside dielectric CMP process to expose Cu for backside RDL connection is different from front-side CMP applications. For the front-side metal, Cu CMP is to remove the thick blanket Cu with dimples or recesses on the wafer surface after Cu plating, where the flat surface topography and no metal residue are the important indicators. For the backside dielectric CMP, on the contrary, the isolating dielectrics covering the Cu studs of the TSVs need to be removed to reveal the TSV Cu surface and minimize dielectric loss on the silicon surface, where the shapes of studs and dielectric removal on the tip of the TSV Cu surface are the indicators (which will be discussed in Sec. 2.9.1).

In this section, the first two CMP applications (see Fig. 2.41) related to TSVs are studied, including front-side Cu CMP for TSVs and RDLs. We propose a two-step Cu CMP procedure with two different Cu slurries to alleviate the Cu dishing for Cu-plated TSV structures and RDLs. The effect of Cu overburden on CMP performance is discussed as well. Based on the results, we provide a set of useful process guidelines for the optimization of Cu topography of TSVs and RDLs in terms of CMP slurry and Cu overburden [11].

CMP for TSVs with Residues

Test Structures   For Cu CMP on TSVs and RDLs, choosing the proper slurry is the key to minimizing metal dishing and oxide erosion. First, TSVs of 10 μm diameter (thickness = 60 μm) on a 300-mm wafer, as shown in Fig. 2.42*a*, is used to study the slurry effect on Cu dishing. The Cu overburden thickness in this case is about 8 μm, calculated through the electric current and deposition time of Cu plating. The barrier metal is Ta (80 nm thick). Second, both TSVs and RDLs, focusing on the impact of Cu overburden thickness, are studied. The TSV interposer test structure (15 μm diameter and 100 μm thick) is shown in Fig. 2.42*b*. Four Cu plating thicknesses (7.2, 7.8, 8.6, and 9.0 μm) are prepared to examine the overburden effect on metal dishing. The Ta barrier is 80 nm thick. Moreover, the test structure of RDLs, having different pattern density from TSVs owing to metal lines and pads, is shown in Fig. 2.42*c*. The Cu overburden thicknesses are 1.2, 2, 4, and 5 μm above the recess location of the 40-μm-wide (thickness = 1 μm) RDLs.

Experimental Setup   The CMP applications are performed by AMAT Reflexion, as shown in Figs. 2.43 and 2.44. The tool consists of three polishing platens. Platen 1 and platen 2 are for Cu polishing, and platen 3 is for barrier polishing (e.g., Ta) and oxide buffing. The slurries used for polishing Cu are studied, and the slurry used for barrier removal, pads, and disks are available commercially. The slurry flow rate is set at

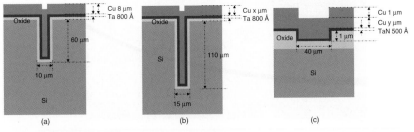

Test samples:
(a) 10-μm diameter (depth = 60 μm) TSV with Cu 8 μm-thick overburden;
(b) 15-μm diameter (depth = 110 μm) TSV with different Cu overburden thicknesses (x = 7.2 μm, 7.8 μm, 8.6 μm and 9.0 μm);
(c) RDL patterns with different Cu overburden thicknesses at the recess location of the 40-μm-linewidth RDL (y = 1.2 μm, 2 μm, 4 μm and 5 μm)

FIGURE 2.42   Schematic diagrams of the test samples: (*a*) 10-μm-diameter (thickness = 60 μm) TSV with an 8-μm-thick Cu overburden; (*b*) 15-μm-diameter (thickness = 110 μm) TSV with different Cu overburden thicknesses (*x* = 7.2, 7.8, 8.6, and 9.0 μm); (*c*) RDL patterns with different Cu overburden thicknesses at the recess location of the 40-μm-linewidth RDL (*y* = 1.2, 2, 4, and 5 μm).

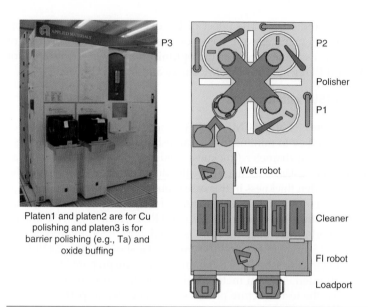

Platen1 and platen2 are for Cu polishing and platen3 is for barrier polishing (e.g., Ta) and oxide buffing

FIGURE 2.43 Applied Materials' chemical mechanical polisher (CMP).

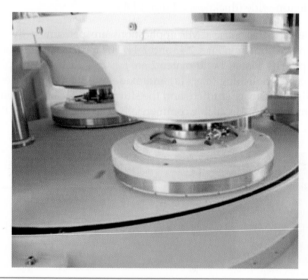

FIGURE 2.44 Platens at work.

300 mL/min. The high-resolution profiler (HRP), optical microscope (OM), and SEM are employed to characterize the Cu surface topography.

Materials and Process We use two Cu slurries for different polishing purposes: One is a slurry with a high Cu removal rate (slurry 1), and the other has a high Cu passivation capability (slurry 2). The removal

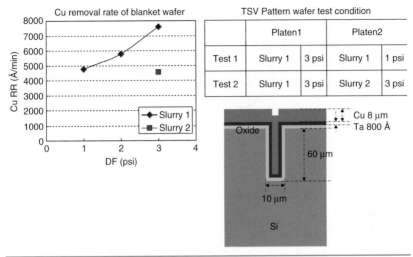

FIGURE 2.45   Cu removal rates of slurry 1 and slurry 2.

rates of blanket Cu with increasing downward forces for both slurry 1 and slurry 2 are shown in Fig. 2.45. The removal rate for slurry 1 at 3 psi is about 7500 Å/min. When the downward force is decreased to 1 psi, the removal rate is reduced to 4700 Å/min, which is close to the removal rate (4500 Å/min) for slurry 2 at 3 psi. The table in Fig. 2.45 shows the test conditions for the effect of the slurries on CMP metal dishing. Slurry 2 has higher Cu passivation capability than slurry 1. Therefore, we use slurry 1 with a high Cu removal rate to remove a thick Cu overburden at 3 psi on platen 1. On the other hand, slurry 1 at 1 psi and slurry 2 at 3 psi (low Cu removal rate) are studied on platen 2.

Results and Discussions

Cu Slurry Selection   Figure 2.46 shows the dishing results of 10-μm-diameter TSVs with an 8-μm-thick Cu overburden after CMP using the test structure shown in Fig. 2.42a. When the same slurry 1 is applied to platen 1 at 3 psi and platen 2 at 1 psi, the metal dishing is 1.2 μm, as shown on the left-hand side of Fig. 2.46. By changing slurry 1 to slurry 2 for platen 2 and increasing the downward force to 3 psi, as shown in Fig. 2.46, the metal dishing is reduced to 250 Å, as shown on the right-hand side of Fig. 2.46. These results imply that even if the removal rates of Cu are similarly low for slurry 1 at 1 psi and for slurry 2 at 3 psi on platen 2, the dishing results are significantly different. In addition, it shows that the slurry with a high Cu passivation capability (slurry 2) applied to platen 2 is very helpful in mitigating the Cu dishing. The present result (250 Å) is better than those (1000 to 4000 Å) obtained in Refs. 27 and 28 with various test conditions on a 10-μm-diameter TSV.

TSV Overburden Effect   Based on the results of metal dishing in the preceding section, slurry 1 for platen 1 at 3 psi and slurry 2 for platen 2

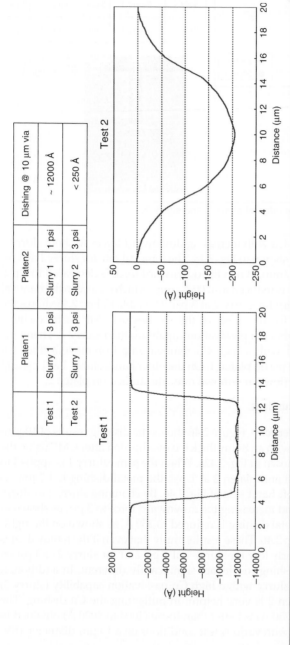

FIGURE 2.46 The dishing at the 10-μm-diameter TSV with (a) slurry 1 on platen 1 with 3 psi and slurry 1 on platen 2 with 1 psi and (b) slurry 1 on platen 1 with 3 psi and slurry 2 on platen 2 with 3 psi.

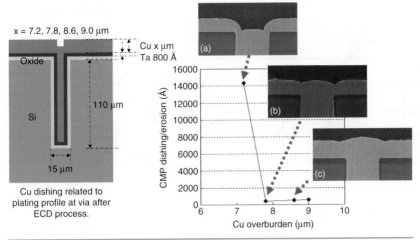

FIGURE 2.47 The TSV dishing/erosion results of different thicknesses of Cu overburden on 15-μm-diameter TSVs after CMP.

at 3 psi are used for study of the impacts of Cu overburden thicknesses on both TSVs and RDLs. The TSV test structure is shown in Fig. 2.42*b*, which includes different Cu overburden thicknesses (i.e., 7.2, 7.8, 8.6, and 9.0 μm). The Cu polishing will stop as soon as it reaches the Ta layer (by means of a laser endpoint detector). Another 60 to 100 seconds will remove the Ta layer. The surface topographies after CMP are measured by HRP. Figure 2.47 shows the dishing results (500 Å) after CMP with different thicknesses of Cu overburden. The dishing data for the 7.2-μm-thick Cu overburden is larger than 1.4 μm and will be explained later. The HRP results of dishing/erosion scanning through 10 vias (pitch = 30 μm) with different thicknesses of Cu overburden are shown in Fig. 2.48. The data imply that thicker Cu overburdens do not reduce the dishing/erosion further; however, recess increases with the Cu overburden. The slurry applied on platen 1 has low Cu passivation capability, and thus thicker Cu overburdens prolong polishing time on platen 1 to remove thick Cu on the field. In addition, since the slurry on platen 1 cannot provide good passivation in the dimple/recess regions of the wafer, the longer polishing time on platen 1 worsens the recess areas so that dishing may increase slightly.

The reason why the 7.2-μm-thick Cu overburden leads to large metal dishing is explained by the insufficient plating time and improper final plating profile (dimple), as shown in Fig. 2.49*a*. The Cu dishing after CMP in Fig. 2.47 happens when the Cu plating profile and overburden look like Fig. 2.49*a*, where there is a dimple over the TSV. Low Cu passivation capability slurry occurs on most areas of the Cu field, so Cu protection is low on the recess metal, resulting in wet etching/erosion behavior on the recess metal right

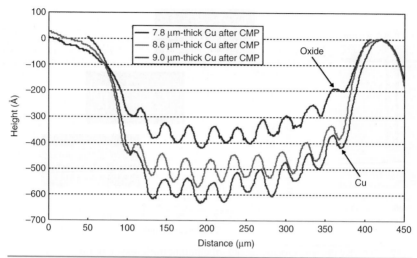

FIGURE 2.48 The TSV surface topographies using an HRP scan through 10 vias after CMP.

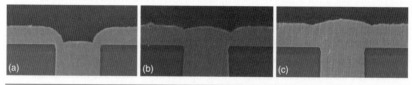

FIGURE 2.49 Different Cu plating times at the end of TSV plating [plating time (a) < (b) < (c) from recess to protrusion].

above the TSV. The etching/erosion becomes more serious after the Cu overburden on the field is removed, which leads to a large Cu dishing.

Moreover, concerning the overburden effect, the overall wafer nonuniformity makes the Cu overburden thicker (12 to 14 μm) near the wafer's edge and thinner near the center (7 to 10 μm) for TSVs with a diameter of 50 μm and a thickness of 150 μm [29]. The Cu via recess after CMP is from 0.25 to 1.5 μm deep depending on the location over the wafer because adjustments of the downward force and speed of CMP are unable to eliminate this overburden nonuniformity [29]. According to our Cu plating tests using blanket wafers, the thickness at the center is 8.5 μm, while it is about 10.1 μm near the edge. Fortunately, the tool used in this study has several pressure zones to control the downward force. Higher pressure is set at the wafer edge to obtain a high removal rate. By tuning the CMP removal-rate profile, the overall wafer nonuniformity effect on the metal recess can be minimized.

CMP for TSV RDLs For RDLs, since the metal pattern density of the metal line and metal pad in RDLs is much higher than in TSVs, study of the effect of Cu overburden on the dishing/erosion for RDLs is necessary. The test structure of RDLs is shown in Fig. 2.42c. Since the Cu plating of RDLs is no longer a bottom-up process, as it is in filling a TSV structure, the plating is more like a conformal behavior. Therefore, overburden in this case is defined as the extra thickness, as indicated by y in Fig. 2.42c, excluding the original RDL step height (1 μm). Optical microscope (OM) images of 1.2- and 2-μm-thick Cu overburdens on 40-μm-linewidth RDLs after CMP are shown in Fig. 2.50. The metal recess on the RDL is found in Fig. 2.50a (from the darker color near the edges of the lines), whereas the flat surface on the RDL is indicated by the bright color shown in Fig. 2.50b.

Figure 2.51a shows the dishing/erosion results of various RDL metal line widths/oxide spaces. The dishing/erosion sizes are, respectively, about 7400 Å for a 1.2-μm overburden and 2700 Å for a 2-μm overburden with 40-μm-linewidth/40-μm-oxide-space RDL. The dishing/erosion results agree with the OM images in Fig. 2.50. The dishing/erosion results of other line-width/oxide-space RDLs, such as 10, 20, and 30 μm, show similar trends as the 40-μm-linewidth/oxide-space RDL. Increasing the Cu overburden to 2 μm significantly improves the metal/oxide recess after CMP. Figures 2.51b and 2.52 show the dishing results of RDL metal pads. A similar trend as that seen with TSVs indicates that thicker Cu overburdens do not reduce the metal dishing, especially at the 200-μm metal pad. The overall results indicate that an optimized thickness of Cu plating overburden should be considered for subsequent Cu CMP performance.

The impact of dishing due to CMP for TSV with RDLs on the downstream 3D packaging processes, such as under-bump metallurgy (UBM), and wafer bumping is very important. Usually, the maximum allowable nonuniformity (tolerance) of the total microbump height is

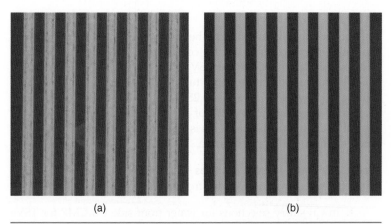

(a)   (b)

FIGURE 2.50   OM images of different thicknesses of Cu plating overburden on 40-μm-linewidth RDL after CMP: (a) 1.2 μm thick and (b) 2 μm thick.

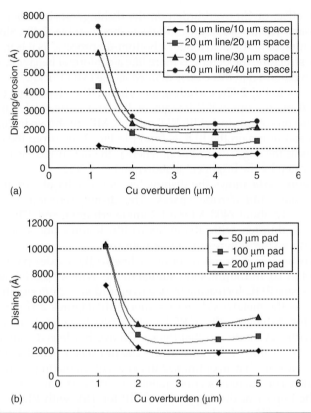

FIGURE 2.51 Effect of different thicknesses of Cu plating overburden on RDL dishing/erosion results after CMP: (a) the linewidth/oxide-space patterns and (b) the metal pads.

5%. Thus, if the total microbump height is 20 μm, then the maximum allowable tolerance is 1 μm. By considering the average tolerance contributed from the CMP for TSV with RDLs, UBM, masking, and wafer bumping, it is reasonable to define the maximum allowable dishing of TSV with RDLs as 0.25 μm. On the other hand, if the acceptable specification is 10% of the total bump height, then the allowable dishing is 0.5 μm. This is not affected by the TSV via size, aspect ratio, and RDL pattern size.

Straight TSV CMP Process Guidelines   Two important points are summarized concerning Cu CMP for front-side TSVs with RDLs.

   1. In Cu slurry selections for wafer front side Cu CMP for TSVs and RDLs, the slurries with high removal rates should go along with those of high passivation capability to reduce the metal

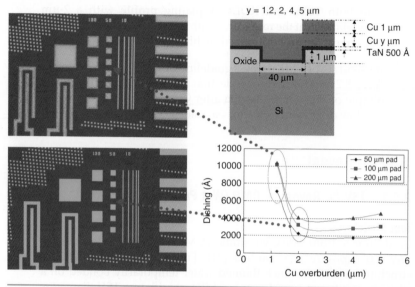

Figure 2.52   Cu overburden effects on CMP of RDLs (Cu overburden affects dishing results).

dishing. Using a slurry with a high Cu removal rate to remove the thick Cu overburden on the field first and then changing to a slurry with a high Cu passivation capability to clean the remaining Cu tends to result in much less metal dishing.

2. The Cu plating result affects the metal/oxide dishing/erosion after CMP. The metal recess or dimple owing to Cu plating is an important indicator of dishing/erosion after CMP and should be minimized. For TSV plating, the overburden profiles from Cu recess to Cu protrusion will lead to much smaller post-CMP metal dishing.

The effects of dishing due to CMP for TSV overburden Cu removal and front-side metallization RDL on the downstream 3D packaging processes, such as UBM and wafer bumping are very important.

2.8.3   Summary and Recommendations

Some important results and recommendations are summarized below:

- It has been demonstrated that proper selection of Cu slurries in the proposed two-step Cu polishing process (from high removal rate to high passivation capability) can reduce metal dishing dramatically.

- An optimized Cu plating overburden has been determined for the dishing/erosion of the metal/oxide that is applicable to both TSVs and RDLs. A plating profile with a 2-μm overburden at the recess area with the step-height smaller than 1 μm can ensure optimized dishing performance.

A useful set of CMP process guidelines for TSV was provided in Sec. 2.8.2 regarding slurry selection, minimizing metal recess or dimple on the pattern region after Cu plating, edge trimming, and low downward force for backside CMP.

2.9   TSV Cu Reveal

2.9.1   TSV Cu Reveal by CMP (Wet Process)

Test Structures
In addition to front-side CMP studies, CMP on the backside oxide for TSV Cu exposure was studied [11]. Figure 2.53 shows the TSV structure on a 50-μm-thick thinned wafer temporarily bonded on a silicon wafer for a backside oxide CMP test with two TSV diameters (10 and 30 μm). The shapes of the Cu studs before and after oxide CMP are also demonstrated.

Experimental Setup
This CMP application is performed by the AMAT Reflexion, as shown in Fig. 2.43. Platen 3 is only used to remove 400-nm oxide on the top of the TSV to expose the Cu studs. The slurry, pad, and disk are

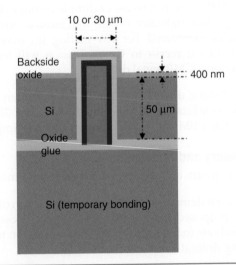

FIGURE 2.53   Backside TSV structure with two TSV diameters ($z = 10$ and 30 μm) on a thinned wafer (50 μm) temporarily bonded on a silicon wafer.

commercially available consumables. The slurry flow rate is set at 300 mL/min. The polishing downward force is set at 1 psi to make the rates of field oxide and Cu removal around 400 and 200 Å/min, respectively. The HRP and OM are employed to characterize the Cu surface topography.

Edge-Trimming Before Front-Side Temporary Bonding

Prior to the study, a wafer transfer test in the CMP tool for the thinned TSV wafer bonded on the silicon wafer was performed to check the possibility of wafer-edge chipping or cracking during the process. With edge trimming (0.5 mm) for the thick blind TSV wafer, no edge chipping was observed after the wafer transfer test, as shown in Fig. 2.54. On the other hand, without wafer edge trimming of the thick blind TSV wafer, edge chipping on the thinned TSV wafer was found after the transfer test. According to the results of thin-wafer transfer tests, all thick blind TSV wafers should go through edge trimming of 0.5 mm before temporary bonding to a support silicon wafer and backside grinding to expose the TSV Cu studs in order to reduce edge chipping of the thinned wafer for the processes that follow.

Results and Discussions

Backside oxide CMP is used for Cu exposure in the TSV region after wafer thinning to 50 plus 2 to 4 μm to expose the TSV Cu stud, and then the Si is etched for 2 to 4 μm. After that, an isolation oxide

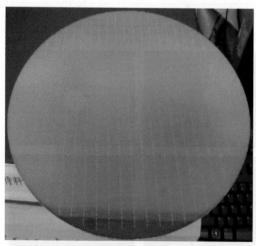

The backside of a 50-μm-thick thin wafer, with edge trimming of 0.5 mm before temporarily bonded on a silicon wafer after transfer test in the CMP tool.

FIGURE 2.54 The backside of a 50-μm-thick thin wafer temporarily bonded on a silicon wafer after a transfer test in the CMP tool.

deposition of 4000 Å is applied to the wafer. Then CMP is used to remove the 4000-Å oxide passivation (~60 seconds) with two indicators: one is little oxide loss on the silicon surface, and the other is the contours of Cu studs after oxide removal on the tip of the Cu surface. Concerning the little oxide loss on the silicon surface after oxide removal on the top of the Cu studs, the polishing action is set to make the rate of field oxide removal around 400 Å/min.

Figure 2.55 shows the two-dimensional (2D) HRP (*above*), OM images (*center*), and 3D HRP contour images (*below*) of a 10-μm-diameter TSV before and after backside oxide CMP for Cu reveal. The surface of the TSV Cu stud obviously shows its shiny color after oxide CMP. Using HRP on the 3D surface topography before and after oxide CMP, the step height of the Cu stud changes from 2.7 to 1.8 μm (see Fig. 2.55). The reduction in step height is almost 9000 Å. Comparing the step-height data before and after CMP (9000 Å) with oxide thickness (4000 Å), the oxide on the top of Cu studs should be removed. Since the density of the

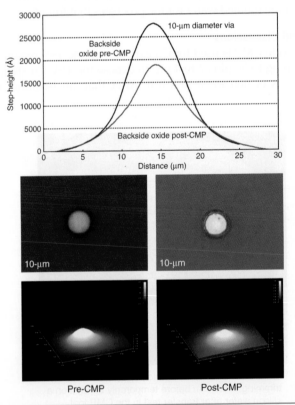

FIGURE 2.55   Results of a 10-μm-diameter TSV before and after backside oxide CMP for Cu exposure. (*Above*) 2D HRP; (*center*) OM images; and (*below*) 3D HRP contour images.

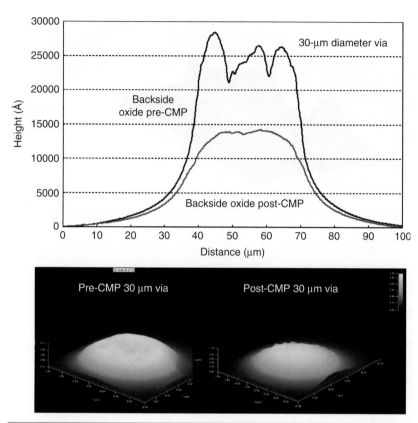

Figure 2.56 Results of a 30-μm-diameter TSV before and after backside oxide CMP for Cu exposure. (*Above*) 2D HRP and (*below*) 3D HRP contour images.

Cu studs over the wafer surface is less than 5 percent and the wafer surface has some Cu studs protruding from it, the CMP downward force basically concentrates on those studs, resulting in efficient reduction in step height. Moreover, Fig. 2.56 (*above*) shows that the shape of the stud before CMP is like a cone, and it keeps the same contour after CMP.

The impact of TSV Cu via diameters on Cu stud shapes is also discussed in this study. Figure 2.56 shows the HRP results of 30-μm-diameter TSVs. The 30-μm-diameter TSV Cu stud has a plateau-like shape, as shown in Fig. 2.56. This is very different from the 10-μm-diameter TSV Cu stud, which has a cone-like shape, as shown in Fig. 2.55. The corresponding 3D surface topographies for the 10- and 30-μm-diameter TSV Cu studs after (60 seconds of) oxide CMP are shown, respectively, in Fig. 2.57. The result implies that the diameter of the via is a key factor for Cu protrusion contour after backside oxide CMP. Moreover, the CMP process may not change the Cu stud

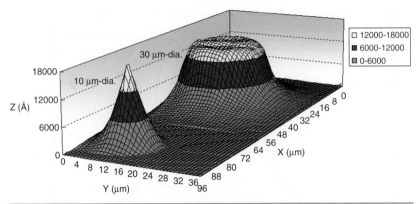

FIGURE 2.57   3D HRP results of 10- and 30-µm-diameter TSVs after oxide CMP for 60 seconds.

contour. Both the 10- and 30-µm TSVs kept the original contours after CMP.

2.9.2   Cu Reveal by Dry Etching Process

Temporary Bonding and Backgrinding

In Ref. 44, IMEC presented a very useful Cu TSV-revealing process, as shown in Fig. 2.58. It can be seen that after temporary bonding to a silicon supporting wafer, the device wafer (designed to be 50 µm thick) is thinned down to 57 µm. The total thickness variation (TTV) of the device wafer after thinning is one of the major parameters to control because it affects the performance of the nail-reveal process. Based on the infrared (IR) time-of-flight measurement, the TTV is in the range of 1.6 µm across a 300-mm blanket wafer as well as a TSV wafer [44].

Isotropic Dry Etching of the Silicon

After thinning, a selective isotropic dry recess etch reveals the TSVs while keeping the Cu protected into the oxide liner (as shown in Fig. 2.59), which prevents Cu oxidation that could occur during subsequent process steps. The differences observed between center-to-edge regions are the results of TSV etch variation, backgrinding, and recess etch variations (2.9 µm center, 1.9 µm edge) [44].

SiN Passivation Deposition

After nail reveal, a 500-nm (SiN) passivation layer is deposited (<200°C) on the backside. This passivation layer is of upmost importance to prevent Cu diffusion through the thin wafer to the FEOL active device layers in case redistribution layers or UBM of microbumps is done on the backside prior to stacking.

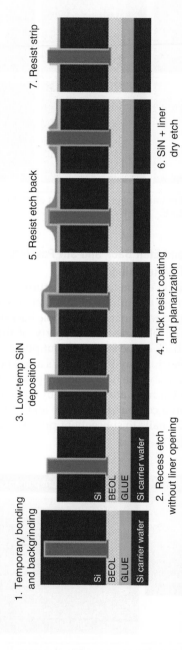

FIGURE 2.58 IMEC's integration of wafer thinning and backside passivation flows after the TSV process.

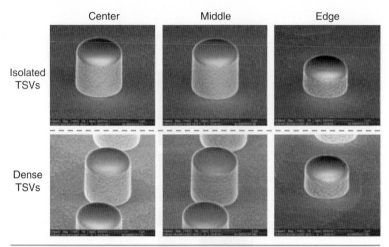

FIGURE 2.59   Top-view SEM image of recessed TSVs after dry recess etch, as shown in step 2 of Fig. 2.58.

Thick Resist Coating and Etch Back and SiN Passivation and Oxide Liner Dry Etching for TSV Revealing

After backside deposition of the low-temperature nitride above the Cu nails (with a thickness uniformity of less than 3 percent), a thick resist layer is spin-coated to planarize the whole backside surface, as shown in Fig. 2.60. The planarization property of this thick resist allows a blanket etch back of the layer without photolithography, as shown in Fig. 2.61, in such a way that all TSVs are exposed, whereas a thin resist layer remains on the field. The SiN passivation layer and the oxide liner on the Cu nails are removed by dry etching. Finally, the resist is stripped in a combination of dry and wet cleaning processes, exposing the Cu nails to subsequent process steps, as shown in Figs. 2.62 and 2.63 [44].

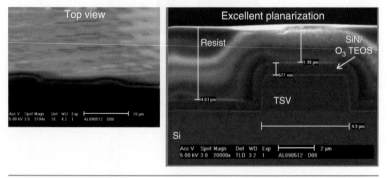

FIGURE 2.60   Tilted SEM view and cross section of dummy TSVs after thick resist planarization.

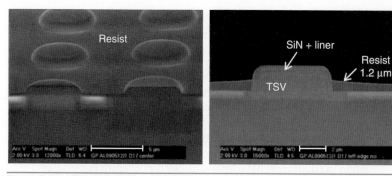

FIGURE 2.61 Top-view SEM image and cross section after thick resist etch back. About 1 μm of resist is left on the field, whereas TSVs covered with a 500-nm SiN passivation are exposed.

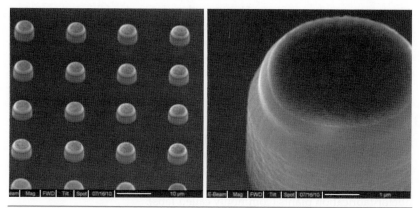

FIGURE 2.62 Top-view SEM image of the Cu nails after passivation layer etch and resist strip.

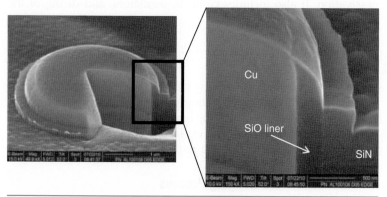

FIGURE 2.63 Focused ion beam (FIB) cross section of a Cu nail after passivation layer etch and resist strip.

2.9.3   Summary and Recommendation

Some important results and recommendation are summarized in the following:

- For TSV Cu reveal by CMP wet process, the materials and process steps are less. However, care must be taken (on the uniformity and coplanarity of the wafer) to prevent Cu-contamination while revealing the Cu. Also, the impact of TSV diameters on the backside isolation oxide CMP for a 50-μm-thick wafer temporarily bonded on a silicon wafer is that Cu studs of larger via diameters have a plateau-like topography. For smaller via diameters, however, the CMP process cannot change the studs' topography from a cone to a plateau, even though the topography stays the same after CMP.

- For TSV Cu reveal by dry etching process, the materials and process step are more. However, the advantage of this process is to avoid the exposure of Cu to bare Si (in the case of Cu reveal by CMP for instance), and together with the application of the diffusion barrier, this prevents Cu-contamination-induced device reliability issues.

2.10   FEOL and BEOL

Front end-of-line (FEOL) is defined as the first portion of IC fabrication where the individual devices such as transistors or resistors are patterned. This process is from a bare wafer to (but not including) the deposition of metal layers. FEOL is usually performed in semiconductor fabrication plants (commonly called fabs).

Back end-of-line (BEOL) is defined as the fabrication in which active devices are interconnected with wiring on the wafer. This process starts from the first layer of metal to bonding pads with passivation. It also includes contacts, insulators, and dicing of the wafer into individual IC chips.

2.11   TSV Processes

For 3D Si integration, usually the vias are fabricated before or after the wafer-to-wafer (W2W) bonding process. On the other hand, for 3D IC integration, the vias can be fabricated by via-first, via-middle, via-last from the front-side, and via-last from the backside processes. For more information, please see Table 2.8.

2.11.1   Via-Before Bonding Process

For W2W bonding with Cu-to-Cu bonding, usually the vias are fabricated before the W2W bonding. Figure 2.64 shows a SEM image

Processes	Methods/Options		
Via forming	Bosch DRIE	Non-Bosch DRIE	Laser
Dielectric deposition	SiO_2	SiN/SiO_2	Polymer
Barrier/seed layers deposition	Ti (or Ta)/Cu	TiW/Cu W/Cu	W/W
Via filling	Cu	W	Conductive polymer, CNT, solder, etc.
Residue removal	CMP	CMP (two-step)	
TSV revealing	Wet etch	Dry etch	
TSV process	TSV before bonding, TSV after bonding	Via first Via middle	Via last (front- or back-side)
Thin-wafer handling	Support (Carrier) wafer	Carrierless	On stacking
Stacking	C2C	C2W	W2W
Micro interconnect	Solder bump	Cu pillar + solder	Au/Cu studs
Bonding	Natural reflow, thermo compression	Direct bonding	Indirect bonding (intermediate layer)

Table 2.8   3D Integration Processes, Methods, and Options

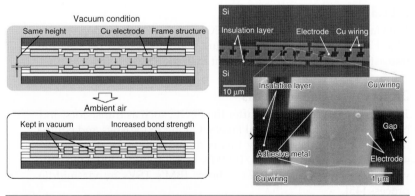

Figure 2.64   W2W with Cu-to-Cu (via-before) bonding by NIMS/AIST/Toshiba/ University of Tokyo for 3D Si integration.

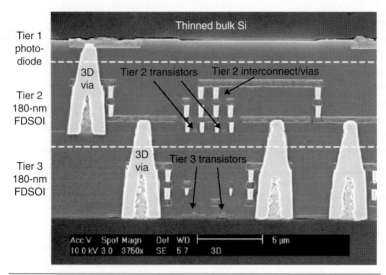

FIGURE 2.65   W2W bonding with SiO_2-to-SiO_2 (via-after) bonding by MIT for 3D Si integration.

of NIMS/AIST/Toshiba/University of Tokyo's cross section with Cu-to-Cu bonding [46].

2.11.2   Via-After Bonding Process

For W2W bonding with SiO_2-to-SiO_2 bonding, usually the vais are fabricated after the W2W bonding. Figure 2.65 shows a SEM image of MIT's cross section with SiO_2-to-SiO_2 bonding [47].

2.11.3   Via-First Process

The vias are fabricated before the FEOL. This can only be done by the fabs. However, even in the fabs, this seldom happens because the transistors are much more important than the TSVs.

2.11.4   Via-Middle Process

The vias are fabricated right after the FEOL, such as transistors, and before the tiny vias and metal layers. Owing to logistics and equipment compatibilities, usually this is done by the fabs. Figure 2.66 shows SEM images of IEEE/IEDM's cross sections with the TSV via-middle process. Figure 2.67 shows IBM's TSVs by via-middle process [45].

2.11.5   Via-Last (From the Front Side) Process

The vias are fabricated after the BEOL, that is, the passivation. Usually this is done from the front side of the wafer and by the fabs with

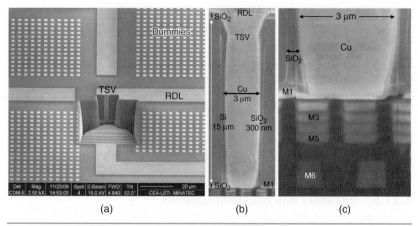

(a)   (b)   (c)

FIGURE 2.66   IEEE/IEDM2010 via-middle process for 3D IC integration.

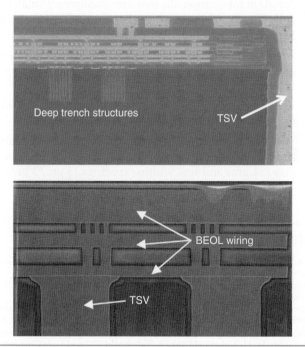

FIGURE 2.67   IBM's TSVs by via-middle process.

very small vias (<1 μm) or the outsourced semiconductor assembly and test (OSAT) with larger vias (≥4 μm). Another factor that needs to be considered is the redistribution layer (RDL). Usually, OSAT's equipments can only make a 3-μm/3-μm line width/space, but the fabs can make submicron widths/spaces.

2.11.6   Via-Last (From the Backside) Process

This is the same as Sec. 2.11.5 except that the vias are fabricated from the backside of the wafer. A couple of examples can be found in Figs. 2.2 (MMIC) and 2.3 (CMOS image sensor), but even speaking straight, they are not 3D IC integration.

2.11.7   How About the Passive Interposers?

When the industry defined the TSV processes for 3D IC integration, there were no passive interposers yet. Also, since there is not active device in the passive interposers, thus they don't fit into any of the preceding!

2.11.8   Summary and Recommendations

Some important results and recommendations are summarized in the following:

- For 3D Si integration, the vias are usually very small (≤ 1 μm), and they are fabricated by either Cu-to-Cu, SiO_2-to-SiO_2, or hybrid (Cu-to-SiO_2) W2W bonding.

- For 3D IC integration, the vias are usually larger (≥ 1 μm), and they are fabricated by either via-first, via-middle, or via-last processes.

- Via-first and via-middle processes are usually performed by the fabs.

- The via-last process is usually performed by the fabs with small vias (<4 μm) and fine RDLs (<3 μm/<3 μm). For larger vias (≥ 4 μm) and line widths/spacings (3 μm/3 μm), OSAT can do the job.

2.12   References

[1] Lau, J. H., *Reliability of RoHS-Compliant 2D and 3D IC Interconnects*, McGraw-Hill, New York, 2011.
[2] Lau, J. H., C. K. Lee, C. S. Premachandran, and A. Yu, *Advanced MEMS Packaging*, McGraw-Hill, New York, 2010.
[3] Shockley, W., "Semiconductive Wafer and Method of Making the Same," U.S. Patent No. 3,044,909, filed on October 23, 1958, and granted on July 17, 1962.
[4] Sekiguchi, M., H. Numata, N. Sato, T. Shirakawa, M. Matsuo, H. Yoshikawa, M. Yanagida, H. Nakayoshi, and K. Takahashi, "Novel Low Cost Integration of Through Chip Interconnection and Application to CMOS Image Sensor," *IEEE/ECTC Proceedings*, San Diego, CA, May 2006, pp. 1367–1374.
[5] Archard, D., K. Giles, A. Price, S. Burgess, and K. Buchanan, "Low Temperature PECVD Of Dielectric Films For TSV Applications," *IEEE/ECTC Proceedings*, 2010, pp. 764–768.
[6] Kumar Praveen, S., Ho Wai Tsan, and R. Nagarajan, "Conformal Low-Temperature Dielectric Deposition Process Below 200°C For TSV Application," *Proceedings of IEEE Electronics Packaging Technology Conference*, 2010, pp. 27–30.
[7] Beica, R., C. Sharbono, and T. Ritzdorf, "Through Silicon Via Copper Electrodeposition For 3D Integration," *IEEE/ECTC Proceedings*, 2008, pp. 577–583.

[8] Lau, J. H., "Evolution, Outlook, and Challenges of 3D IC/Si Integration," *IEEE/ICEP Proceedings* (keynote), Nara, Japan, April 13, 2011, pp. 1–17.

[9] Lau, J. H., "Evolution, Challenge, and Outlook of TSV, 3D IC Integration and 3D Si Integration," *International Wafer Level Packaging Conference* (plenary), San Jose, CA, October 3–7, 2011, pp. 1–18.

[10] Hsin, Y. C., C. Chen, J. H. Lau, P. Tzeng, S. Shen, Y. Hsu, S. Chen, C. Wn, J. Chen, T. Ku, and M. Kao, "Effects of Etch Rate on Scallop of Through-Silicon Vias (TSVs) in 200-mm and 300-mm Wafers," *IEEE/ECTC Proceedings*, Orlando, FL, June 2011, pp. 1130–1135.

[11] Chen, J. C., P. J. Tzeng, S. C. Chen, C. Y. Wu, J. H. Lau, C. C. Chen, C. H. Lin, Y. C. Hsin, T. K. Ku, and M. J. Kao, "Impact of Slurry in Cu CMP (Chemical Mechanical Polishing) on Cu Topography of Through Silicon Vias (TSVs), Re-distributed Layers, and Cu Exposure," *IEEE/ECTC Proceedings*, Orlando, FL, June 2011, pp. 1389–1394.

[12] Selvanayagam, C., J. H. Lau, X. Zhang, S. Seah, K. Vaidyanathan, and T. Chai, "Nonlinear Thermal Stress/Strain Analysis of TSV (Through Silicon Via) and Their Flip-Chip Microbumps," *IEEE Transactions in Advanced Packaging*, Vol. 32, No. 4, November 2009, pp. 720–728.

[13] Tang, G., O. Navas, D. Pinjala, J. H. Lau, A. Yu, and V. Kripesh, "Integrated Liquid Cooling Systems for 3D Stacked TSV Modules," *IEEE Transactions on Components and Packaging Technologies*, Vol. 33, No. 1, 2010, pp. 184–195.

[14] Khan, N., V. Rao, S. Lim, S. Ho, V. Lee, X. Zhang, R. Yang, E. Liao, Ranganathan, T. Chai, V. Kripesh, and J. H. Lau, "Development of 3D Silicon Module with TSV for System in Packaging," *IEEE Transactions on CPMT*, Vol. 33, No. 1, March 2010, pp. 3–9.

[15] Chai, T., X. Zhang, J. H. Lau, C. Selvanayagam, K. Biswas, S. Liu, D. Pinjala, G. Tang, Y. Ong, S. Vempati, E. Wai, H. Li, B. Liao, N. Ranganathan, V. Kripesh, J. Sun, J. Doricko, and C. Vath, "Development of Large Die Fine-Pitch Cu/Low-k FCBGA Package with Through Silicon Via (TSV) Interposer," *IEEE Transactions on Components, Packaging and Manufacturing Technology*, Vol. 1, No. 5, 2011, pp. 660–672.

[16] Ranganathan, N., L. Ebin, L. Linn, V. Lee, O. Navas, V. Kripesh, and N. Balasubramanian, "Integration of High Aspect Ratio Tapered Silicon Via for Through-Silicon Interconnection," *IEEE/ECTC Proceedings*, Orlando, FL, May 2008, pp. 859–865.

[17] Lau, J. H., "TSV Manufacturing Yield and Hidden Costs for 3D IC Integration," *IEEE/ECTC Proceedings*, Las Vegas, NV, June 2010, pp. 1031–1041.

[18] Chaabouni, H., M. Rousseau, P. Leduc, A. Farcy, R. El Farhane, A. Thuaire, G. Haury, A. Valentian, G. Billiot, M. Assous, F. De Crecy, J. Cluzel, A. Toffoli, D. Bouchu, L. Cadix, T. Lacrevaz, P. Ancey, N. Sillon, and B. Flechet, "Investigation on TSV Impact on 65-nm CMOS Devices and Circuits," *Proceedings of IEEE International Electron Devices Meeting*, 2010, pp. 35.1.1–35.1.4.

[19] Reed, J. D., S. Goodwin, C. Gregory, and D. Temple, "Reliability Testing of High Aspect Ratio Through Silicon Vias Fabricated with Atomic Layer Deposition Barrier, Seed Layer and Direct Plating and Material Properties Characterization of Electrografted Insulator, Barrier and Seed Layer for 3D Integration," *Proceedings of IEEE International 3D Systems Integration Conference*, 2010, pp. 1–8.

[20] Hung Y., C. Hsieh, S. Jeng, H. Tao, C. Min, and Y. Mii, "A New Enhancement Layer To Improve Copper Interconnect Performance," *Proceedings of IEEE International Interconnect Technology Conference*, 2010, pp. 1–3.

[21] Powell, K., S. Burgess, T. Wilby, R. Hyndman, and J. Callahan, "3D IC Process Integration Challenges and Solutions," *Proceedings of IEEE International Interconnect Technology Conference*, 2008, pp. 40–42.

[22] Teh, W. H., R. Caramto, S. Arkalgud, T. Saito, K. Maruyama, and K. Maekawa, "Magnetically Enhanced Capacitively Coupled Plasma Etching For 300-Mm Wafer-Scale Fabrication Of Cu Through-Silicon-Vias For 3D Logic Integration," *Proceedings of IEEE International Interconnect Technology Conference*, 2009, pp. 53–55.

[23] Wu, C., S. Chen, P. Tzeng, J. H. Lau, Y. Hsu, J. Chen, Y. Hsin, C. Chen, S. Shen, C. Lin, T. Ku, and M. Kao, "Oxide Liner, Barrier and Seed Layers, and

Cu-Plating of Blind Through Silicon Vias (TSVs) on 300mm Wafers for 3D IC Integration," *Proceedings of IMAPS International Conference*, Long Beach, CA, 2011, pp. 1–7.

[24] Beica, R., C. Sharbono, and T. Ritzdorf, "Through Silicon Via Copper Electrodeposition for 3D Integration," *IEEE/ECTC Proceedings*, 2008, pp. 577–583.

[25] Vempati Srinivasa Rao, Ho Soon Wee, Lee Wen Sheng Vincent, Li Hong Yu, Liao Ebin, Ranganathan Nagarajan, Chai Tai Chong, Xiaowu Zhang, and Pinjala Damaruganath, "TSV Interposer Fabrication for 3D IC Packaging," *IEEE 11th Electronics Packaging Technology Conference*, 2009, pp. 431–437.

[26] Lee Wen Sheng Vincent, Navas Khan, Liao Ebin, S. W. Yoon, and V. Kripesh, "Cu Via Exposure by Backgrinding for TSV Applications," *IEEE 9th Electronics Packaging Technology Conference*, 2007, pp. 233–237.

[27] Takahashi, K., Y. Taguchi, M. Tomisaka, H. Yonemura, M. Hoshino, M. Ueno, Y. Egawa, Y. Nemoto, Y. Yamaji, H. Terao, M. Umemoto, K. Kameyama, A. Suzuki, Y. Okayama, T. Yonezawa, and K. Kondo, "Process Integration of 3D Chip Stack with Vertical Interconnection," *IEEE/ECTC Proceedings*, 2004, pp. 601–609.

[28] Takahashi, K., Y. Taguchi, M. Hoshino, K. Tanida, M. Umemoto, T. Yonezawa, and K. Kondo, "Development of Less Expensive Process Technologies for Three-Dimensional Chip Stacking with Through-Vias," *Electronics and Communications in Japan*, Part 2, Vol. 88, No. 7, pp. 50–60, 2005.

[29] Dean Malta, Christopher Gregory, Dorota Temple, Trevor Knutson, Chen Wang, Thomas Richardson, Yun Zhang, and Robert Rhoades, "Integrated Process for Defect-Free Copper Plating and Chemical-Mechanical Polishing of Through-Silicon Vias for 3D Interconnects," *IEEE/ECTC Proceedings*, 2010, pp. 1769–1775.

[30] Tsai, W., H. H. Chang, C. H. Chien, J. H. Lau, H. C. Fu, C. W. Chiang, T. Y. Kuo, Y. H. Chen, R. Lo, and M. J. Kao, "How to Select Adhesive Materials for Temporary Bonding and De-Bonding of Thin-Wafer Handling in 3D IC Integration?" *IEEE ECTC Proceedings*, Orlando, FL, June 2011, pp. 989–998.

[31] Dang, B., P. Andry, C. Tsang, J. Maria, R. Polastre, R. Trzcinski, A. Prabhakar, and J. Knickerbocker, "CMOS Compatible Thin Wafer Processing Using Temporary Mechanical Wafer, Adhesive and Laser Release of Thin Chips/Wafers for 3D Integration," *IEEE/ECTC Proceedings*, 2010, pp. 1393–1398.

[32] Tamura, K., K. Nakada, N. Taneichi, P. Andry, J. Knickerbocker, and C. Rosenthal, "Novel Adhesive Development for CMOS-Compatible Thin Wafer Handling," *IEEE/ECTC Proceedings*, 2010, pp. 1239–1244.

[33] Zhang, X., A. Kumar, Q. X. Zhang, Y. Y. Ong, S. W. Ho, C. H. Khong, V. Kripesh, J. H. Lau, D.-L. Kwong, V. Sundaram, Rao R. Tummula, and Georg Meyer-Berg, "Application of Piezoresistive Stress Sensors in Ultra Thin Device Handling and Characterization," *Sensors and Actuators A: Physical*, Vol. 156, 2009, pp. 2–7.

[34] Kwon, W., J. Lee, V. Lee, J. Seetoh, Y. Yeo, Y. Khoo, N. Ranganathan, K. Teo, and S. Gao, "Novel Thinning/Backside Passivation for Substrate Coupling Depression of 3D IC," *IEEE/ECTC Proceedings*, Orlando, FL, 2011, pp. 1395–1399.

[35] Lee, J., V. Lee, J. Seetoh, S. Thew, Y. Yeo, H. Li, K. Teo, and S. Gao, "Advanced Wafer Thinning and Handling for Through Silicon Via Technology," *IEEE/ECTC Proceedings*, Orlando, FL, 2011, pp. 1852–1857.

[36] Jourdain, A., T. Buisson, A. Phommahaxay, A. Redolfi, S. Thangaraju, Y. Travaly, E. Beyne, and B. Swinnen, "Integration of TSVs, Wafer Thinning and Backside Passivation on Full 300-mm CMOS Wafers for 3D Applications," *IEEE/ECTC Proceedings*, Orlando, FL, 2011, pp. 1122–1125.

[37] Halder, S., A. Jourdain, M. Claes, I. Wolf, Y. Travaly, E. Beyne, B. Swinnen, V. Pepper, P. Guittet, G. Savage, and L. Markwort, "Metrology and Inspection for Process Control During Bonding and Thinning of Stacked Wafers for Manufacturing 3D SIC's," *IEEE/ECTC Proceedings*, Orlando, FL, 2011, pp. 999–1002.

[38] Bhardwaj, J., H. Ashraf, and A. McQuarrie, "Dry Silicon Etching for MEMS," *Proceedings of Microstructures and Microfabricated Systems*, Montreal, Quebec, Canada, May 4–9, 1997, pp. 1–13.

[39] C. K. Chung, "Geometrical Pattern Effect on Silicon Deep Etching by an Inductively Coupled Plasma System," *Journal of Micromechanics and Microengineering*, Vol. 14, 2004, pp. 656–662.

[40] Liu, H.-C., Y.-H. Lin, and W. Hsu, "Sidewall Roughness Control in Advanced Silicon Etch Process," *Journal of Microsystem Technologies*, Vol. 10, 2003, pp. 29–34.

[41] Ranganathan, N., K. Prasad, N. Balasubramanian, and K. Pey, "A Study of Thermo-Mechanical Stress and Its Impact on Through-Silicon Vias," *Journal of Micromechanics and Microengineering*, Vol. 18, 2008, pp. 1–13.

[42] Chen, K., A. A. Ayon, X. Zhang, and S. M. Spearing, "Effect of Process Parameters on the Surface Morphology and Mechanical Performance of Silicon Surfaces after Deep Reactive Ion Etching (DRIE)," *Journal of Microelectromechanical Systems*, Vol. 11, 2002, pp. 264–275.

[43] Tian, J., and M. Bartek, "Simultaneous Through-Silicon Via and Large Cavity Formation Using Deep Reactive Ion Etching and Aluminum Etch-Stop Layer," *IEEE/ECTC Proceedings*, Lake Buena Vista, FL, May 27–30, 2008, pp. 1787–1792.

[44] Jourdain, A., T. Buisson, A. Phommahaxay, A. Redolfi, S. Thangaraju, Y. Travaly, E. Beyne, and B. Swinnen, "Integration of TSVs, Wafer Thinning and Backside Passivation on Full 300-mm CMOS Wafers for 3D Applications," *IEEE/ECTC Proceedings*, May 2011, pp. 1122–1125.

[45] Farooq, M. G, T. L. Graves-Abe, W. F. Landers, C. Kothandaraman, B. A. Himmel, P. S. Andry, C. K. Tsang, E. Sprogis, R. P. Volant, K. S. Petrarca, K. R. Winstel, J. M. Safran, T. D. Sullivan, F. Chen, M. J. Shapiro, R. Hannon, R. Liptak, D. Berger, and S. S. Iyer, "3D Copper TSV Integration, Testing and Reliability," *Proceedings of IEEE/IEDM*, Washington, DC, 2011, pp. 7.1.1–7.1.4.

[46] Shigetou, A., T. Itoh, M. Matsuo, N. Hayasaka, K. Okumura, and T. Suga, "Bumpless Interconnect Through Ultrafine Cu Electrodes by Mans of Surface-Activated Bonding (SAB) Method," *IEEE Transaction on Advanced Packaging*, Vol. 29, No. 2, May 2006, pp. 218–226.

[47] Burns, J., B. Aull, C. Keast, C. Chen, C. Chen, J. Knecht, V. Suntharalingam, K. Warner, P. Wyatt, and D. Yost, "A Wafer-Scale 3-D Circuit Integration Technology," *IEEE Transactions on Electron Devices*, Vol. 53, No. 10, October 2006, pp. 2507–2516.

CHAPTER 3

Through-Silicon Vias: Mechanical, Thermal, and Electrical Behaviors

3.1 Introduction

In this chapter, the mechanical, thermal, and electrical behaviors of Cu-filled through-silicon vias (TSVs) are considered. Useful simulation results and test results are presented. In addition, some design guidelines and test methods for Cu-filled TSVs are provided and discussed. TSV technology covers multidisciplinary areas of science and engineering (see, e.g., Refs. 1–50).

3.2 Mechanical Behavior of TSVs in System-in-Package

The copper-filled TSV is one of the critical aspects of three-dimensional (3D) integrated-circuit (IC) integration and 3D Si integration. This is so because (1) the thermal coefficient of expansion (TCE) of silicon is approximately $2.5 \times 10^{-6}/°C$, (2) the TCE of copper is approximately $17.5 \times 10^{-6}/°C$, (3) there is a very large local thermal expansion mismatch between the silicon chip and the filled copper that could create failures (delaminations) near the interface between the filled copper and the silicon and between the copper and the dielectric liner, and (4) Cu pumping occurs when the 3D structure is subjected to heating.

3.2.1 Mechanical Behavior of TSVs for Active/Passive Interposers

In this section, design for reliability (DFR) of the high-performance 3D IC integration system-in-package (SiP) (shown in Fig. 3.1) with

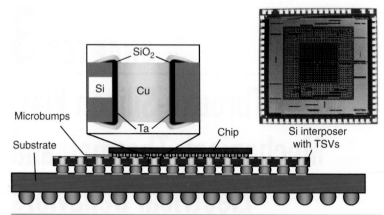

Figure 3.1   Schematic of the Cu–low-*k* chip (21 × 21 mm) on a TSV interposer (25 × 25 mm) (2.5D IC integration).

copper-filled TSVs is examined using the finite-element method. A very large chip (21 × 21 mm) with nine metal Cu–low-*k* layers is supported by a silicon interposer (25 × 25 × 300 mm) with copper-filled TSVs. This TSV interposer is made with multilayer, double-sided metallization, CuNiAu underbump metallurgy (UBM) on the top side, and SnAgCu (150-μm-pitch) solder bumps on the bottom side. The bismaleimide triazene (BT) substrate size is 45 × 45 mm square with 1-2-1 layer configuration (core thickness = 0.8 mm) [1–3] and a plastic ball-grid array (PBGA) pad pitch of 1 mm [4]. The package specification is shown in Table 3.1.

Nonlinear thermal stress owing to the local thermal expansion mismatch at the interfaces between the copper, silicon, and dielectric is determined for a wide range of aspect ratios (of the TSV thickness and diameter). Figure 3.2*a* shows a simplified diagram of the TSV for finite-element analysis [2, 3]. The assumptions are (1) since the layer of tantalum (1 kÅ) is much thinner than the dielectric layer (1 μm), its effect is negligible, (2) the redistribution layer is very thin and therefore negligible, (3) silicon and silicon dioxide are not stressed beyond their elastic zone, (4) copper undergoes elastic deformation, followed by plastic deformation, and (5) the silicon interposer with the TSV is stress-free at 125°C.

One-quarter of the TSV is modeled in an axisymmetric simulation. The boundary conditions used and the mesh at the critical interfaces are shown in Fig. 3.2*b, c*, respectively. Note that directions 1 and 2 represent the radial and axial directions, respectively. A static temperature ramp-down analysis from 125 to −40°C (and vice versa) is carried out to simulate the maximum stress state. The modeling matrix and material properties used are shown in Tables 3.2 and 3.3, respectively.

Overall package	Body size	45 × 45 mm
Top chip	Thickness	1.8 mm
	chip size	21 × 21 mm
	IC technology	Cu/low-k, 65 nm technology
	I/O	11,000
	Pitch/solder	150 µm/SnAg
TSV interposer	Outer/inner pitch	300 µm/600 µm
	Tapered shape	100 µm/50 µm
Cu	Layer	2
	Thickness	3 µm
	Via diameter	40 µm
	UBM/solder	CuNiAu/SnAgCu
Organic substrate	Layer	4
	Core thickness	800 µm
	Pitch	1 mm
	I/O	2000

TABLE 3.1   Package Configurations and Dimensions

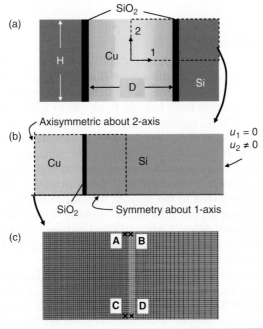

FIGURE 3.2   (a) Schematic of simplified TSV. (b) Quarter model of TSV with applied boundary conditions. (c) Mesh of critical regions.

Diameter (D)	25 µm, 50 µm, 75 µm
Aspect ratio (H/D)	1, 2, 3, 5, 7, 10, 12 for D = 25 µm 1, 2, 3, 5, 7 for D = 50 µm 1, 2, 3, 5 for D = 75 µm

TABLE 3.2   Simulation Cases

Material	Silicon	Silicon Oxide	Electroplated Copper
Young's modulus (GPa)	129.617 @ 25°C 128.425 @ 150°C	70	70
Poisson ratio	0.28	0.16	0.34
TCE (ppm/°C)	2.813 @ 25°C 3.107 @ 150°C	0.6	18
Plastic properties (MPa)			240 @ 0 ε 250 @ 0.003 ε 255 @ 0.007 ε 255 @ 0.009 ε 250 @ 0.017 ε

TABLE 3.3   Material Properties for Simulations

3.2.2 DFR Results

During the temperature cycling of the TSV, each temperature ramp-up results in copper expanding more than 5 times as much as silicon and more than 10 times as much as silicon dioxide. The deformation resulting from heating from −40 to 125°C and cooling from 125 to −40°C (exaggerated 100 times) is shown in Fig. 3.3a, b. It can be seen in Fig. 3.3a that the silicon dioxide layer is highly strained as it is dragged along by the axial expansion of the copper (the Cu wants to pump up and out of the via) and compressed by the radial expansion of the copper. This deformation, however, is largely not transferred to the bulk of the silicon because silicon is much stiffer than copper and silicon dioxide. Conversely, during temperature ramp-down, a high strain is induced on the material near the interface indicated by points A and B, which are the potential failure locations.

Failure may occur at two critical points. First, failure could occur owing to the tearing action during contraction at the interface between copper and silicon dioxide (points A and B in Fig. 3.2c). Second, cracking of copper or silicon dioxide could occur at the midplane of the TSV (points C and D in Fig. 3.2c). For the simulations

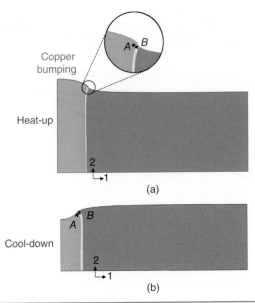

Copper
bumping

Heat-up

2
1

(a)

Cool-down

A   B

2
1

(b)

FIGURE 3.3   Deformation of the TSV at critical regions exaggerated 100 times by heat-up (a) and cool-down (b).

carried out, the radial stress at points A and B and the axial strains at points C and D were monitored to give an indication of the stress and strains causing delamination and cracking in those regions.

The radial stresses in the copper and silicon dioxide at the critical corner plotted against the diameter of the TSV for different aspect ratios are shown in Fig. 3.4. The radial stress increases linearly with the diameter of the TSV, with a steeper increase in the TSV with larger aspect ratios. This is indicated by the slopes of the graphs marked in Fig. 3.4. Although a larger diameter of copper in the TSV induces a similar thermal strain, the deformation in copper is increased, resulting in a larger strain and therefore larger stresses. The stress in silicon dioxide is greater than that in copper for all cases owing to its higher elastic modulus.

In addition, the change in stress owing to varying of the aspect-ratio effect is most evident in TSVs with smaller diameters—approximately 15 percent increase in radial stress is observed in a TSV of $D = 25$ μm compared with 2 percent increase for $D = 75$ μm. The axial strains at points C and D in copper and silicon dioxide are plotted in Fig. 3.5. As expected, owing to their close proximity, the strains in copper and silicon dioxide are identical for all aspect ratios and diameters. At an aspect ratio of 1, the axial strains are largest. As the aspect ratio increases, the strain decays. Beyond an aspect ratio of 5, axial strain no longer depends on aspect ratio. Since the axial strain values are

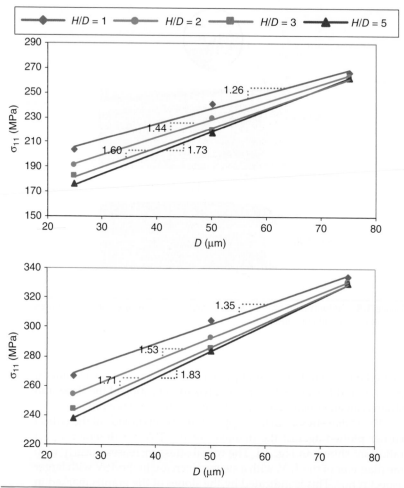

FIGURE 3.4   Radial stress (σ_{11}) at critical corner for (a) copper (point A as shown in Fig. 3.3) and (b) silicon dioxide (point B as shown in Fig. 3.3).

below the elongation values of copper (2 percent) and silicon dioxide (30 percent), cracking on the midplane is not expected for the copper-filled TSVs.

The widened range of radial stress in the case of an aspect ratio of 1 (Fig. 3.4) and the convergence of axial strain beyond an aspect ratio of 5 (Fig. 3.5) are caused by the proximity of the free edge on top of the TSV to the fixed edges (the axis and the midplane), which constraints the deformation. Figure 3.6 shows the axial strain contours of vias of various aspect ratios. Vias with an aspect ratio larger than 7 experience large strains at the top instead of the more distributed strains seen in vias with smaller aspect ratios.

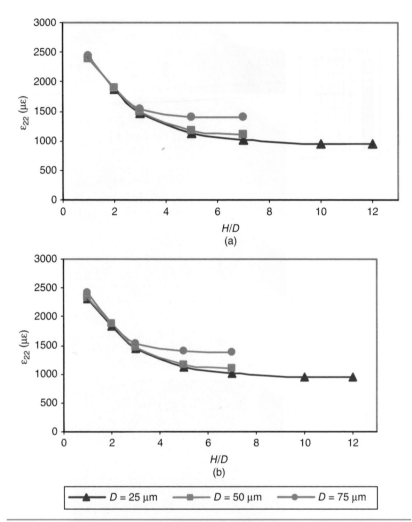

FIGURE 3.5 Axial strain (ε_{22}) at midplane for (a) copper and (b) silicon dioxide.

3.2.3   TSVs with a Redistribution Layer

In TSV fabrication, once the via is made by deep reactive-ion etching (DRIE; see Sec. 2.4), a thin passivation layer of silicon oxide is deposited on the silicon, as shown in Sec. 2.5. This is followed by barrier/seed layers (Sec. 2.6) and then via filling through electroplating (Sec. 2.7). Next, the wafer undergoes chemical-mechanical polishing (CMP; see Sec. 2.8), and then another thin passivation layer of silicon oxide is deposited.

The copper in the redistribution layer (RDL) then is sputtered at room temperature. Finally, the entire structure undergoes annealing

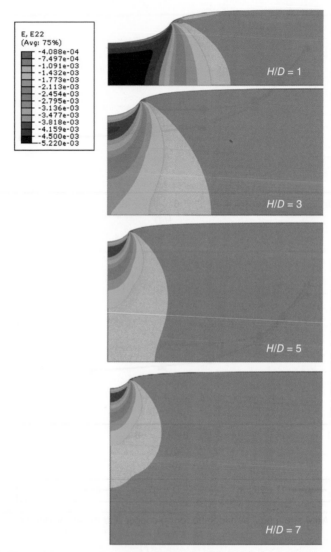

Figure 3.6 Axial strain (ε_{22}) contours of a filled TSV of $D = 25$ μm at various aspect ratios.

to remove any residual stress induced by the deposition processes. As a result, residual stress is assumed to be negligible in modeling work.

The redistribution layer in the model would simulate the real situation to a larger extent because it would include the possibility of cracking of the RDL. A schematic drawing of the cross section of the filled via, the one-quarter finite-element model, and the mesh at the critical location is shown in Fig. 3.7. For the model including the RDL

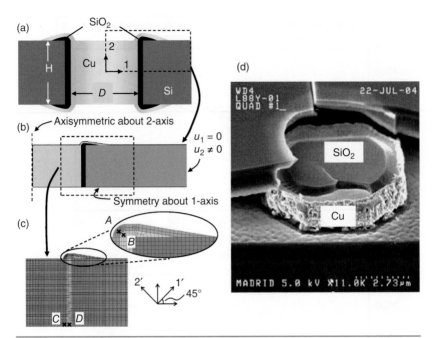

FIGURE 3.7 (a) Diagram of a simplified TSV with RDL. (b) Quarter model of a TSV with applied boundary conditions. (c) Mesh of critical region. (d) Failure mode. (The SEM image is from Tezzaron.)

layer, failure is expected to occur in the RDL layer at 45 degrees from the surface of the copper via. Hence the strains along directions 1 and 2 have been resolved (transformed) to determine the strain in the diagonal direction, ε_D along the 1'-direction (as indicated in Fig. 3.7).

The diagonal strains at points A and B are plotted in Fig. 3.8. It can be seen that there is a linear relationship between the diagonal strain and the diameter of the via. As expected, the strain in the silicon oxide is larger than that in the copper, and the silicon oxide could fail, as shown in Fig. 3.7d. This is due to the local thermal expansion mismatch between the Cu ($17.5 \times 10^{-6}/°C$) and Si ($2.5 \times 10^{-6}/°C$) in the z direction reported in May at IEEE/ECTC2008 [3], and the "mismatched" Cu pushes the silicon oxide to crack (see also Fig. 3.3). This phenomenon, called *copper pumping* (see Fig. 3.7d), was reported in the fall of 2008 by Patti of Tezzaron (according to Philip Garrou, contributing editor of *Semiconductor International*, on December 3, 2009).

The axial strain at point C is shown in Fig. 3.9 for the model with RDL. The axial strain curves at point D are not shown because they had identical axial strains as the copper. Note that these curves are similar to those in Fig. 3.5a, which shows the axial strains in copper in the model without the RDL. Hence, for the purpose of determining axial strain, the RDL may be ignored.

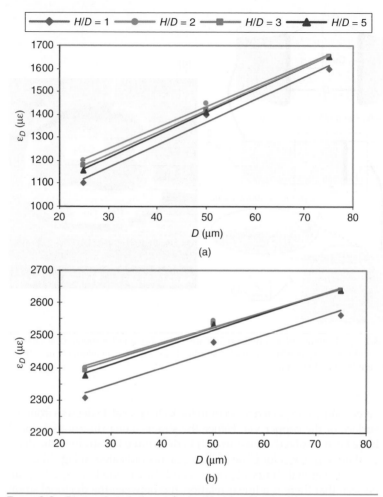

FIGURE **3.8**   Diagonal strain ($\varepsilon_D = \varepsilon_1'$) at critical location for (a) copper (point A) and (b) silicon oxide (point B) in filled vias with RDL.

3.2.4   Summary and Recommendations

Some important results are summarized as follows:

- In general, above an aspect ratio (thickness/diameter) of 5, there is little dependence of stress and strain on the aspect ratio of the TSV.

- For the perfect TSV structures modeled, failures at the interfaces of the TSV are unlikely because the strains in these elements are not large enough to cause failures.

- The imperfections that could arise during fabrication (e.g., poor bonding between interfaces and the presence of asperities,

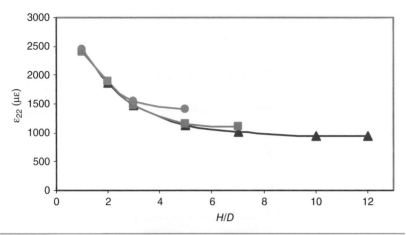

<small>FIGURE 3.9</small>   Axial strain (ε_{22}) at midplane for copper (point C).

which act as points of stress concentration) compounded with the stresses owing to the local thermal expansion mismatch could result in failure. Thus the TSV structure should be fabricated as perfectly as possible.

- The silicon dioxide dielectric layer is in a higher state of stress than the copper as a result of its lower TCE compared with copper. Hence extra care should be taken during the dielectric-layer fabrication to ensure a high-quality TSV.

3.3   Mechanical Behavior of TSVs in Memory-Chip Stacking

Homogeneous memory-chip stacking is different from active/passive interposer. Unlike the active/passive interposers, the homogeneous memory-chip stacks (like the memory cube) do not have RDLs. The TSVs are used to connect straight through the memory chips with very small vias to increase the bandwidth, as shown in Fig. 3.10. In this section, the mechanical behavior of the structure shown in Fig. 3.10 is determined via the nonlinear finite-element method as well as fracture mechanics. Specifically, the location and magnetite of maximum stress and strain are determined. Then the interfacial energy-release rates are estimated at the critical stress/strain areas [23] by using a modified virtual crack-closure (MVCC) technique [15].

3.3.1   Boundary-Value Problem

The driving forces of TSV mechanical failures (delaminations) are maximum stresses and energy release rate (ERR) at interfaces between the Cu and SiO$_2$ and between Cu and Si owing to the thermal-expansion mismatch among the Cu, Si, and SiO$_2$. If the maximum interfacial thermal

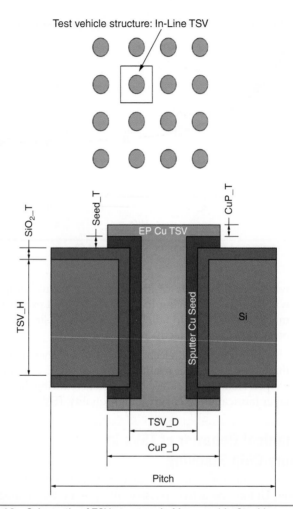

FIGURE 3.10   Schematic of TSV structure in Memory-chip Stacking

stress is greater than the allowable stress or the interfacial ERR exceeds the critical ERR, then interfacial delamination will take place.

For the purpose of realizing thermal stress distributions and the interfacial ERR in TSVs, a 3D finite-element model of a symmetric single in-line TSV is constructed to be simulated. Two kinds of horizontal cracks that embed in the interface of SiO_2 passivation and the Cu seed layer (Cu pad and TSV wall delamination cases) are also introduced to estimate the interfacial ERR by using a MVCC technique [15–17]. All the material properties are assumed to be linear elastic except for Cu TSVs which are treated as nonlinear materials to have precisely thermal stresses in this test vehicle. To capture the most important geometric parameters in TSVs, the design of experiment (DoE) analysis is employed to emphasize the significance of crack length,

TSV diameter, TSV pitch, depth of TSV, SiO_2 thickness, and Cu seed-layer thickness. The proposed results will be useful if design optimization for keeping the delamination off of TSVs in 3D IC integration is needed.

3.3.2 Nonlinear Thermal Stress Analyses for TSVs

Because thermal effects are always seen in 3D IC integration and have a prominent influence not only on temperature distribution but also on stress distribution, the problems of thermal-induced stress become critical reliability issues in the industry. For the sake of examining thermal stress distributions in TSV structures in 3D IC integration, 3D finite-element modeling (FEM) with the computational finite-element software ANSYS is employed. An electroplated copper TSV (EP-Cu TSV) is simulated to evaluate the thermal stress distribution and interfacial delamination behavior. The EP-Cu TSV FEM is designed as a TSV blanketed with an SiO_2 passivation layer and filled with EP-Cu, which is sputtered Cu as a seed layer covered on SiO_2. Figure 3.10 shows a schematic of a symmetric single in-line TSV and Fig. 3.11 shows its finite-element model, which is pure hexahedral element-meshed (a one-fourth element-meshed model is

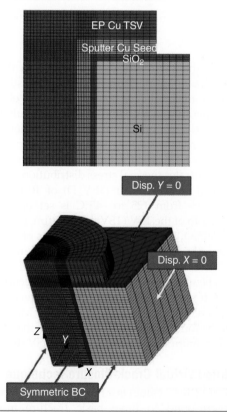

FIGURE 3.11   Finite-element model of TSV structure in 3D IC integration.

Material	E (GPa)	ν	TCE ($\times 10^{-6}$/°C)
Die	129.62 (@25°C) 128.43 (@150°C)	0.28	2.18 (@25°C) 3.11 (@150°C)
TSV [18] (electroplated Cu)	70	0.34	18.0
SiO$_2$	70	0.16	0.6
Cu seed layer	110	0.35	17

TABLE 3.4   Material Properties of TSVs

Strain	Stress (MPa) @ 25°C	Stress (MPa) @ 125°C
0.001	120	110
0.004	186	179
0.01	217	214
0.12	234	231
0.04	248	245

TABLE 3.5   Stress-Strain Relationship for Copper

shown owing to the symmetry characteristic). In the TSV model, copper is treated as a temperature-dependent plastic material [18], whereas all other materials are considered to be elastic materials, including a temperature-dependent Young's modulus and TCE for silicon. The material properties of each component are listed in Tables 3.4 and 3.5. In Table 3.4, E, v, and TCE are, respectively, Young's modulus, Poisson's ratio, and the thermal coefficient of expansion.

Figure 3.12 illustrates thermal stress distributions of TSVs in 3D IC integration with a TSV diameter (TSV_D) of 10 μm. The uniform temperature change from 125 to –45°C is set up for the loading condition. The depth of the TSV (TSV_H) is 50 μm, the thickness of the SiO$_2$ passivation (SiO$_2$_T) is 0.5 μm, the thickness of the Cu seed layer (Seed_T) is 1 μm, the Cu pad size (CuP_D) is 20 μm, and the pitch is 40 μm (see Fig. 3.10). The thickness of the Cu pad (CuP_T) is 2 μm. From the figure it can be seen that the maximum stress occurs in the circumference of the Cu seed layer and the SiO$_2$ passivation surface. The corresponding von Mises stress is found to be 488 MPa. The results for more cases can be found in Ref. 23, which shows that if TSV diameter increases, the maximum von Mises stress in the TSV structure increases.

3.3.3   Modified Virtual Crack-Closure Technique

Since the critical ERR (G value) is a function of mix mode G_I, G_{II}, and G_{III}, which correspond to three basic fracture modes—mode I

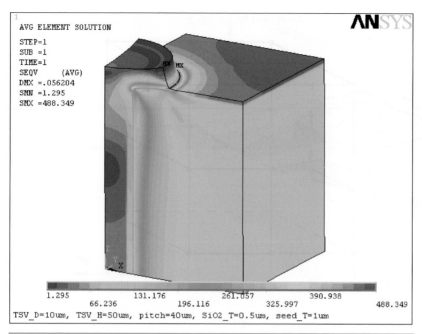

FIGURE 3.12   Thermal stress contours of TSVs in 3D IC integration
(TSV diameter = 10 μm).

(opening mode), mode II (sliding mode), and mode III (tearing mode)—which are determined separately. For the eight-node solid elements shown in Fig. 3.13, the three components of ERR, G_I, G_{II}, and G_{III}, can be obtained by the following equations [15]:

$$G_I = -\frac{1}{2\Delta A} Z_{Li}(w_{Ll} - w_{Ll^*})$$

$$G_{II} = -\frac{1}{2\Delta A} X_{Li}(u_{Ll} - u_{Ll^*})$$

(3.1)

$$G_{III} = -\frac{1}{2\Delta A} Y_{Li}(v_{Ll} - v_{Ll^*})$$

where $\Delta A = \Delta a \times b$, as shown in Fig. 3.13. ΔA is the area virtually closed, Δa is the length of the elements at the delamination front, and b is the width of the elements. In Figure 3.13, X_{Li}, Y_{Li}, and Z_{Li} denote the forces at the delamination front in column L, row i. The corresponding displacement behind the delamination at the upper-face node row l are, respectively, denoted as u_{Ll}, v_{Ll}, and w_{Ll}, and the corresponding displacement at the lower-face node row l^* are, respectively, denoted as u_{Ll^*}, v_{Ll^*}, and w_{Ll^*}.

In finite-element analysis, the values of G_I/G_T, G_{II}/G_T, and G_{III}/G_T always may change with a change in the element size around the

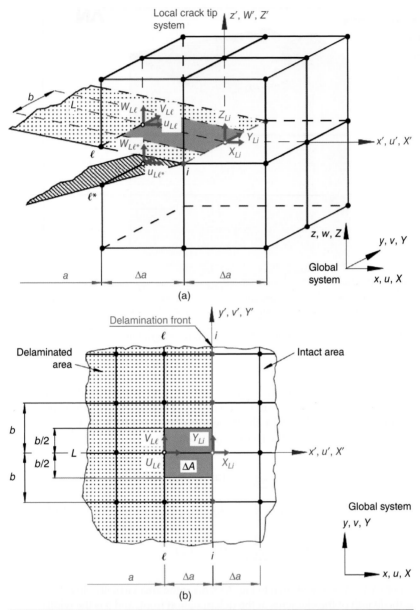

FIGURE 3.13 Modified virtual crack-closure technique: (a) 3D view; (b) see-through view of upper surface [15].

crack tip, especially when the element size is very small. (Here, G_T is the total ERR and is the summation of three components of ERR, that is, $G_T = G_I + G_{II} + G_{III}$.) However, in our finite-element simulation, the element size is relatively large; hence the values of G_I/G_T, G_{II}/G_T, and G_{III}/G_T are insensitive to the element size.

In general, the interface fracture criterion can be expressed as

$$\left(\frac{G_{\mathrm{I}}}{G_{\mathrm{IC}}}\right)^{l} + \left(\frac{G_{\mathrm{II}}}{G_{\mathrm{IIC}}}\right)^{m} + \left(\frac{G_{\mathrm{III}}}{G_{\mathrm{IIIC}}}\right)^{n} = 1 \qquad (3.2)$$

where l, m, and n are constants, and G_{IC}, G_{IIC}, and G_{IIIC} are, respectively, the critical ERRs for pure fracture modes I, II, and III. In the present study, G_{III} is neglected in the following ERR discussion of TSVs.

3.3.4   Energy Release Rate Estimation for TSVs

In this section, the thermal loading condition of –45 to 125°C (stress-free temperature is set to be 25°C) is used to investigate the fundamental delamination behavior in the proposed TSV structure. A circular arc of delamination that embedded between the Cu seed layer and the SiO_2 layer is applied to simulate the delamination behavior. The delamination is modeled as two independent coincided (two close but separated) surfaces, and a contact pair is set between each surface of the delamination to avoid penetration between the nodes. The geometry of the baseline model that was chosen for the investigation of crack-length effect is $CuP\_T = 2\ \mu m$; $Seed\_T = 1\ \mu m$; $SiO_2\_T = 0.5\ \mu m$; $CuP\_D = 20\ \mu m$; $TSV\_D = 10\ \mu m$; $TSV\_H = 20\ \mu m$; and pitch = 25 μm.

Two kinds of delamination propagation cases (a horizontal crack that embeds in the interface of the SiO_2 passivation and the Cu seed layer) are assigned (Cu pad inbound and Cu pad outbound), as shown in Fig. 3.14. The different crack length for each delamination

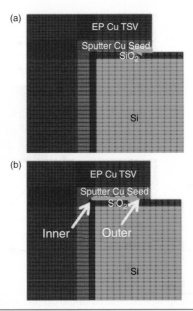

FIGURE 3.14   Two kinds of delamination propagation cases, including (a) Cu pad inbound and (b) Cu pad outbound.

propagation case is simulated at first to investigate critical crack length, which resulted in a higher ERR at the crack front, and the critical crack length is used to perform the parametric study.

Figure 3.15 illustrates the results of the Cu pad inbound case at a temperature of –45°C. The results indicate that the G values are inversely proportional to the crack length in the Cu pad inbound case. Besides, a mixed mode exists in the Cu pad inbound case when the crack front is close to the pad edge. Figure 3.16 illustrates the results of the Cu pad outbound case under –45 to 125°C thermal loading conditions. It is observed that an opening or mixed mode is seen at 125°C for the inner and outer crack fronts, whereas a sliding mode exists at –45°C.

Moreover, for the inner crack front, as the crack length became longer, the G_I value increased, whereas G_{II} reached a relative maximum value around the midpoint of the assigned crack-length range (in this case, $G_I = G_{II}$ when the crack length equals 3 μm for the inner crack front). However, the outer crack front shows an opposite trend from the inner crack front. This delamination behavior can be ascribed to the TCE mismatch between the materials in the TSV structure, where both EP-Cu and sputtered Cu have higher TCEs than silicon and SiO_2.

3.3.5   Parametric Study of Energy Release Rate for TSVs

According to the preceding simulation results, the thermal loading condition that induces an opening or mixed delamination mode and the crack length that results in higher G values are defined in the

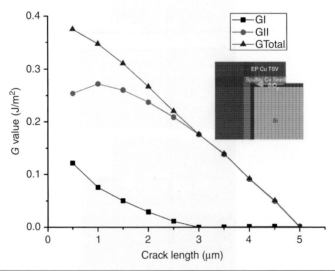

FIGURE 3.15   Crack-length-effect results of the Cu pad inbound case at –45°C under –45 to 125°C thermal loading conditions.

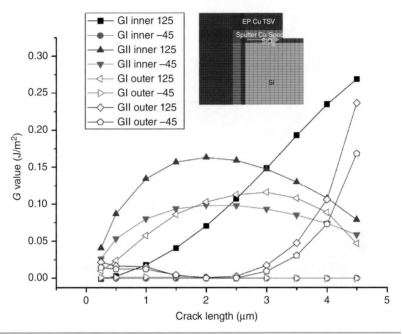

Figure 3.16 Crack-length-effect results of the Cu pad outbound case under −45 to 125°C thermal loading conditions.

parametric study, as shown in Table 3.6. In the parametric study, the SiO_2_T effect is introduced first in the Cu pad inbound case to simplify the discussion of the other parameters, and the results are shown in Fig. 3.17. It can be seen that the G values are proportional to all TSV_D and, for TSV_D = 10 μm and 20 μm, inversely proportional to TSV_H, pitch, and SiO_2_T. This shows that the volume ratio of copper and silicon plays an important role in the G values (because the TCEs of both are significantly different). Moreover, the G value becomes stable as the TSV_H gets larger than 50 μm. If the structure's liner (SiO_2_T) changes to 1.0 μm, the G value does not change because the TSV_H is larger than 60 μm. On the other hand, if the structure's liner (SiO_2_T) changes to 0.5 μm, the G value is steady because the TSV_H is larger than 70 μm. Moreover, the Young's modulus of SiO_2 is relatively soft compared with other materials, which suggests that the

Delamination Case	Cu Pad Outbound	Cu Pad Inbound
Thermal loading condition	−45 to 125°C	125 to −45°C
Crack length	1 μm	4 μm

Table 3.6 Thermal Loading Condition and Crack Length for the Parametric Study

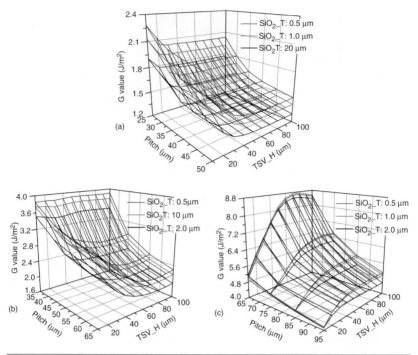

FIGURE 3.17   (a) SiO_2_T effect on $TSV\_D = 10$ μm model in the Cu pad inbound case. (b) $TSV\_D = 20$ μm. (c) $TSV\_D = 50$ μm.

SiO_2 can be viewed as a stress buffer layer between the Cu seed layer and the silicon and results in the aforementioned SiO_2_T trend. Given that the trend of SiO_2_T is clear, the thinnest $SiO_2\_T = 0.5$ μm is used in the next parametric study for the other delamination cases.

Figure 3.18 shows the parameteric study results of the Cu pad inbound and Cu pad outbound (outer crack front and inner crack front) cases. The SiO_2 thickness of 0.5 μm and the Cu seed layer thickness of 1.0 μm are used in these cases. In the Cu pad inbound and Cu pad outbound cases, the G values are proportional to TSV_D. In addition, the volume-ratio effect of copper and silicon affects the trend of TSV_H, which can be observed from the variation in pitch. While TSV_D increases, the volume ratio of copper and silicon also rises as the pitch decreases. A more rapid increase occurs for each TSV_D cases because the pitch is shorter. The pitch only makes a significant contribution to G values in the $TSV\_D = 50$ μm model, which shows that if a larger TSV_D exists, the effects of pitch need to be focused.

Moreover, the Cu pad inbound case has a higher G value than the the Cu pad outbound case, where the outer crack front is larger overall than the inner crack front. The G value in the Cu pad inbound case with an inner crack front is the smallest of all the TSV_D cases. This result indicates that if the crack front is close to the edge of the pad, a higher G value is induced under the proposed thermal loading

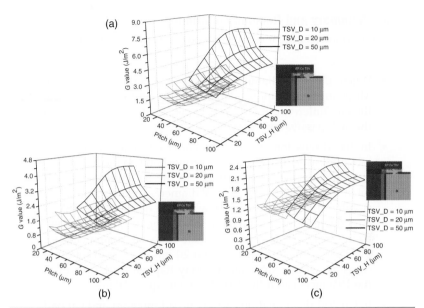

FIGURE 3.18 (a) Parametric study of Cu pad inbound case. (b) Parametric study of Cu pad outbound case (outer crack front). (c) Parametric study of the Cu pad inbound case (inner crack front).

conditions. From Fig. 3.18 it is seen that the G value is less than $1.5\,\mathrm{J/m^2}$ in the TSV_D = 10 μm model and less than $2.5\,\mathrm{J/m^2}$ in the TSV_D = 20 μm model in Cu pad delamination cases. In TSV_D = 50 μm model, the G values are less than $8.0\,\mathrm{J/m^2}$ in the Cu pad inbound case and less than $4.2\,\mathrm{J/m^2}$ in the Cu pad outbound case with the outer crack front, as well as less than $2.4\,\mathrm{J/m^2}$ in the Cu pad inbound case with the inner crack front.

From the literatures [12, 20–22], it is known that the critical strain energy release rate (Gc) value for copper is 10 J/m² and for SiO$_2$ is only about 8.5 J/m². (Hence the Gc value of 8.5 J/m² is used as the index value to determine whether the delamination occurs or not in this chapter.) Thus some results for or against delamination are summarized as follows:

1. In the discussion of the SiO$_2$ thickness effect in the Cu pad inbound case, it is safe to avoid delamination if the TSV diameter equals 10 and 20 μm. However, delamination may occur as the depth of the TSV increases if the TSV diameter equals 50 μm.

2. The G values are less than 1.5 J/m² if the TSV diameter is 10 μm and less than 2.5 J/m² if the TSV diameter is 20 μm in Cu pad delamination cases (Cu pad inbound and Cu pad outbound). Moreover, the G values are less than 8.0 J/m² if the TSV diameter is 50 μm. Thus one shouldn't worry about the crack growth in Cu pad delamination cases.

3.3.6   Summary and Recommendations

Some important results and recommendations are summarized as follows:

- When the TSV diameter increases, larger maximum von Mises stresses are induced in the TSV structure and located at the circumference interface between the Cu seed layer and the SiO_2 passivation surface.

- In the discussion of SiO_2 thickness effect in the Cu pad inbound case, the G values are proportional to the TSV diameter and inversely proportional to the SiO_2 thickness, and pitch, except for the case of TSV diameter equal to 50 μm. The G value increases no further when the depth of the TSV is larger than 50 and 80 μm if the TSV diameter equals to 10 and 20 μm, respectively. Thus it is safe to avoid delamination if the TSV diameter equals 10 and 20 μm in Cu pad inbound cases. However, delamination may happen as the depth of the TSV increases if the TSV diameter equals 50 μm.

- In Cu pad delamination cases (Cu pad inbound and Cu pad outbound), the G values are proportional to the TSV diameter. For variations in pitch, it is observed that the volume-ratio effect of copper and silicon affects the trend in depth of the TSV and the pitch only plays a significant role in G values if the TSV diameter is larger. Higher G values occur in the Cu pad inbound case compared with the Cu pad outbound case.

- The G values are less than 1.5 J/m² if the TSV diameter is 10 μm and less than 2.5 J/m² if the TSV diameter is 20 μm in Cu pad delamination cases. Moreover, it is seen that the G value is less than 8.0 J/m² in the Cu pad inbound case and less than 4.2 J/m² in the Cu pad outbound case without a crack front as well as less than 2.4 J/m² in the Cu pad inbound case with an inner crack front if the TSV diameter equals to 50 μm. Thus one shouldn't worry about crack growth in Cu pad delamination cases (an SiO_2 thickness of 0.5 μm and a Cu seed layer thickness of 1.0 μm are adopted in these cases).

3.4   Thermal Behaviors of TSVs

3.4.1   Equivalent Thermal Conductivity of TSV Chip/Interposer

Thermal management of 3D integration systems is very important, and low-cost and effective thermal management design guidelines and solutions are desperately needed for the widespread use of 3D integrations. Even with the most advanced software and high-speed hardware, it is impossible to model all the TSVs in a 3D integration system. In this section, based on the theory of heat transfer and computational fluid dynamics (CFD), the equivalent thermal conductivities of a TSV interposer/chip with various TSV diameters, pitches, and aspect ratios are developed. These equivalent thermal conductivities then are used (with the TSV chip/interposer as a block) to perform all the simulations [8].

Figure 3.19 provides a schematic of high-performance packaging systems. The high-performance chip is connected to the TSV interposer (or chip) through microbumps, and the bottom of the TSV interposer/chip is connected to the organic substrate by ordinary solder bumps. Given the large thermal expansion mismatch between the silicon interposer/chip and the organic substrate, underfill is needed so that the solder joints remain reliable. Underfill also may be necessary between the high-performance chip and the copper-filled TSV chip (or interposer) because, depending on the number (pitch) of vias and the via diameter, these two chips may create a very large thermal expansion mismatch, and the microbump solder joints may not be able to survive. This is also true for the case of memory-chip stacking.

As mentioned earlier, the equivalent thermal conductivity of the TSV interposer/chip is determined via detailed 3D CFD analyses. The TSV chip/interposer is shown schematically in the upper-right corner of Fig. 3.20. It can be seen that unlike the thermal conductivity of an ordinary silicon chip, which is isotropic because of the copper-filled TSVs, the thermal conductivity of a TSV silicon chip is anisotropic; that is, the thermal conductivity in the xy-planar directions ($k_{eq,x} = k_{eq,y}$) is not equal to that in the z-normal direction ($k_{eq,z}$).

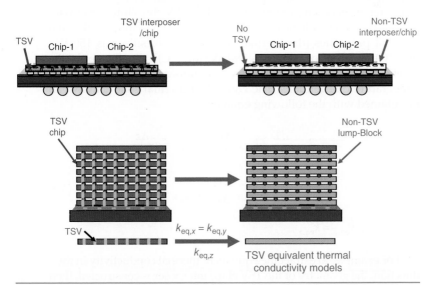

FIGURE 3.19   Equivalent thermal conductivity model for TSV chips and interposer.

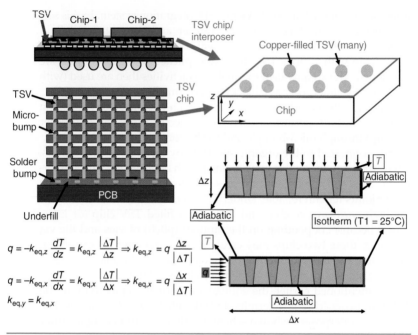

FIGURE 3.20   Equivalent thermal conductive models for computational fluid dynamics.

The approaches for extracting the equivalent thermal conductive in the z-direction ($k_{eq,z}$) and x- and y-directions ($k_{eq,x} = k_{eq,y}$) are shown, respectively, in the center right and bottom right of Fig. 3.20. First, construct the geometry of the copper-filled TSV chip/interposer with various diameters, pitches, and aspect ratios. Then input the thermal material properties [the thermal conductivity of silicon is 150 W/(m · °C) and that of copper is 390 W/(m · °C)]. Finally, apply the kinetic and kinematic boundary conditions, and calculate the temperature distributions. The equivalent thermal conductivity can be obtained with the following equations:

$$q = -k_{eq,z}\frac{dT}{dz} = k_{eq,z}\frac{|\Delta T|}{\Delta z} \Rightarrow k_{eq,z} = q\frac{\Delta z}{|\Delta T|}$$

$$q = -k_{eq,x}\frac{dT}{dx} = k_{eq,x}\frac{|\Delta T|}{\Delta x} \Rightarrow k_{eq,x} = q\frac{\Delta x}{|\Delta T|} \tag{3.3}$$

$$k_{eq,y} = k_{eq,x}$$

For example, to extract the equivalent thermal conductivity in the z direction, the geometry of the TSV chip/interposer is constructed, then a uniform heat flux q is imposed on the top surface of the TSV chip, and the bottom surface is set as an isotherm boundary (i.e., 25°C), whereas the four surrounding boundaries are set as adiabatic boundaries. By using the Flowtherm software, the average temperature on the top surface of the TSV chip can be calculated, and consequently, $k_{eq,z}$ can be obtained.

Figures 3.21 and 3.22, respectively, show the equivalent thermal conductivity for $k_{eq,z}$ and $k_{eq,x} = k_{eq,y}$ of a copper-filled TSV chip/interposer with various diameters, pitches, and aspect ratios. It can be seen that (1) the equivalent thermal conductivity in all directions of the TSV chip is larger than that of a pure silicon chip, (2) the equivalent thermal conductivity in all directions is larger for smaller pitches of the TSV chip, and (3) the equivalent thermal conductivity in all directions is larger for larger diameters of the TSV chip. For engineering convenience, the results in Figs. 3.21 and 3.22 have been curve-fitted into the following empirical equations for equivalent thermal conductivity:

$$k_{eq,z} = 150 + 188D^2/P^2 \qquad (3.4)$$

$$k_{eq,x} = k_{eq,y} = 150 + 105D^2/P^2 \qquad (3.5)$$

where P is the pitch, $D = (D_1 + D_2)/2$ is the diameter, and D_1 and D_2 are the diameters of a tapered TSV chip. The accuracy of these equations has been demonstrated by showing the correlation between the empirical equations (for the case of $P = 0.3$ mm) and the detailed

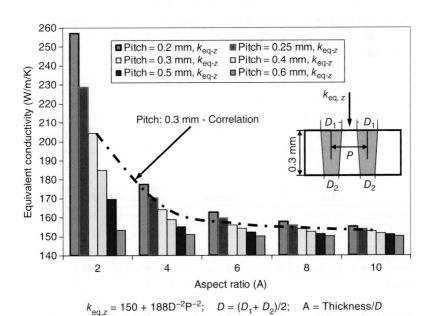

FIGURE 3.21 The equivalent thermal conductivity (in the vertical direction) for copper-filled TSV chips with various diameters, pitches, and aspect ratios.

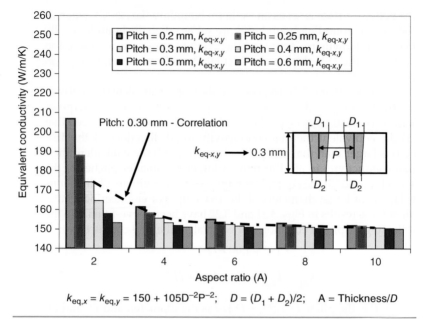

$$k_{eq,x} = k_{eq,y} = 150 + 105D^{-2}P^{-2}; \quad D = (D_1 + D_2)/2; \quad A = \text{Thickness}/D$$

FIGURE 3.22   The equivalent thermal conductivity (in the horizontal direction) for copper-filled TSV chips with various diameters, pitches, and aspect ratios.

3D CFD analyses, as shown in Figs. 3.21 and 3.22. Consequently, these empirical equations will be used for TSV chip/interposer as a lumped block without any vias for analysis of the 3D SiP. Table 3.7 shows the material properties for the simulations.

3.4.2   Effect of TSV Pitch on Equivalent Thermal Conductivity of Chip/Interposer

Figure 3.23 shows the pitch effect of the Cu-filled TSV on the equivalent thermal conductivity of the TSV interposer/chip with a thickness of 300 mm and an average diameter of 75 μm. It can be seen

	Chip	TSV	Bumps	Underfill	PCB
Material	Si (TSV)	Cu	SnAg	Polymer	FR4
K **(W/m/°C)**	Empirical equation	390	57	0.5	// 0.8 ⊥ 0.3
Dimension **(mm)**	5 × 5	Ø 0.05	Ø 0.20 Height = 0.15	5 × 5 × 0.15	76 × 114 × 1.6
Power (W)	0.2 W	N.A	N.A	N.A	N.A

TABLE 3.7   Package Configuration, Materials, and Power for Simulation

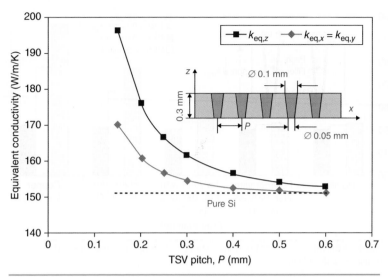

Figure 3.23 Thermal conductivity of copper-filled TSV chips/interposers with various pitches.

that (1) the equivalent thermal conductivity of the TSV interposer/ chip increases when the TSV pitch decreases (i.e., more vias for a given chip size), (2) the equivalent thermal conductivity is larger than that of the pure silicon material [the thermal conductivity of silicon is $150 \, \text{W}/(\text{m} \cdot {}^\circ\text{C})$ and that of copper is $390 \, \text{W}/(\text{m} \cdot {}^\circ\text{C})$], (3) the equivalent thermal conductivity in the z direction ($k_{eq,z}$) is larger than that in the x and y directions ($k_{eq,x}$ and $k_{eq,y}$), and (4) the equivalent thermal conductivity is more sensitive to smaller TSV pitches; that is, there are lots of Cu-filled vias.

3.4.3 Effect of TSV Filler on Equivalent Thermal Conductivity of Chip/Interposer

Figures 3.24 and 3.25 show the equivalent thermal conductivity of the TSV interposer/chip with different filler materials in the normal direction and in the planar direction, respectively. The wafer is assumed to be 0.3 mm thick, the copper plating thickness is 10 μm, and the aspect ratio is 4. It can be seen that the overall trend is that the equivalent thermal conductivity increases with the thermal conductivity of the filler materials. However, the filler material has negligible effect on the equivalent thermal conductivity for TSVs with larger pitches (e.g., pitch = 0.6 mm). Furthermore, when the thermal conductivity of the filler materials [e.g., 4 and 25 $\text{W}/(\text{m} \cdot \text{K})$] is much less than that of the pure silicon interposer, the filler material has a much greater effect on the equivalent thermal conductivity in the x and y directions than that in the z direction. This is due to the fact that the

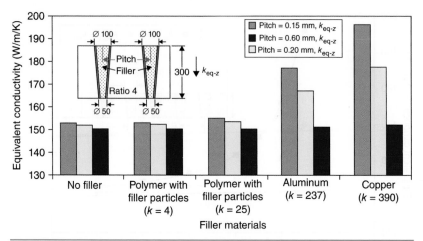

FIGURE 3.24   Equivalent Thermal conductivity of TSV chips/interposers with different fillers (Z-direction).

thermal path in the x or y direction is partially isolated by the low thermally conductive material, whereas the thermal path goes smoothly through the silicon in the z direction.

3.4.4   Effect of Plating Thickness of a Partially Cu-Filled TSV Interposer/Chip on the Equivalent Thermal Conductivity

Figure 3.26 shows a TSV with five different partially filled Cu thicknesses ($t = 5$ μm, $t = 10$ μm, $t = 15$ μm, $t = 20$ μm, and $t = 25$ μm.) The wafer is 0.3 mm thick, and there is no filler. It is assumed that there is no free convection in the vias owing to the tiny spaces.

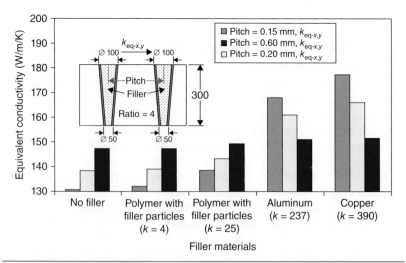

FIGURE 3.25   Equivalent thermal conductivity of TSV chips/interposers with different fillers (x and y directions).

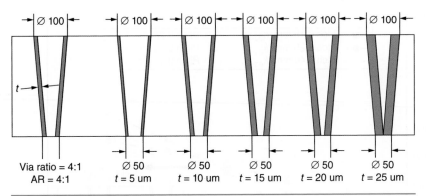

FIGURE 3.26   TSV with five different partially filled Cu thicknesses ($t = 5$, 10, 15, 20, and 25 µm).

Figures 3.27 and 3.28 show the variation of equivalent thermal conductivity of TSV interposer/chips with the partially plated copper thickness for different TSV pitches, respectively, (*a*) in the normal direction (*z* direction) and (*b*) in the planar directions (*x* and *y* directions). It can be seen that equivalent thermal conductivities of the TSV interposer/chip increase with plating thickness.

In addition, the equivalent thermal conductivities of the TSV interposer/chip are more sensitive to the copper plating thickness for smaller pitches of vias. For example, in the case of a TSV pitch of

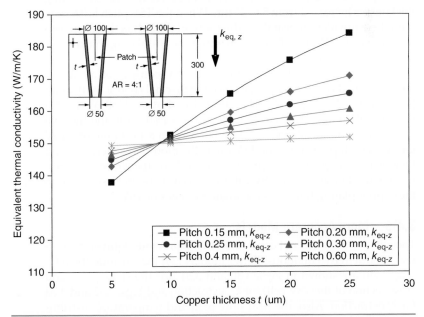

FIGURE 3.27   Equivalent thermal conductivity of TSV chips/interposers with different partially filled Cu thicknesses (*z* direction).

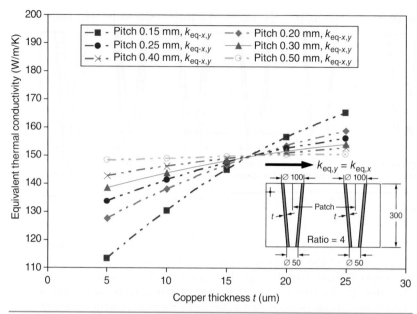

FIGURE 3.28 Equivalent thermal conductivity of TSV chips/interposers with different partially filled Cu thicknesses (x and y directions).

0.15 mm, the equivalent thermal conductivity in the z direction increases from 138 to 187 W/(m · K) (~40 percent) if the Cu plating thickness is increased from 5 to 25 µm. When the TSV pitch is 0.6 mm, then the equivalent thermal conductivity in the z direction increases from 149 to 153 W/(m · K) (<3 percent) if the Cu plating thickness is increased from 5 to 25 µm.

Why? Physically, the partially filled TSV consists of air and a metal plating layer (i.e., copper). Since the thermal conductivity of air [0.026 W/(m · K)], copper [390 W/(m · K)], and silicon [150 W/(m · K)] is quite different, various TSVs induce the variation in the thermal conductivity of the interposer/chip. For a given aspect ratio of a TSV interposer/chip, the finer the TSV pitch, the more TSVs there are, and there is more copper or air but less silicon. Thus the equivalent thermal conductivities of the TSV interposer/chip are more sensitive to the copper plating thickness for smaller pitches of TSV.

3.4.5 More Accurate Models

The effect of SiO$_2$ was not considered in the previous equivalent thermal conductivity equations. Better models [24] that include the SiO$_2$ (Fig. 3.29) are shown in Figs. 3.30 and 3.31 (for the stereogram and cross section of the TSV cell) for k_{xy} extraction and Figs. 3.32 and 3.33 for k_z extraction. Also, the heat-flow path and temperature distribution in the lateral and vertical directions of a TSV cell are shown, respectively, in Figs. 3.31 and 3.33. Based on parametric study and curve fitting [24], we have the following:

For equivalent k_{xy} the empirical equation is

$$k_{xy} = \left(90t_{SiO_2}^{-0.33} - 148\right)\left(\frac{D_{TSV}}{P}\right)H^{0.1} + 160t_{SiO_2}^{0.07} \qquad (3.6)$$

For equivalent k_z, the empirical equation is

$$0.002 \leq \frac{t_{SiO_2}}{H} \leq 0.01 \Rightarrow k_z = 128 \cdot \exp\left(\frac{D_{TSV}}{P}\right) \qquad (3.7)$$

where $0.2\ \mu m \leq t_{SiO_2} \leq 0.5\ \mu m$
  $10\ \mu m \leq D_{TSV} \leq 50\ \mu m$
  $H \geq 20\ \mu m$
  $0.1 \leq D_{TSV}/P \leq 0.77$

Also, P = TSV pitch; D_{TSV} = TSV diameter; H = chip thickness; and t_{SiO_2} = SiO$_2$ thickness, which can be found in Figs. 3.29 through 3.33.

3.4.6 Summary and Recommendations

- For thermal performance of 3D IC integration, the SiO$_2$ layer serves as a thermal barrier in a TSV, which reduces the value of k_{xy} and k_z. On the other hand, the filled Cu increases the value of k_{xy} and k_z.

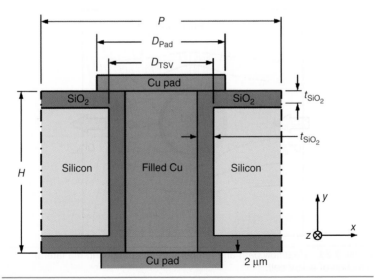

FIGURE 3.29   Schematic diagrams of a TSV cell.

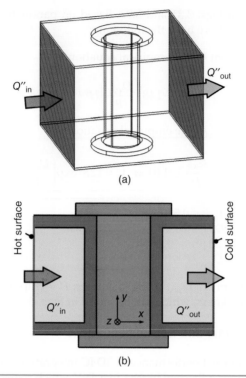

(a)

(b)

FIGURE 3.30   TSV cell for k_{xy} extraction: (a) the stereogram of the cell; (b) the cross section of the cell.

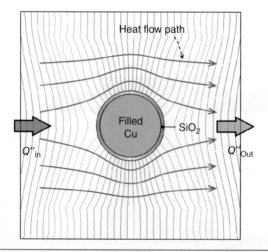

FIGURE 3.31   Heat-flow path and temperature distribution in the lateral direction of a TSV cell.

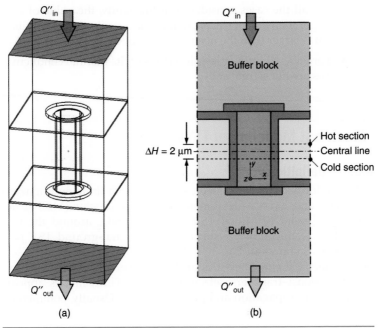

FIGURE 3.32   TSV cell for k_z extraction: (a) the stereogram of the cell; (b) the cross section of the cell.

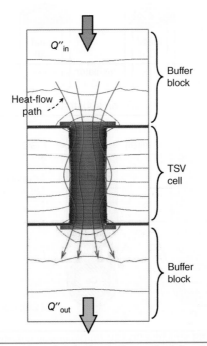

FIGURE 3.33   Temperature distribution in the vertical direction of a TSV cell.

- For all the cases considered in this study, the value of k_z is larger than that of silicon; that is, the contribution of the filled Cu is larger than the negative effect of the SiO_2 layer.
- For $SiO_2 = 0.1\ \mu m$ or less, its effects on its thermal conductivities are not significant.

3.5 Electrical Modeling of TSVs

3.5.1 Definition

TSVs are formed inside a silicon substrate with finite and nonzero resistivity and have a metal-insulator-semiconductor (MIS) structure [28]: [metal (Cu-filled TSV)–insulator (SiO_2 around the Cu-filled TSV)–semiconductor substrate (Si)] signal-ground TSV pair (SG-TSV), as shown in Fig. 3.34. There are silicon substrate effects on the electrical characteristics of the TSV, such as slow-wave and dielectric quasi–transverse electromagnetic (TEM) modes appearing at signal propagation and power delivery. Usually, commercial

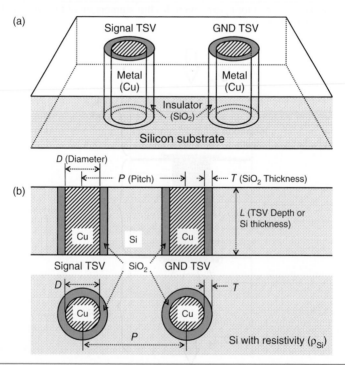

FIGURE 3.34   (a) MIS structure SG-TSV. (b) Dimension definition of (a) for the proposed MIS structure SG-TSV model.

silicon substrate has finite resistivity in the range of 5 to 40 $\Omega \cdot$ cm, and the electrical characteristics of MIS structure TSVs are more strongly affected by the two modes.

3.5.2 The Model and Equations

Figure 3.35 shows the proposed MIS structure SG-TSV model and parameters [28]. The model parameters are defined by dimensions such as D [TSV diameter (μm)], L [TSV length (μm)], P [pitch (μm); distance between two TSV centers], T [SiO$_2$ insulator thickness (μm)], and ρ_{Si} [Si substrate resistivity ($\Omega \cdot$ cm)] as follows (see Figs. 3.34 and 3.35).

$$R_T = \frac{L\sqrt{1 + 0.25 \times \text{frequency}}}{5.8\text{E}7\,(D/2)^2\,\pi} \tag{3.8}$$

$$L_s = \left(\frac{0.8P}{50} + 1\right)\frac{\mu_0 L}{2\pi}\left[\ln\left(\frac{2L}{D/2}\right) - 1\right] \tag{3.9}$$

$$M = \left(\frac{0.8P}{50} + 1\right)\frac{\mu_0 L}{2\pi}\left\{\ln\left[\frac{L}{P} + \sqrt{1 + \left(\frac{L}{P}\right)^2}\right] - \sqrt{1 + \left(\frac{P}{L}\right)^2} + \frac{P}{L}\right\} \tag{3.10}$$

$$C_{SiO_2} = \frac{3.9\varepsilon_0 L}{T}(2\pi D/2) \tag{3.11}$$

$$C_{Si} = \frac{11.9\varepsilon_0 \pi L}{\ln\left[\frac{P}{D} + \sqrt{\left(\frac{P}{D}\right)^2 - 1}\right]} \tag{3.12}$$

$$R_{Si} = \frac{11.9\varepsilon_0 \rho_{Si}}{100C_{Si}} = \frac{1}{G_{Si}} \tag{3.13}$$

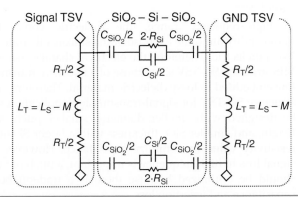

FIGURE 3.35   Proposed MIS structure SG-TSV model and parameters defined in Eqs. (3.8) through (3.13).

where R is resistance, L is inductance, C is capacitance, and M is mutual partial inductance between two TSVs. Also, μ_0 and ε_0 are permeability and permittivity of free space, respectively. In Eq. (3.8), frequency is in gigahertz. Equations (3.8), (3.9), and (3.11) and Eqs. (3.10), (3.12), and (3.13) have units of (Ω/TSV), (H/TSV), (F/TSV), (H/SG-TSV), (F/SG-TSV), and (Ω/SG-TSV), respectively.

In Eq. (3.8), $\sqrt{1+0.25\times}$ frequency reflects the skin depth effect of the TSV pillar. In Eqs. (3.9) and (3.10), $(1 + 0.8P/50)$ corrects itself and mutual inductances of the TSV pillar depending on the TSV pitch. Equation (3.11) can be obtained from metal-insulator-metal (MIM) capacitance between the TSV pillar and the Si substrate, considering concentrated electrical field distributions in SiO_2 when waves are in slow-wave mode. Equations (3.12) and (3.13) come from capacitance of a two-wire transmission line in dielectric quasi-TSM mode and the relation of $G_{Si} = 2\pi f_e C_{Si}$, respectively [50].

3.5.3   Summary and Recommendations

Some important results and recommendations are summarized as follows [28]:

- In the low-frequency range, when a signal propagates through SG-TSV, the effective dielectric constant and permeability determining the signal propagation velocity are affected by the thickness of the SiO_2 around the two TSVs and the pitch between the two TSVs, respectively. When the SiO_2 thickness is smaller and the pitch is larger, the effective dielectric constant becomes larger, and slow-wave mode effect appears strongly, showing enlarged capacitance and severely distorted flatness of signal loss.

- In high-frequency range, the dielectric constant and permeability are same as those of the silicon substrate acting like a dielectric. Therefore, the electrical characteristics of an MIS structure SG-TSV show those of a two-wire transmission line embedded in lossy dielectric material. Therefore, an MIS structure SG-TSV for signal-transmission application should be designed with smaller diameter, smaller pitch, shorter length, and smaller SiO_2 thickness with a larger Si substrate resistivity value to minimize capacitance. However, if the signal loss in high frequency is important, a pitch of SG-TSV should be optimized because there is a tradeoff between capacitance and insertion loss.

3.6   Electrical Test of Blind TSVs

3.6.1   Motivation

As mentioned in Chap. 2, during manufacture, all TSVs are blind vias, and the TSV components produced require additional processing before completion. However, if the manufactured TSVs are of poor quality, the processing tasks that follow will be wasted. As a result, TSV manufacturing requires a method for evaluating the quality of manufactured TSVs to determine whether processing can continue, thereby reducing costs. In this section, the main objective is to develop a method to measure the electrical quality of manufactured TSVs. Another objective is to integrate the TSV testing apparatus and high-frequency coupling testing into a quality testing circuit that can measure TSV coupling parameters and achieve the objective of monitoring the conditions of TSV manufacture [46].

3.6.2   Testing Principle and Apparatus

Figure 3.36 shows vertical and cross-sectional views of an example implementation of the two-TSV 3D IC integration TSV testing apparatus, in which there are two completed TSVs with a certain design distance in between. The TSVs are connected to different signal input and output pathways. In this method, although the wafer substrate is not connected to the lowest operational potential, mutual coupling occurs between the parasitic parameters R_{sub} and C_{sub}.

Once a low-frequency signal is inputted in either TSV, it is not transmitted to the other TSV owing to the insulation of the parasitic capacitance between the TSVs. The two TSVs are similar to the two

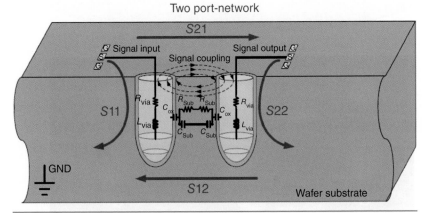

FIGURE 3.36   Two-TSV structure of a 3D IC integration blind TSV testing apparatus.

ends of a broken circuit. However, when a high-frequency signal is inputted in one of the TSVs, the signal can reach the other TSV through capacitance coupling. Therefore, the signal energy of the coupling can be measured in the other TSV. Since the signal energy is related to the coupling parameter between the two TSVs, and since the coupling parameter is comprised of parasitic parameters, the TSV parasitic parameters thus can be derived using the signal-measuring apparatus and coordinating method. By comparing the parasitic parameters with those of normal TSVs, the status of the manufactured TSV can be determined. High-frequency S-parameter measurements can be employed for precision. The $S11$, $S21$, $S22$, and $S12$ parameters can be determined using ground-signal-ground (GSG) probing to quantify the four S-parameters of the TSV precisely.

Figure 3.37 shows the blind TSV's floor plan of GSG pattern. With this pattern, we can analyze and simulate the frequency of the $S11$ and $S12$ parameters.

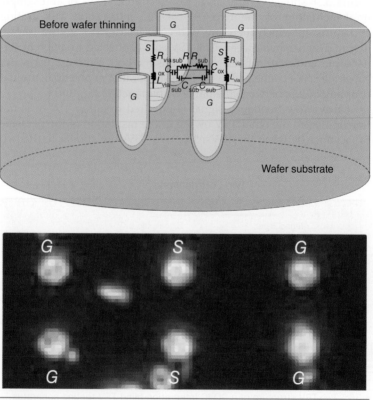

FIGURE 3.37 The TSV's floor plan of GSG pattern. Photograph of the GSG pattern.

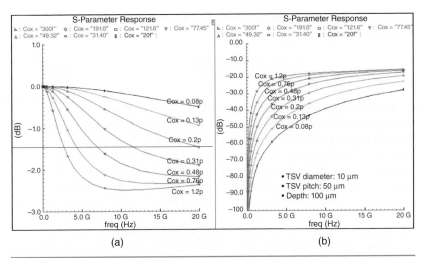

FIGURE 3.38   Equivalent capacitance distributions of varying S-parameters: (a) S11 (reflection coefficients); (b) S12 (transmission coefficients).

The results of simulating changes in the status of TSV sidewall manufacturing are shown in Fig. 3.38. If the manufactured sidewall is unstable, C_{ox} may increase or decrease depending on the thickness of the sidewall. Consequently, this study simulated an increase in changes in C_{ox}. As seen in the figure, when C_{ox} increases (sidewalls are thinner), the signal enters from the input end. With the coupling effect, the strength of the signal at the output end is greater. When C_{ox} decreases (sidewalls are thicker), the signal passed through the coupling effect is weaker. Moreover, $C_{ox} = 0f$ (sidewall manufacturing fails, causing the TSV and the substrate to short circuit) may lead to an infinite coupling frequency. Therefore, this proves that using this 3D IC TSV testing apparatus in conjunction with high-frequency coupling effects can test the status of the TSV directly after manufacture.

Furthermore, to enable clients to quickly judge the quality of the manufacturing, this study converted an electrical property, the C_{ox} parameter, back to relations in physical structure, in other words, the thickness of the manufactured sidewalls. First, assume that $H \gg D$, and only consider the cylinder capacitor formula, which is applied to the TSVs, as shown in Eq. (3.14) [49]:

$$Cox = \frac{2\pi\varepsilon H}{\ln\left(\dfrac{D}{D-2d}\right)} \tag{3.14}$$

where $C_{ox}(f)$ is the greatest capacitance of the sidewalls, which is a function of frequency; H is the height of the TSV; D is the diameter of the TSV; ε is the dielectric constant; and d is the thickness of the sidewalls.

The parameters are illustrated in Fig. 3.36. Equation (3.14) can be rewritten to Eq. (3.15):

$$d > \frac{D}{2}\left(1 - e^{-\frac{2\pi\varepsilon H}{\mathrm{Cox}(f)}}\right) \tag{3.15}$$

For example, in terms of electrical properties to determine whether the sidewall thickness meets specifications, limitations are defined in two parts. The first is the minimal limit of sidewall manufacturing that depends on direct-current (dc) conditions. This uses the perspective of leakage currents, in which leakage must be less than a certain value. Generally, the silica insulator layer equivalent impedance on the sidewalls must be greater than 1 GΩ [47]. To ensure that these conditions are within a safe range, the safety factor recommended here is 2. In other words, equivalent impedance must be greater than 2 GΩ, thereby establishing that minimal sidewall thickness is greater than 0.2 μm.

With the limitations of dc conditions, the sidewall thickness required for different operation frequencies must be obtained. Therefore, the concept of energy transfer is incorporated into the alternating-current (ac) portion. According to different system applications, the signal transfer energy through the TSVs can be defined. In this study, we define the signal as less than half the energy transferred through the TSVs may be lost, which is 3 dB [48]. The return loss (represents the reflection of energy) must be smaller than 3 dB. With a safety factor of 2, then the return loss must be smaller than 1.5 dB. The insertion loss (represents the transmission of energy) must be greater than 20 dB, and a larger value represents less loss [48]. These specifications determine the thickness of the SiO$_2$. According to these thickness specifications and a pair of GSG pattern measurements, the user can easily diagnose whether the TSV has an obvious electrical defect or not. Consequently, further TSV processes can be decided as go (good or pass) or no-go (not good or do not pass).

Figure 3.39 shows a flowchart of electrical testing of the blind TSV wafer. It consists of three major parts. The first part is to design the electrical GSG test pattern, TSV diameter, and SiO$_2$ thickness based on the application frequency and Eqs. (3.14) and (3.15). The test location is recommended at the scrape areas of the wafer. The second part is to process the wafer until the end of CMP and before metallization and then measure the S-parameter of the GSG pattern. The last part is to compare the measurement results with the specifications and decide if it is good or no-good for further processes.

3.6.3   Experimental Procedures, Measurements, and Results

As mentioned earlier, first we need to choose a TSV diameter D and a TSV thickness H for the GSG test pattern. By using the

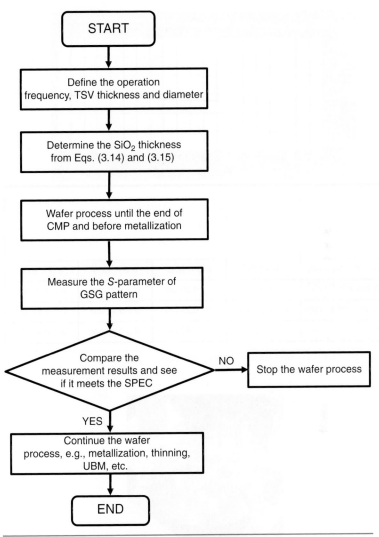

FIGURE 3.39   Flowchart for blind TSV electrical testing.

Resistance-Capacitance (RC) model for the simulations, the C_{ox} values with different frequencies can be obtained, as shown in Fig. 3.40 (*above*). For the specific TSV operating (application) frequency, we can determine the C_{ox} value, and then, by using Eqs. (3.14) and (3.15) with the specific TSV operating frequency, we can determine the TSV SiO_2 thickness, as shown in Fig. 3.40 (*below*).

Figure 3.37 (*below*) shows a top-view photograph of the GSG test pattern. These TSVs are fabricated on a 300-mm back-end-of-line (BEOL) process (right before front-side metallization.) Figure 3.41

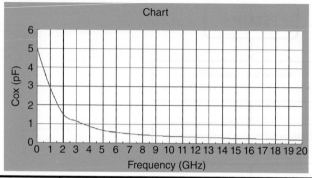

SiO$_2$ Thickness	$d > \dfrac{D}{2} \cdot (1 - e^{-\frac{2\pi \varepsilon H}{Cox\,(f)}})$ & $d > 0.2\ um$			
Operation Frequency	**0 ~ 5 G**	**0 G ~ 10 G**	**0 G ~ 15 G**	**0 G ~ 20 G**
$D = 10\ um, H = 50\ um$	$d > 0.20\ um$	$d > 0.20\ um$	$d > 0.21\ um$	$d > 0.26\ um$
$D = 10\ um, H = 100\ um$	$d > 0.20\ um$	$d > 0.29\ um$	$d > 0.42\ um$	$d > 0.51\ um$
$D = 10\ um, H = 200\ um$	$d > 0.30\ um$	$d > 0.57\ um$	$d > 0.80\ um$	$d > 0.98\ um$

FIGURE 3.40   TSV SiO$_2$ liner thickness specification.

FIGURE 3.41   Measurement setup for 12-in (300-mm) wafers.

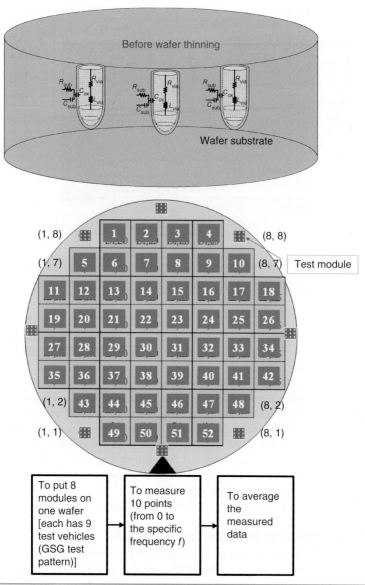

FIGURE 3.42   Blind TSV measurement method.

shows the measurement setup for the 12-in (300-mm) wafer. Figure 3.42 shows the blind TSVs electrical measurement method. The eight test modules are located at the scrape areas of the wafer, as shown in Fig. 3.42. Each test module has nine test vehicles (GSG test pattern). At each test vehicle, 10 measurements are made with frequencies ranging from 0 Hz to the specific frequency (20 GHz in this case). Then an average of all these measured data is made. Finally, the measured return

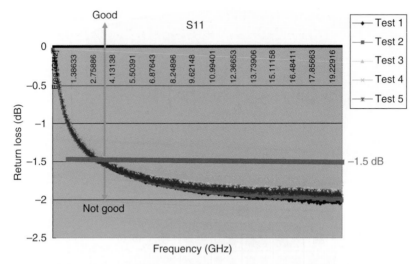

PASS Condition: S11 > −1.5 dB @ 3.5 GHz and TSV operation frequency < 3.5 GHz

FIGURE 3.43   Pass/fail conditions of blind TSVs based on S11.

loss and calculated specification from the equations are compared, and the user easily can determine whether the TSV has an obvious electrical defect or not.

Figures 3.43 and 3.44 show the measurement results of the two TSV structures with S11 and S12 parameters, respectively. It can be seen that the values of (1) return losses are smaller than −1.5 dB at operating frequencies smaller than 3.5 GHz and (2) insertion losses are larger than −20 dB for up to 20 GHz. This means that this TSV process is dominated at reflection parameter S11, which can work at the lower 3.5 GHz. If we want it to work at a higher frequency (>3.5 GHz), this TSV process needs to be improved. The 3D IC process can be monitored easily during the blind TSV process by the present method.

3.6.4   Blind TSV Electrical Test Guidelines
Figure 3.45 shows the electrical test guideline for blind TSVs. The first step (left-hand side) defines the TSV modules and the specification of TSV SiO_2 liner thickness. The second step measures the S-parameters of the TSV modules on the wafer and averages these measurement data. The third step compares the measured return loss and calculated specification from the equations. Based on these three steps, the user can easily determine whether the TSV has an obvious electrical defect or not.

3.6.5   Summary and Recommendations
A novel test method and set of specifications for determining the electrical integrity of blind TSVs have been provided and discussed. Since the

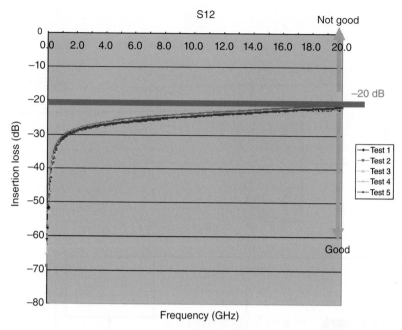

PASS Condition: S12 < −20 dB @ 20 GHz and TSV operation frequency < 20 GHz

FIGURE 3.44   Pass/fail conditions of blind TSVs based on S12.

electrical test is performed and judgment (such as good or no good) can be made right after the CMP and before metallization/UBM of the blind TSVs, it is very useful for avoiding an unnecessary waste of manufacturing cost and time by taking out the disqualified wafers in advance. Some important results and recommendations are summarized as follows [46]:

- It has been found that the TSV liner thickness can be determined by Eqs. (3.14) and (3.15) and depends on the application frequency and aspect ratio of the TSV.

- A pair of TSVs with GSG structure is recommended for the key test design for TSV liner thickness.

- By using the S-parameter analysis technique with the vector network analyzer (including a pair of GSG structure patterns), a measurement method for the return and transmission losses of each key test pattern has been proposed.

- Based on the return loss (which must be less than 1.5 dB) and the insertion loss (which must be larger than 20 dB), the thickness of the TSV liner can be determined.

- Based on these thickness specifications, a pair of GSG pattern measurements, and comparison of the measured return loss

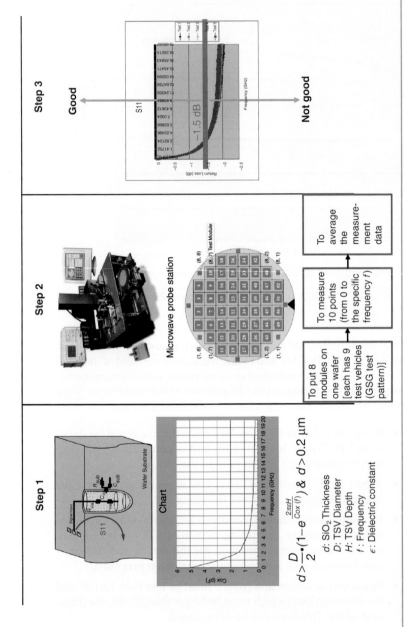

FIGURE 3.45 Blind TSV test guidelines.

with the calculated value from Eqs. (3.14) and (3.15), it is easily diagnosed whether the TSV liner has an obvious electrical defect or not.

- A set of useful electrical test guidelines for blind TSVs for 3D IC integration has been provided.

3.7   References

[1] Lau, J. H., *Reliability of RoHS-Compliant 2D and 3D IC Interconnects*, McGraw-Hill, New York, 2011.

[2] Chai, T., X. Zhang, J. H. Lau, C. Selvanayagam, K. Biswas, S. Liu, D. Pinjala, G. Tang, Y. Ong, S. Vempati, E. Wai, H. Li, B. Liao, N. Ranganathan, V. Kripesh, J. Sun, J. Doricko, and C. Vath, "Development of Large Die Fine-Pitch Cu/Low-k FCBGA Package with Through-Silicon Via (TSV) Interposer," *IEEE Transactions on Components, Packaging and Manufacturing Technology*, Vol. 1, No. 5, 2011, pp. 660–672.

[3] Selvanayagam, C., J. H. Lau, X. Zhang, S. Seah, K. Vaidyanathan, and T. Chai, "Nonlinear Thermal Stress/Strain Analysis of Copper Filled TSV (Through-Silicon Via) and Their Flip-Chip Microbumps," *IEEE Transactions in Advanced Packaging*, Vol. 32, No. 4, November 2009, pp. 720–728.

[4] Lau, J. H., *Ball Grid Array Technology*, McGraw-Hill, New York, 1996.

[5] Shen, Y. L., "Thermomechanical Stresses in Copper Interconnects: A Modeling Analysis," *Microelectronic Engineering*, Vol. 83, 2006, pp. 446–459.

[6] Vandevelde, B., C. Okoro, M. Gonzalez, B. Swinnen, and E. Beyne, "Thermomechanics of 3D-Wafer Level and 3D Stacked IC Packaging Technologies," *International Conference on Thermal, Mechanical and Multi-Physics Simulation and Experiments in Microelectronics and Micro-Systems*, 2008, pp. 1–7.

[7] Kuo, T. Y., S. Chang, Y. Shih, C. Chiang, C. Hsu, C. Lee, C. Lin, Y. Chen, and W. Lo, "Reliability Tests for a Three-Dimensional Chip Stacking Structure with Through Silicon Via Connections and Low Cost," *IEEE/ECTC Proceedings*, 2008, pp. 853–858.

[8] Lau, J. H., and G. Y. Tang, "Thermal Management of 3D-IC Integration with TSV (Through-Silicon Via),", *IEEE/ECTC Proceedings*, 2009, pp. 635–640.

[9] Hsieh, M. C., C. K. Yu, and S. T. Wu, "Thermomechanical Simulative Study for 3D Vertical Stacked IC Packages with Spacer Structures," *Semiconductor Thermal Measurement, Modeling and Management Symposium*, 2010.

[10] Hsieh, M. C., C. K. Yu, and W. Lee, "Effects of Geometry and Material Properties for Stacked IC Package with Spacer Structure," *International Conference on Thermal, Mechanical and Multi-Physics Simulation and Experiments in Microelectronics and Micro-Systems*, 2009, pp. 239–244.

[11] Ranganathan, N., K. Prasad, N. Balasubramanian, and K. L. Pey, "A Study of Thermomechanical Stress and Its Impact on Through-Silicon Vias," *Journal of Micromechanics and Microengineering*, Vol. 18, 2008.

[12] Liu, X., Q. Chen, P. Dixit, R. Chatterjee, R. R. Tummala, and S. K. Sitaraman, "Failure Mechanisms and Optimum Design for Electroplated Copper Through-Silicon Vias (TSVs)," *IEEE/ECTC Proceedings*, 2009, pp. 624–629.

[13] Hsieh, M. C., and C. K. Yu, "Thermomechanical Simulations for 4-Layer Stacked IC Packages," *International Conference on Thermal, Mechanical and Multi-Physics Simulation and Experiments in Microelectronics and Micro-Systems*, 2008, pp. 254–260.

[14] Ladani, L. J., "Stress Analysis of Three-Dimensional IC Package as Function of Structural Design Parameters," *Microelectronic Engineering*, Vol. 87, 2010, pp. 208–215.

[15] Krueger, R., "The Virtual Crack Closure Technique: History, Approach and Application," ICASE Report No. 2002–10, NASA, April 2002.

[16] Wu, C. J., M. C. Hsieh, C. C. Chiu, M. C. Yew, and K. N. Chiang, "Interfacial Delamination Investigation Between Copper Bumps in 3D Chip Stacking Package by Using the Modified Virtual Crack Closure Technique," *Microelectronic Engineering*, 2011 (in press).

[17] Hsieh, M. C., S. T. Wu, R. M. Tain, C. H. Chen, and C. K. Lin, "Thermomechanical Analysis and Interfacial Energy Release Rate Estimation for Metal-Insulator-Metal Capacitor Device," *Microelectronic Engineering*, 2011 (in press).

[18] Read, D. T., Y. W. Cheng, and R. Geiss, "Morphology, Microstructure, and Mechanical Properties of a Copper Electrodeposit," *Microelectronic Engineering*, Vol. 75, 2004, pp. 63–70.

[19] Iannuzzelli, R., "Predicting Plated-Through-Hole Reliability in High-Temperature Manufacturing Process," *IEEE/ECTC Proceedings*, 1991, pp. 410–421.

[20] Wang, H. W., and H. W. Ji, "Size Effect and Microscopic Experimental Analysis of Copper Foils in Fracture," *Key Engineering Materials*, Vol. 324–325, 2006, pp. 253–266.

[21] *AZoM: The A to Z of Materials and AZojomo.*

[22] Volinsky, A. A., N. R. Moody, and W. W. Gerberich, "Interfacial Toughness Measurements for Thin Films on Substrates," *Acta Materialia*, Vol. 50, 2002, pp. 441–466.

[23] Hsieh, M. C., S. Wu, C. Wu, J. H. Lau, R. Tain, and R. Lo, "Investigation of Energy Release Rate for Through-Silicon Vias (TSVs) in 3D IC Integration," *EuroSimE Proceedings*, April 2011, pp. 1/7–7/7.

[24] Chien, J., Y. Chao, J. H. Lau, M. Dai, R. Tain, M. Dai, P. Tzeng, C. Lin, Y. Hsin, S. Chen, J. Chen, C. Chen, C. Ho, R. Lo, T. Ku, and M. Kao, "A Thermal Performance Measurement Method for Blind Through-Silicon Vias (TSVs) in a 300-mm Wafer," *IEEE ECTC Proceedings*, Orlando, FL, June 2011, pp. 1204–1210.

[25] Akasaka, Y., "Three-Dimensional IC Trends," *Proceedings of the IEEE*, Vol. 74, No. 12, December 1986, pp. 1703–1714.

[26] Akasaka, Y., and T. Nishimura, "Concept and Basic Technologies for 3D IC Structure," *IEEE Proceedings of International Electron Devices Meetings*, Vol. 32, 1986, pp. 488–491.

[27] Gat, A., L. Gerzberg, J. Gibbons, T. Mages, J. Peng, and J. Hong, "CW Laser of Polycrystalline Silicon: Crystalline Structure and Electrical Properties," *Applied Physics Letters*, Vol. 33, No. 8, October 1978, pp. 775–778.

[28] Pak, J., J. Cho, J. Kim, J. Lee, H. Lee, K. Park, and J. Kim, "Slow Wave and Dielectric Quasi-TEM Modes of Metal-Insulator-Semiconductor (MIS) Structure Through-Silicon Via (TSV) in Signal Propagation and Power Delivery in 3D Chip Package," *IEEE/ECTC Proceedings*, May 2011, pp. 667–672.

[29] Lau, J. H., C. K. Lee, C. S. Premachandran, and A. Yu, *Advanced MEMS Packaging*, McGraw-Hill, New York, 2010.

[30] Lau, J. H., "Evolution, Outlook, and Challenges of 3D IC/Si Integration," *IEEE/ICEP Proceedings* (keynote), Nara, Japan, April 13, 2011, pp. 1–17.

[31] Lau, J. H., "Evolution, Challenge, and Outlook of TSV, 3D IC Integration and 3D Si Integration," *International Wafer Level Packaging Conference* (plenary), San Jose, CA, October 3–7, 2011, pp. 1–18.

[32] Roullard, J., S. Capraro, A. Farcy, T. Lacrevaz, C. Bermond, P. Leduc, J. Charbonnier, C. Ferrandon, C. Fuchs, and B. Flechet, "Electrical Characterization And Impact On Signal Integrity Of New Basic Interconnection Elements Inside 3D Integrated Circuits," *IEEE/ECTC Proceedings*, 2011, pp. 1176–1182.

[33] Cadix, L., M. Rousseau, C. Fuchs, P. Leduc, A. Thuaire, R. El Farhane, H. Chaabouni, R. Anciant, J.-L. Huguenin, P. Coudrain, A. Farcy, C. Bermond, N. Sillon, B. Flechet, and P. Ancey, "Integration and Frequency Dependent Electrical Modeling of Through-Silicon Vias (TSV) for High-Density 3DICs," *IEEE/ECTC Proceedings*, 2010, pp. 1–3.

[34] Sheng, F., S. Chakravarty, and D. Jiao, "An Efficient 3D-to-2D Reduction Technique for Frequency-Domain Layered Finite Element Analysis of

Large-Scale High-Frequency Integrated Circuits," *IEEE Proceedings of Electrical Performance of Electronic Packaging*, 2007, pp. 295–298.

[35] Thorolfsson, T., and P. D. Franzon, "System Design for 3D Multi-FPGA Packaging," *IEEE Proceedings of Electrical Performance of Electronic Packaging*, 2007, pp. 171–174.

[36] Sun, J., J. Lu, D. Giuliano, P. Chow, and J. Gutmann, "3D Power Delivery for Microprocessors and High-Performance ASICs," *IEEE Proceedings of Applied Power Electronics Conference*, 2007, pp. 127–133.

[37] Khan, M. A., and A. Q. Ansari, "Quadrant-Based XYZ Dimension Order Routing Algorithm for 3D Asymmetric Torus Routing Chip (ATRC)," *Proceedings of International Conference on Emerging Trends in Networks and Computer Communications*, 2011, pp. 121–124.

[38] Geng, F., D. Xiaoyun, and L. Luo, "Trends in Networks and Computer Communications," *International Conference on Electronic Packaging Technology & High Density Packaging*, 2009, pp. 85–90.

[39] Cadix, L., M. Rousseau, C. Fuchs, P. Leduc, A. Thuaire, R. El Farhane, H. Chaabouni, R. Anciant, J.-L. Huguenin, P. Coudrain, A. Farcy, C. Bermond, N. Sillon, B. Flechet, and P. Ancey, "Integration and Frequency Dependent Electrical Modeling of Through-Silicon Vias (TSV) for High-Density 3DICs," *International Interconnect Technology Conference*, 2010, pp. 1–3.

[40] Gu, X., B. Wu, M. Ritter, and L. Tsang, "Efficient Full-Wave Modeling of High Density TSVs for 3D Integration," *IEEE/ECTC Proceedings*, 2010, pp. 663–666.

[41] Soon, W., W. Seung, Z. Qiaoer, K. Pasad, V. Kripesh, and J. H. Lau, "High RF Performance TSV Silicon Carrier for High-Frequency Application," *IEEE/ ECTC Proceedings*, 2008, pp. 1946–1952.

[42] Lamy, Y. P. R., K. B. Jinesh, F. Roozeboom, D. J. Gravesteijn, and W. F. A. Besling, "RF Characterization and Analytical Modeling of Through-Silicon Vias and Coplanar Waveguides for 3D Integration," *Transactions on Advanced Packaging*, Vol. 33, No. 4, 2010, pp. 1072–1079.

[43] Chaabouni, H., M. Rousseau, P. Leduc, A. Farcy, R. El Farhane, A. Thuaire, G. Haury, A. Valentian, G. Billiot, M. Assous, F. De Crecy, J. Cluzel, A. Toffoli, D. Bouchu, L. Cadix, T. Lacrevaz, P. Ancey, N. Sillon, and B. Flechet, "Investigation on TSV Impact on 65-nm CMOS Devices and Circuits," *IEEE Proceedings of International Electron Devices Meeting*, 2010, pp. 35.1.1–35.1.4.

[44] Wu, J. H., J. Scholvin, and J. A. del Alamo, "Through-Wafer Interconnect in Silicon for RFICs," *IEEE Transactions on Electronic Devices*, Vol. 51, No. 11, November 2004, pp. 1765–1771.

[45] Leung, L. L. W., and K. J. Chen, "Microwave Characterization and Modeling of High Aspect Ratio Through-Wafer Interconnect Vias in Silicon Substrates," *IEEE Transactions of Microwave Theory and Techniques*, Vol. 53, No. 8, August 2005, pp. 2472–2480.

[46] Sheu, S., Z. Lin, J. Hung, J. H. Lau, P. Chen, S. Wu, K. Su, C. Lin, S. Lai, T. Ku, W. Lo, and M. Kao, "An Electrical Testing Method for Blind Through Silicon Vias (TSVs) for 3D IC Integration", *Proceedings of IMAPS International Conference*, Long Beach, CA, October 2011, pp. 208–214.

[47] Sunohara, M., T. Tokunaga, T. Kurihara, and M. Higashi, "Silicon Interposer with TSVs (Through-Silicon Vias) and Fine Multilayer Wiring," *IEEE/ECTC Proceedings*, 2008, pp 847–852.

[48] Ziolkowski, R. W., "An Efficient, Electrically Small Antenna Designed for VHF and UHF Applications," *IEEE Antennas and Wireless Propagation Letters*, Vol. 7, August 2008, pp. 217–220.

[49] Pozar, D., *Microwave Engineering*, 3rd ed., Wiley, New York, 2004.

[50] Shibata, T., and E. Sano, "Characterization of MIS Structure Coplanar Transmission Lines for Investigation of Signal Propagation in Integrated Circuits," *IEEE Transactions on Microwave Theory and Techniques*, Vol. 38, No. 7, July 1990, pp. 881–890.

CHAPTER 4

Thin-Wafer Strength Measurement

4.1 Introduction

Wafer thinning and thin-wafer handling are the second most important key enabling technologies for three-dimensional (3D) integrated circuit (IC) and Si integrations. One of the reasons why 3D IC and Si integrations is so popular is because after stacking of the thin wafers/chips, the total thickness of the stack is still thinner than that of an ordinary wafer/chip. Low profile is the most desirable feature of mobile products such as smart phones and tablets. The drawbacks of thin wafers are warpage and low strength. The thinner the wafer, the larger is the potential warpage, and the lower is the strength. In this chapter, piezoresistive stress sensors are used to measure the strength and warpage of thin wafers. In addition, the effects of the wafer back-grinding process on the nanomechanical behavior of multilayered Cu–low-k stacks are presented.

4.2 Piezoresistive Stress Sensors for Thin-Wafer Strength Measurement

4.2.1 Problem Definition

Silicon piezoresistive stress sensors are powerful tools for in situ stress measurement [1–10]. Piezoresistive stress sensors are known to be a potential device for measuring stress in semiconductor device packages [1–4]. Edwards et al. [4] used n-type piezoresistive stress sensors to qualitatively evaluate stress states in plastic packages. For quantitative measurement of stress using piezoresistive stress sensors, various calibration methods have been developed to determine piezoresistive coefficients [3, 5–9]. Calibrated piezoresistive silicon stress sensors have been used widely to evaluate stress at the silicon die surface after several packaging steps, such as die attach, underfill, and encapsulation [1, 3, 10]. The large coefficient of thermal expansion (CTE) mismatch between silicon chips and other packaging materials, such as epoxy and substrate or board, has been reported to generate large residual stress, ranging from 100 to 200 MPa at the silicon die surface after different packaging steps [1, 3, 10].

For 3D integration, piezoresistive silicon stress sensors can be used for measuring the strength of the device and interposer wafers during and after all processes, such as wafer thinning, SiO_2 deposition, metallization, and electroplating. This section discusses the design, fabrication, calibration, and use of a piezoresistive stress sensor to measure the strength and warpage of wafers after mounting on a dicing tape and wafer thinning.

4.2.2 Design and Fabrication of Piezoresistive Stress Sensors

The layout of an n-type stress-sensor design is shown in Figs. 4.1 and 4.2. The sensors R_1, R_2, R_3, and R_4 are fabricated along the [110], [$\bar{1}$00], [$\bar{1}$10], and [010] directions of a p-type [100] silicon wafer. The resistance of the R_1 and R_3 sensors is about 0.625 kΩ, and that of the R_2 and R_4 resistors is about 0.932 kΩ. The sheet resistance of all the sensors is about 178 Ω. The difference in resistances of different sensors is due to their size, which is 350×100 μm for R_1 and R_3 and 420×80 μm for R_2 and R_4.

These n-type stress sensors are fabricated on a silicon wafer using the conventional process. First, a 100-Å thermal silicon dioxide (SiO_2) layer was grown on the p-type [100] silicon wafer. After that, a photoresist (PR) was coated on the wafer, and then it was removed from the area where the sensors were to be fabricated. After removing the PR, arsenic was implanted on the p-type silicon wafer to form

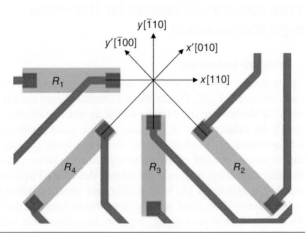

FIGURE 4.1 Layout of a piezoresistive stress sensor rosette.

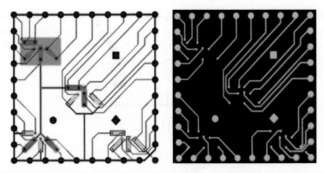

FIGURE 4.2   Optical image of a piezoresistive stress sensor test chip showing the metallization lines with probing pads.

n-type resistor, and then PR was stripped from the wafer. After PR stripping, a 4000-Å plasma-enhanced chemical-vapor deposition (PECVD) of an SiO_2 layer was deposited on the wafer, then contact windows were opened in the oxide for resistors, and finally, TaN and Al metallizations were deposited and patterned to form metal contacts for resistors, metallization lines, and probing pads. After metallization patterning, a 5000-Å PECVD SiO_2 layer was deposited on the wafer and patterned to open the probing pads. Figure 4.2 is an optical image of sensor chip showing the metallization lines and probing pads. Four sensor rosettes were fabricated at different locations of a 5- × 5-mm chip.

4.2.3   Calibration of Stress Sensors

The two basic requirements for piezoresistive stress sensors to perform stress measurements are (1) a measurable resistance change in the sensors with the application of loads and (2) a sensor calibration to determine the piezoresistive coefficients. Sensor calibration involves measuring the resistance change in the sensors with the application of a known stress and then determining the piezoresistive coefficients using the theory of piezoresistivity [11]. Based on the theory of piezoresistivity, if an uniaxial stress σ_x is applied in the [110] direction of the sensor rosette shown in Fig. 4.1, then the piezoresistive coefficients are given by

$$\Pi_{11} + \Pi_{12} = \frac{1}{\sigma_x}\left(\frac{\Delta R_1}{R_{10}} + \frac{\Delta R_3}{R_{30}}\right) \tag{4.1}$$

$$\Pi_{44} = \frac{1}{\sigma_x}\left(\frac{\Delta R_1}{R_{10}} - \frac{\Delta R_3}{R_{30}}\right) \tag{4.2}$$

where ΔR_i and R_{i0}, respectively, are the stress-induced resistance change and the initial resistance of the ith sensor. Once the values of the piezoresistive coefficients are known (and assuming the case of plane stress $\sigma_z = 0$), piezoresistive sensors can be used to determine the unknown in-plane stress components (σ_x and σ_y) by measuring the change in resistance of sensors ΔR_i and R_{i0}

$$\sigma_x = \frac{\Pi_{44}\left(\dfrac{\Delta R_1}{R_{10}} + \dfrac{\Delta R_3}{R_{30}}\right) + (\Pi_{11} + \Pi_{12})\left(\dfrac{\Delta R_1}{R_{10}} - \dfrac{\Delta R_3}{R_{30}}\right)}{2\Pi_{44}(\Pi_{11} + \Pi_{12})} \qquad (4.3)$$

$$\sigma_y = \frac{\Pi_{44}\left(\dfrac{\Delta R_1}{R_{10}} + \dfrac{\Delta R_3}{R_{30}}\right) - (\Pi_{11} + \Pi_{12})\left(\dfrac{\Delta R_1}{R_{10}} - \dfrac{\Delta R_3}{R_{30}}\right)}{2\Pi_{44}(\Pi_{11} + \Pi_{12})} \qquad (4.4)$$

Various methods such as four-point bending (4PB) of a strip of sensor chips [3, 5–10], 4PB using a sensor chip [8], wafer-level vacuum chuck [9], and a hydrostatic load [3] have been used to calibrate piezoresistive stress sensors. In this section, the 4PB method with a strip of sensor chips is used for calibration owing to its simple setup and the availability of more literature data [3, 5–10] for reference. Figure 4.3 shows the strip of sensor chips for calibration of the piezoresistive stress sensors. It can be seen that the size is 70×10 mm and that there are 28 sensor chips in two rows. However, only the sensor rosette that is located at the center of the strip is used for calibration.

In order to perform the 4PB of the strip of sensor chips, a 4PB fixture and a loading machine to apply the load on the strip, a microscope to visualize the probing pads, two microprobes to probe the

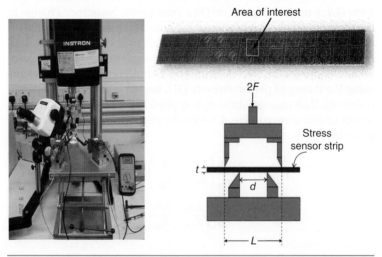

FIGURE 4.3   Four-point bending (4PB) setup and sample for determination of the piezoresistive coefficients.

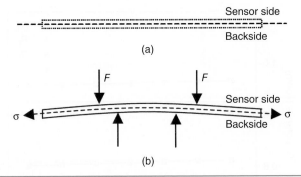

FIGURE 4.4 Schematic illustration of the bending of the sensor strip showing (a) the zero-stress state and (b) the tensile-stress state at the sensor plane.

sensor pads, and a multimeter to measure the sensor's resistance, as shown in Fig. 4.3, are needed. Here, an Instron microtester with a 100-N load cell is used to apply a given force (2F) on the strip. A schematic representation of the 4PB fixture is also shown in Fig. 4.3. A uniform uniaxial tensile stress σ, as shown in Fig. 4.4, can be applied on the sensor plane by using this type of 4PB fixture, and the applied stress can be calculated from the applied load 2F based on standard 4PB theory [5,8] as

$$\sigma = \frac{3F(L-d)}{t^2 h} \tag{4.5}$$

where L is the loading span, d is the supporting span, h is the width of the strip, and t is the thickness of the strip (chip). This formula works well if the deflection of the strip owing to the load is small and the t and h are small compared with the d and L [8]. Thus, in this chapter, L, d, h, and t are chosen to be of 50, 20, 10, and 0.73 mm, respectively.

In the calibration process, first the resistance of the stress sensors is measured without any applied load. Then a given load (with an increment of 4 N up to 24 N) is applied on the strip using the 4PB fixture and Instron microtester, and the corresponding resistances of the stress sensors are measured with the help of the microprobe and multimeter. Considering the variation in the readings, at least eight strips of sensors from two different wafers are tested to calculate the average values.

Figure 4.5 shows the resistances of sensors as a function of the applied stress. It can be seen that the resistance of n-type sensors decreases with an increase in applied stress. To determine the piezoresistive coefficients from the resistance versus applied-stress data, according to Eqs. (4.1) and (4.2), the slopes of plots between normalized resistance changes (i.e., $\Delta R_1/R_{10} \pm \Delta R_3/R_{30}$) and applied stress, as shown in Fig. 4.6, are measured and tabulated in Table 4.1.

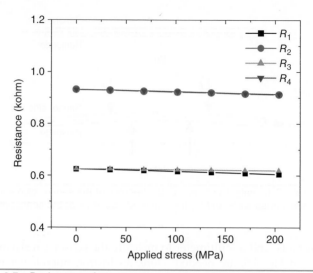

FIGURE 4.5   Resistance of sensors as a function of applied stress.

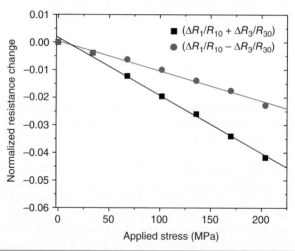

FIGURE 4.6   Normalized resistance change of sensors as a function of applied stress.

| $\Pi_{11} + \Pi_{12}$ | -1.98×10^{-4} MPa^{-1} |
| Π_{44} | -1.06×10^{-4} MPa^{-1} |

TABLE 4.1   Piezoresistive Coefficients of an n-Type Stress Sensor

The piezoresistive coefficients $\Pi_{11} + \Pi_{12}$ and Π_{44} of the n-type sensor are found to be -1.98×10^{-4} and -1.06×10^{-4} MPa^{-1}, respectively. These values for the piezoresistive coefficients are in agreement with previously reported work [2].

Once the piezoresistive coefficients $\Pi_{11} + \Pi_{12}$ and Π_{44} of the n-type stress sensor are determined, the stresses σ_x and σ_y can be calculated from Eqs. (4.3) and (4.4) with the measured change in resistance (ΔR_i) and initial resistance (R_{i0}). In this chapter, the piezoresistive stress sensors will be used to evaluate the stresses in wafers after thinning (back-grinding) and mounting of a dicing tape.

4.2.4 Stresses in Wafers after Thinning (Back-Grinding)

In order to study the effects of wafer thinning on the strength of the wafer, three original sensor wafers (730 μm) were thinned down to three different thicknesses, namely, 400, 200, and 100 μm [12]. A commercial back-grinding machine (Okamoto GNX 200) was used for the thinning process, which consisted of three major tasks: rough grinding, fine grinding, and polishing. The process parameters for the mesh size of the grinding wheel, spindle speed, chuck table speed, and feed rate are, respectively, No. 600, 3000 rpm, 230 rpm, and 180 to 190 μm/min for rough grinding and No. 2000, 3400 rpm, 230 rpm, and 12 to 16 μm/min for fine grinding. The final polishing is with a 200 cc/min slurry flow, 150 g/cm^2 load, 230 rpm pad speed, and 210 rpm chuck speed.

After thinning of the wafers, a resistance measurement is done at the same nine locations (as shown in Fig. 4.7) on the sensor wafer, and the stress components σ_x and σ_y are determined by Eqs. (4.3) and (4.4) and are shown in Table 4.2 for the case of a wafer thinned down to 100 μm. It can be seen that a large amount of compressive stress has been

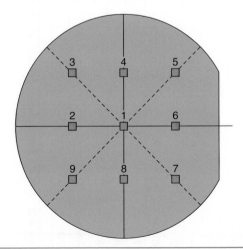

FIGURE 4.7   Stress measurement locations on a wafer.

TABLE 4.2 n-Plane Stress Components σ_x and σ_y Measured at Different Locations on a Device Wafer after Thinning to Different Thicknesses

Wafer Thickness	100 μm		200 μm		400 μm	
Location No.	Stress σ_x (MPa)	Stress σ_y (MPa)	Stress σ_x (MPa)	Stress σ_y (MPa)	Stress σ_x (MPa)	Stress σ_y (MPa)
1	-16.0	-23.6	0.1	-7.4	12.8	20.3
2	-47.3	-24.6	-31.1	-8.5	-11.5	-3.9
3	-112	-89.2	-60	-52.3	-35.7	-28.1
4	-71.6	-48.8	-27.6	-20.0	-35.7	-28.1
5	-51.9	-44.3	-19.5	-12.0	-11.5	-3.9
6	-47.3	-24.6	-11.5	-3.9	-15.0	7.7
7	-51.9	-44.3	0.1	-7.4	12.8	20.3
8	-79.7	-56.9	-50.8	-13.0	1.2	23.8
9	-27.6	-20.0	-3.4	4.2	9.3	31.9
Average	-56.1	-41.8	-22.6	-13.4	-8.1	4.4
SD	28.6	22.2	21.9	16.0	18.9	22.1

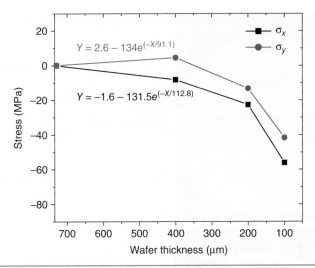

$Y = 2.6 - 134e^{(-X/91.1)}$

$Y = -1.6 - 131.5e^{(-X/112.8)}$

FIGURE 4.8 Variation in in-plane stresses as a function of wafer thickness.

generated in the sensor wafer after thinning. Figure 4.8 shows the average stress as a function of wafer thickness, and it is apparent that the amount of in-plane stress (compressive) increases exponentially with a decrease in wafer thickness.

The degree of compressive stress on the sensor wafer is further confirmed by measuring the wafer bending profile and the warpage (bow) of the original and back-grinded wafers. The wafer bow was measured using 33 point measurements, and the results are shown in Fig. 4.9. The measured values are listed in Table 4.3. Transition of the wafer bow from positive to negative indicates that the stress state at the sensor wafer plane changed from tensile to compressive. This transition in stress state can be understood with the help of sensor wafer processing materials and parameters.

Basically, the sensor wafer consists of two layers of materials. The first layer (bottom layer) is a single-crystal silicon layer, and the second layer (top layer) is mainly 9000 Å of plasma-enhanced chemical vapor deposition (PECVD) silicon dioxide. This PECVD SiO_2 is deposited on the silicon wafer at approximately 400°C. Thus, when the sensor wafer cools down to room temperature because of the thermal-expansion mismatch between the thin silicon dioxide (CTE = 0.54×10^{-6}/K [13]) layer and the silicon substrate (CTE = 2.33×10^{-6}/K [13]), it creates a bending (bowing upward) of the composite sensor wafer that generates tensile stress on top of the wafer.

Figure 4.10 shows a schematic illustration of the bending force that is developed in the sensor device wafer as a result of wafer processing. From the figure it can be seen that the stress at the surface of the original device wafer at room temperature must be tensile as a

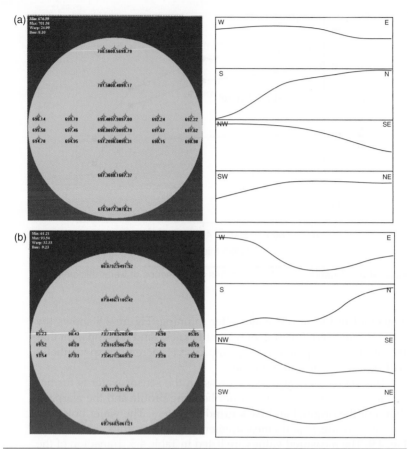

FIGURE 4.9 Warpage data and bending profile of (a) original wafer and (b) 100-μm thinned wafer.

result of wafer processing, and this was confirmed with the bending profile of the original device wafer shown in Fig. 4.9a. However, the stress state at the surface of original device wafer at room temperature was assumed as "no stress" for the stress sensor, and this was the reference point for stress measurement. Thus, if the tensile stress at

Wafer Thickness (μm)	Bow (μm)
730 (original)	8.1
400	9.9
200	−3.9
100	−9.2

TABLE 4.3   Warpage of Sensor Device Wafers Showing a Transition from Positive Bow to Negative Bow with a Decrease in Wafer Thickness

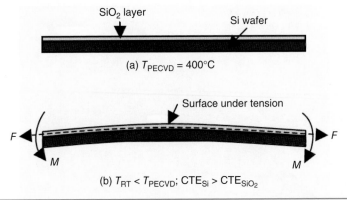

FIGURE 4.10 Schematic illustration of the bending force developed in the wafer (a) during PECVD silicon oxide deposition at 400°C and (b) at room temperature after PECVD deposition.

the device wafer surface decreases, then the stress sensor will detect an increase in compressive stress, and the same was observed in this chapter. The increase in compressive stress (decrease in tensile stress) at the thinned device wafer surface can be attributed to the decrease in stiffness of the Si layer owing to the decrease in Si layer thickness, the subsurface damage that occurred at the backside of the Si wafer, and the thermomechanical stresses imposed during the wafer back-grinding process.

The occurrence of subsurface damage at the backside of the Si wafer was confirmed via transmission electron microscopic (TEM) analysis of the back-grinded Si wafer, as shown in Fig. 4.11. From the TEM analysis, it was found that a thin (<200-nm) layer of amorphous and polycrystalline silicon with heavy subsurface damage, such as dislocations and staking faults, forms at the surface of the ground Si owing to the back-grinding process. This subsurface damage can result in the generation of intrinsic stress in the back-grinded Si wafer [13].

To further understand the generation of residual stress as a result of the wafer back-grinding process, a simple quantitative analysis was carried out using the well-known Stoney formula [14]. Based on this formula, residual stress in a circular composite film-substrate plate having a deflection of δ can be expressed by [13]

$$\sigma = \frac{\delta E d^2}{3(1-\mu)t l^2} \tag{4.6}$$

where E is the Young's modulus of the substrate material, μ is Poisson's ratio of the substrate material, d is the substrate thickness, t is the film thickness, and l is the plate radius. After putting the wafer radius,

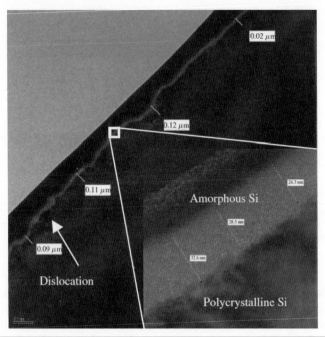

FIGURE 4.11   TEM image of a back-grinded Si wafer showing the formation of a thin (<200-nm) layer of amorphous and polycrystalline Si with many dislocations.

thickness, and bow data and the physical constants (E = 10,890 kg/ mm^2 and μ = 0.42) [15] in Eq. (4.6), stresses in back-grinded device wafers were calculated and the values tabulated in Table 4.4. The calculated stress values confirm that, as discussed previously, the stress at the surface of original device wafer is tensile, and it decreases and changes to compressive after wafer back-grinding.

The calculated stress is compared with the average stress measured using the piezoresistive sensor. For the comparison, the calculated stress

Wafer Thickness (μm)	Calculated Stress (MPa)	Normalized Calculated Stress (MPa)
730 (original)	29.4	0
400	10.8	−18.6
200	−1.1	−30.5
100	−0.6	−30.0

TABLE 4.4   Stress in Back-Grinded Device Wafers Calculated Using the Stoney Formula and Wafer Warpage Data

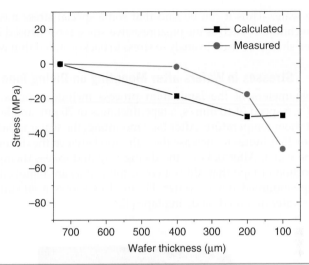

FIGURE 4.12   Calculated and measured stress values as a function of wafer thickness.

is normalized based on the fact that in sensor's measurement, stress in the original device wafer was considered to be zero. Figure 4.12 shows a comparison between calculated and measured stress values. From the figure, it is clear that up to a device wafer thickness of 200 μm, both the measured and the calculated stresses show nearly the same trend of an increase in compressive stress with a decrease in device wafer thickness. However, after 200 μm, the calculated stress deviates from the measured stress.

In addition, a variation is also observed between the calculated and measured stress values. This variation in stress values and deviation in trend could occur for two reasons. The first is that the Stoney formula measures the stress that generates in thin film, whereas the sensor measures the stress that develops at the top layer of the Si wafer where the piezoresistive sensor is fabricated. Although these two layers are very close, they are different in terms of location as well as material. The second reason could be the result of subsurface damage that occurs at the ground surface of the Si wafer (see Fig. 4.11) as a result of the back-grinding process. Since this subsurface damage can generate intrinsic stress in the back-grinded Si wafer, its exact influence on overall residual stress may not be revealed by the thin-film Stoney formula. From this analysis, it is clear that although the Stoney formula can provide some information about the stress in a back-grinded wafer, it may not reveal the exact influence of the back-grinding process on stress in a thin device wafer. This is especially true for device wafers that are thinner than 100 μm because in such wafers, subsurface damage is much closer to the front side of the device wafer. Thus it can be said that some special stress measurement techniques, such as the piezoresistive stress sensors used in this chapter, should be used to analyze stress in back-grinded thin wafers.

4.2.5   Stresses in Wafers after Mounting on Dicing Tape

The parameters of the lamination process include a pressure of 0.3 MPa, a speed of 10 mm/s, a tape thickness of 70 μm, and lamination at room temperature. After tape mounting, the wafer is baked at 70°C for 15 minutes to increase the adhesion between the dicing tape and the wafer. After baking, the dicing ring and excess dicing tape (the portion of tape that did not cover the wafer) are removed from the tape-mounted device wafer. Figure 4.13 shows a 400-μm thin device wafer mounted on dicing tape [12].

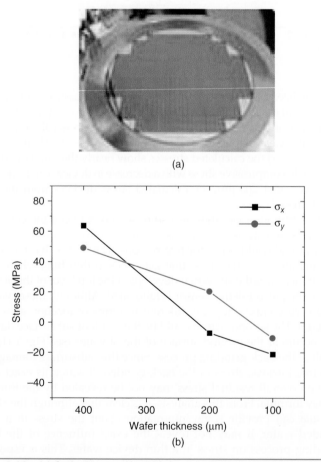

(a)

(b)

FIGURE 4.13   (a) Optical image of a 400-μm thin device wafer mounted on dicing tape. (b) In-plane stresses at the surface of thin device wafers after mounting on dicing tape.

The resistance measurement is performed at nine locations on the wafer, as shown in Fig. 4.14. The in-plane stress components σ_x and σ_y are determined by Eqs. (4.3) and (4.4), and their average values are shown in Fig. 4.13. It can be seen that dicing-tape mounting increases the tensile stress at the surface of the sensor wafer. Given this increase in tensile stress, the stress in the 400-µm-thick wafer is in tension, whereas the stress in the 100-µm-thick wafer is still in compression, but its magnitude decreases significantly. The presence of these tensile and compressive stresses in the dicing-tape-mounted wafer is further confirmed by the wafer bending-profile and warpage values (by using 33-point measurements), as shown in Fig. 4.14 and Table 4.5, respectively.

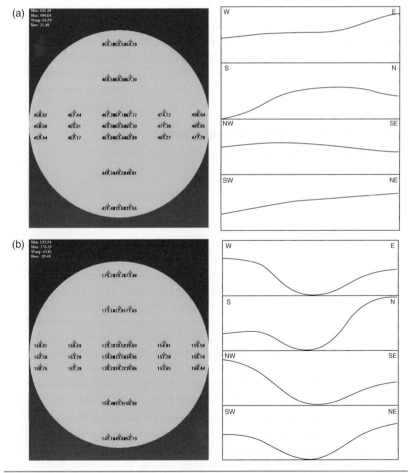

FIGURE 4.14 Warpage data and bending profile of (a) 400-µm and (b) 100-µm thin device wafers after dicing tape mounting.

Wafer Thickness (μm)	Bow (μm)
400	21.5
200	−20.8
100	−29.5

TABLE 4.5   Warpage of Sensor Device Wafers Having Different Thicknesses after Dicing Tape Mounting

The transition in stress state in the sensor wafer after mounting with dicing tape can be explained by means of the dicing-tape mounting process. The dicing-tape mounting process is actually a lamination process, in which a polymer tape is stretched and laminated on the backside of the wafer. After that, the dicing tape attempts to return to its original shape owing to its viscoelasticity. Thus, after the lamination process, the shrinkage in the dicing tape results in the development of a downward bending force on the sensor wafer, and therefore, the top surface of the wafer experiences a tensile stress. The effect of this phenomenon is similar to that of CTE mismatch, except in this phenomenon stress develops due to the shrinkage in starched polymeric tape.

4.2.6   Summary and Recommendations

Some important results and recommendations are summarized as follows:

- Piezoresistive stress sensors have been designed, fabricated, and calibrated successfully. The measured piezoresistive coefficients are in good agreement with previously reported studies.

- Stress sensor revealed that wafer back-grinding generates compressive stress at the surface of thin device wafers and that the compressive stress increases exponentially with decreases in wafer thickness. In 100-μm thin wafers, compressive stress increased to 112 MPa at some locations with an average value of 56 MPa. Thus it is suggested that stress sensors can be used to monitor the stress that develops as a result of wafer back-grinding and can be used to select appropriate process parameters and materials, such as plasma etching for stress relaxation and low-temperature PECVD oxide as a dielectric material.

- A transition in stress from compressive to tensile has been found to occur in thin device wafers when mounting them on a dicing tape. In a 400-μm thin wafer, stress changed from 8 MPa (compressive) to 64 MPa (tensile) owing to the viscoelestic behavior of the dicing tape. This indicates that thin devices undergo a severe stress transition during dicing-tape mounting and that appropriate handling methods, such as the use of a support wafer, are needed to overcome this problem.

4.3   Effects of Wafer Back-Grinding on the Mechanical Behavior of Cu–Low-*k* Chips

4.3.1   Problem Definition

Poor die strength of thinned wafers is mainly due to the scratches, crystal defects, and stresses that form during mechanical back-grinding [16]. So far, many researchers have studied wafer thinning/back-grinding processes extensively in terms of die strength by assessing the quality of the ground surface [16–21]. Besides the degradation of die strength owing to the small thickness of thinned wafers, some researchers have found subsurface damage that results from back-grinding [22]. Blech et al. have studied the effect of mechanical wafer back-grinding on wafer deformation [23]. Hoh Huey Jiun et al. studied the die fracture strength by considering the surface roughness of the thinned surface and the amount of thickness removed by stress-relief methods [21]. In addition to surface roughness of the thinned wafer, grinding direction has a significant effect on the die strength [24]. Singulation of thinned wafers always poses problem because it induces chipping and rough edges during the dicing process. Chen et al. have found that dicing before grinding is a good solution for ultrathin chip applications and that it enhances chip strength around 10 to 15 percent [20]. In the back-grinding process, it is impractical to retain native die strength, but it can be controlled to some extent by employing stress-relief methods after the grinding process. The commonly used stress-relief methods include wet chemical etching, dry or plasma etching, dry polishing, and chemical and mechanical polishing (CMP) [25, 26].

In recent years, many researchers have assessed the quality of the back-grinding process extensively with the help of die strength evaluation. The die strength of the thinned wafer can be evaluated by using different mechanical testing techniques, such as the three-point bend test, the four-point bend test, the ball-on-ring test, ball-breaker tests, and ring-on-ring tests. These mechanical tests are greatly influenced by several process and material parameters, such as surface roughness or finish, degree of thinning, stress-relief process, and the quality of the dicing edge [20–23, 25, 26]. However, the available literature related to the effect of grinding processes on the active side of a die/chip is limited, and this necessitates a focused study of the effect of wafer back-grinding on the active side of a chip.

In this section, the effect of the back-grinding process on the active side of a Cu–low-*k* stack using the nanoindentation and nanoscratch techniques combined with TEM analysis is presented. Usually, the active side of the chip is a few micrometers thick, and it cannot be studied using the methods that are used conventionally for die strength evaluation. Therefore, we have chosen sophisticated methods such as nanoindentation and nanoscratch techniques for this study [17].

Nanoindentation and nanoscratch tests have been carried out on both normal (no back-grinding) and back-grinded stacks to study fracture strength, modulus, hardness, and adhesive/cohesive strength. Generally, after back-grinding, the die strength of the thinned chip decreases significantly. However, in this study, we noticed an improvement in the mechanical behavior of the back-grinded Cu–low-k stack. Thus we can say that, on one hand, the back-grinding process lowers die strength in one way but, on the other hand, improves the mechanical integrity of the Cu–low-k stack.

4.3.2   Experiments

Test-Vehicle Fabrication

The test vehicle [17] used in this study is a multilayer low-k stack of 15 different thin films comprising SiN, USG, Blok (SiC), and black diamond (BD low-k), as shown in Fig. 4.15. All samples were prepared on 8-in Si [100] wafers in a semiconductor-processing clean room of class 1000 environment. The test structure employed in this study was designed exclusively to study the BD (low-k) integrity, and it resembles the Cu–low-k structure of the three metallizations. Thus the structure has three BD low-k layers at different levels according to the back-end-of-the-line (BEOL) interconnect design specifications, and it does not have any copper metal lines. The total thickness of the multilayer low-k stack in the test vehicle is around 3400 nm. The main intention in preparing the test vehicle without metal line was that in the reality, low-k stack regions are more vulnerable than regions with metal lines. Hence this study provides an outlook about the response of the low-k test structures during the back-grinding process [17].

Back-Grinding Process

Wafer thinning of the low-k stacked wafers was carried out using a commercial back-grinding system (Disco Corporation, Japan). In the thinning process, first, the coarse grinding was done using a No. 300 grit wheel, then fine grinding was done using No. 2000 grit, and finally, dry polishing was carried out to remove any subsurface damage. The dry polishing was a special stress-relief process developed by Disco Corporation using No. 8000 grit without using any chemicals. The main objective of employing this method was to offer superlative stability to the back-grinded wafers [27].

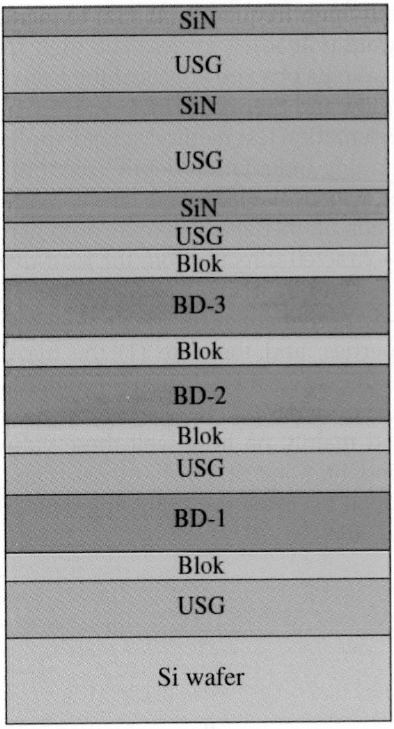

SiN	
USG	
SiN	
USG	
SiN	
USG	
Blok	
BD-3	
Blok	
BD-2	
Blok	
USG	
BD-1	
Blok	
USG	
Si wafer	

Figure 4.15 Multilayered (15-layer) Cu–low-*k* stack test vehicle.

Nanoindentation Testing

Nanoindentation is a powerful mechanical characterization technique that is similar to conventional hardness-testing methods but is performed on much smaller scale using special equipment called a *nanoindenter*. This method involves indenting a material of interest whose mechanical properties are unknown with another material (called an *indenter*) whose properties are known. The unique advantage of the nanoindentation technique over conventional hardness testing is that both hardness and the modulus can be easily extracted from the nanoindentation curve [28]. Besides these properties, the technique also can determine residual stress, elastic-plastic behavior, creep, relaxation properties, fatigue, and fracture toughness.

In this study, nanoindentation tests were performed on the Nano Indenter XP (MTS Corporation, United States) with a continuous-stiffness-measurement (CSM) attachment and a Berkovich indenter. This CSM attachment has the unique advantage of providing mechanical properties as a function of penetration depth. In nanoindentation experiments, CSM is carried out by applying a harmonic force at a relatively high frequency (45 Hz) to increase load without completely separate unloading cycles. The high frequency used in the CSM method avoids obscure effects of the tested samples such as creep, viscoelasticity, and thermal drift.

In the nanoindentation test method, a load applied on the sample, and the corresponding indentation depth is continuously monitored and recorded as a load-displacement curve. Without imaging the hardness impression on the tested sample, both hardness and elastic modulus can be extracted directly from the load-displacement curve by using the Oliver and Pharr method [28]. Three important parameters of the load-displacement curve play a vital role in assessing the mechanical properties, and they are (1) the maximum load P_{max}, (2) the maximum displacement h_{max}, and (3) contact stiffness from the initial loading curve $S = dP/dh$. The accuracy and repeatability of the properties depend mainly on how well these values are measured during the nanoindentation test. The hardness H and elastic modulus E can be determined by using the following equation developed by Oliver and Pharr [28]:

$$\frac{E}{1-\upsilon^2} = \frac{\sqrt{\pi}}{2} \frac{1}{\sqrt{A_{max}}} \frac{dp}{dh} \tag{4.7}$$

and

$$H = \frac{P_{max}}{A_{max}} \tag{4.8}$$

where υ is Poisson's ratio, dP/dh is the initial slope of the indentation unloading curve, P_{max} is the maximum load, and A_{max} is the projected contact area at the maximum load. For a perfect Berkovich indenter, projected area A can be calculated by $A = 24.56h_c^2$, where h_c is true contact depth. The load at which system fails is termed as the critical load (Lc) from nanoindentation tests.

Nanoscratch Testing

The adhesion strength of a film-substrate system is defined mainly by the interfacial bond strength and depends solely on the interfacial properties. The interactions at the interface of a film-substrate system may be chemical, electrostatic, or van der Waals type. Empirically, adhesion strength is the stress or load required to detach the thin film from the substrate. Up to the present, numerous techniques have been developed for measuring the adhesion of a film-substrate system [29], but among them, the nanoscratch technique is being used widely to determine adhesion or cohesion strength of thin film-substrate systems [30]. In this section, all nanoscratch tests were performed on the Nano Indenter XP using the nanoscratch attachment. A diamond stylus was used as the scratch indenter, and it was scratched over the low-k stack system under ramp loading conditions until some well-defined failure was observed. In the current study, adhesion/cohesion strength is reported in terms of the critical normal load (LC) from nanoscratch tests. For each sample, at least five tests were conducted, and all tests showed good repeatability. For the scratch test, a conical diamond indenter of 5 μm was used because it has a uniform face in all directions. The main advantage of using the conical diamond indenter is that it is symmetric at all alignments, which removes the directionality effect imposed by pyramid-type tips [31]. The diamond stylus is scratched over the low-k stack for a length of 500 μm under ramp loading condition from 0 to 250 mN. All tests were performed with a constant scratch velocity of 1 μm/s.

4.3.3 Results and Discussion

Failure Load, Hardness H, and Elastic Modulus E by Nanoindentation

In this study, failure load, hardness H, and elastic modulus E are computed by analyzing the nanoindentation load-displacement curves. Figure 4.16 shows the nanoindentation load-displacement curves of normal (no back-grinding) and back-grinded low-k stacks (BG 500, 300, 150, and 75 μm) and their comparison using the CSM attachment. At least 10 nanoindentation tests were performed on each sample system, and all exhibited good repeatability. Poisson's ratio of the low-k stack systems was maintained around 0.2. From the figure, it is

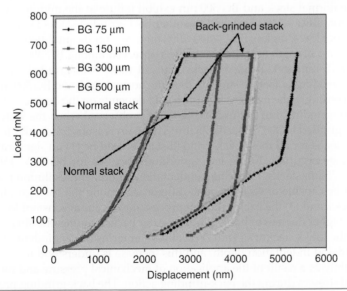

FIGURE 4.16 Typical load-displacement curves of normal and back-grinded samples and their comparisons.

obvious that the normal stack shows a pop-in event (failure load/ fracture strength of the stack) at lower loads and indentation depths compared with back-grinded stacks. These pop-in events in the nanoindentation curve result from film cracking and delamination of the stack in the form of blisters [32]. Normal stacks failed at a load of 456.25 ± 21.22 mN and an indentation depth of 2422.41 ± 58.53 nm, whereas back-grinded stacks failed at loads in the range of 482 to 661 mN and indentation depths of 2405 to 2979 nm. The fracture load and depth values of all types of samples are summarized in Table 4.6 and Fig. 4.17. Figure 4.18 shows optical images of the residual nanoindentation impressions of the normal and back-grinded stacks. From the nanoindentation curves and optical imaging analysis, it is clear that the nanoindentation responses of the normal and back-grinded stacks are different in terms of failure load and depth. The normal stack and BG 500 μm show extensive delamination and chipping, whereas the other back-grinded stacks (BG 300, 150, and 75 μm) show delamination blisters, and this behavior is in good agreement with the nanoindentation pop-in events. BG 500 μm exhibits a mixed response because it shows chipping during nanoindentation and moderate pop-in failure load at 482.17 mN, and this might be due to the moderate degree of back-grinding. In the case of the other back-grinded stacks (BG 300, 150, and 75 μm), even higher nanoindentation loads were not able to damage/chip-off the low-*k* stack and caused interfacial delamination only.

One common feature observed among the normal and back-grinded stacks is that the failure of the stack occurs in the low-*k* region. The normal stack and BG 500 μm exhibit failure at the middle low-*k* region, BD2 (~2400 nm), and the other back-grinded stacks fail at the bottom low-*k* region, BD1 (2800–2900 nm). No significant difference in fracture strength (pop-in events) was observed among BG 300-, 150-, and 75-μm back-grinded stacks, and all these back-grinded stacks showed higher facture strength than the normal stack and BG 500 μm. Accordingly, the increase in failure load depends on the degree of the back-grinding, but not much difference is observed when the wafers are grinded to 300, 150, and 75 μm. The main reasons for similar failure loads for all these back-grinded stacks could be (1) no significant change occurred during back-grinding processes for BG 300-, 150-, and 75-μm stacks, (2) the nanoindentation equipment resolution may not be capable of sensing small differences among the back-grinded stacks, and (3) the back-grinding process may occur at constant loads for higher degrees of back-grinding. After back-grinding, the strength of the low-*k* stack was enhanced mainly in terms of nanoindentation load and indentation depth, and this increase is understood to occur mainly as a result of the application of mechanical pressure and thermal stresses during the back-grinding action. The back-grinding pressure or load may improve the adhesion, especially the van der Waals forces at the multilayered interfaces, and cause densification of the

| Sample | Failure Load and Displacement | | Properties Corresponding to Middle of the Low-*k* Region (GPa) | | | Critial Load, Lc (mN) |
	Load (MN)	Indentaion Depth (nm)	Hardness @2000 nm	Elastic Modulus @500 nm		
Normal stack	456.25 ± 21.22	2422.41 ± 58.53	4.27 ± 0.1	41.81 ± 2.10		72.80 ± 3.98
BG-500 μm	482.17 ± 25.25	2405.86 ± 70.47	4.94 ± 0.36	51.55 ± 3.25		98.65 ± 7.50
BG-300 μm	661.20 ± 7.57	2979.79 ± 21.58	5.10 ± 0.27	53.01 ± 2.47		100.73 ± 3.56
BG-150 μm	658.45 ± 4.74	2809.01 ± 30.60	5.10 ± 0.64	53.64 ± 1.95		96.00 ± 3.76
BG-75 μm	658.60 ± 12.21	2942.52 ± 71.20	5.06 ± 0.71	49.75 ± 2.62		96.30 ± 2.22

TABLE 4.6   Summary of Nanomechanical Properties of Normal and Back-Grinded Samples

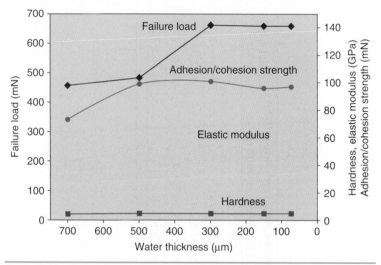

FIGURE 4.17   Failure load, adhesion/cohesion strength, elastic modulus *E*, and hardness *H* as a function of wafer thickness.

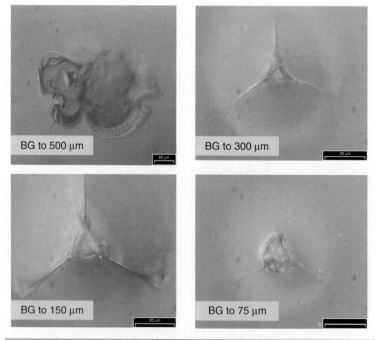

FIGURE 4.18   Optical images of residual nanoindentation impressions of back-grinded samples.

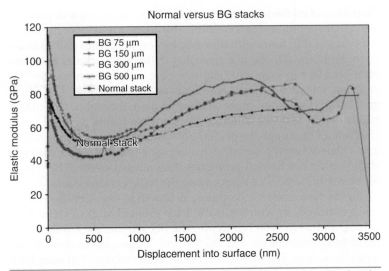

FIGURE 4.19 Elastic modulus as a function of displacement for normal and back-grinded samples.

individual films of the stack. Many researchers in the packaging field are investigating whether back-grinding processes deteriorate die strength, but this is not the same phenomenon with the active side of a chip stack.

Figures 4.19 and 4.20 show, respectively the elastic modulus and hardness as a function of the indentation depth for normal and back-grinded stacks measured using the nanoindentation CSM technique.

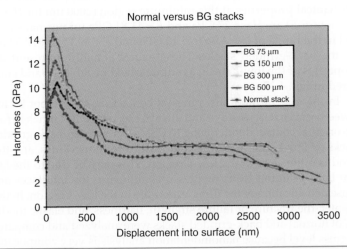

FIGURE 4.20 Hardness as a function of displacement for normal and back-grinded samples.

Properties of all the samples are not constant but strongly depend on the contact depth, and this is mainly due to the presence of the different types of thin films with diverse physical properties and 15 interfaces. From the figures, it is clear that initially all samples exhibited high hardness and modulus values owing to the presence of the SiN layer on top of the stack. However, as shown in Fig. 4.20, back-grinded stacks showed higher hardness values throughout the indentation depth, and this difference is significant until approximately 1500 nm. Even though the BG 500-μm samples show moderate failure load values when compared with other back-grinded stacks, they still show higher mechanical properties than normal stacks. There is no difference in hardness values among the back-grinded stacks.

In the case of elastic modulus, as shown in Fig. 4.19, the overall trend is mixed, in which initially back-grinded stacks exhibit high values, and from 1000 nm depth, the BG 150-μm stack follows the trend of normal stacks, and BG 75- and 300-μm stacks show lower modulus values. The mixed trend of modulus values of all samples is due mainly to the fact that the elastic modulus is a highly sensitive and intrinsic property. Elastic modulus is greatly influenced by the layers beneath the testing films and substrate.

In the case of hardness, the difference between normal stacks and back-grinded stacks is very clear when compared with the modulus values. The minimum values of hardness and elastic modulus from the property plots can represent the low-k BD region because it is known that the BD (low-k) has lower mechanical strength than the other constituents of the stack (namely, SiN, USG, and Blok). Thus the minimum region for hardness is around 2000 nm, which is in the middle of the low-k region, and for modulus, the minimum region is found to be at around 500 nm. The mechanical properties at this minimum region (2000 nm for H and 500 nm for E) are $H = 4.27$ GPa and $E = 41.81$ GPa for normal stacks and in the range of $H = 4.94$ to 5.10 GPa and $E = 49.75$ to 51.55 GPa for the back-grinded stacks. These values actually represent the properties of the low-k stack, as summarized in Table 4.6 and compared in Fig. 4.17.

In the case of modulus, physical contact depths cannot be followed because it is sensitive and always exerts its effect much before the original contact depth owing to CSMs. As a whole, back-grinded stacks exhibit higher hardness and elastic modulus values, and this trend is quite clear in the low-k region. This may be due to the fact that back-grinding loads or pressures influence the interfaces and cause densification of the films, especially in the low-k region. In this chapter, the fracture or failure strength, hardness, and elastic modulus of normal and back-grinded stacks were analyzed and compared at gross level because nanoindentation analysis is very complicated and not well established for the multilayered stacks.

Adhesion/Cohesion Strength by Nanoscratch

Adhesion or cohesion strength of normal and back-grinded stacks is reported in terms of the critical normal load (LC). The critical normal load is the load where the first abrupt decrease in scratch depth is observed. Figure 4.21 shows the nanoscratch profiles as scratch depth versus normal load for all samples, and corresponding optical images are given in Fig. 4.22. From Fig. 4.21, LC of the normal and back-grinded stacks is well defined as the load where the stack is fully chipped off (delaminated) from the substrate, and it can be verified from the corresponding optical images (see Fig. 4.22). In Fig. 4.22, optical images of the both normal and back-grinded stacks are captured at three different locations, beginning, middle, and end of the scratch track profile. Before reaching the LC, minor damage to the stacks is observed in terms of cracks, and it can be seen on the scratch profile as a small kink just before reaching the LC point. Arrows over the scratch track of all samples indicate the LC where significant delamination or chip-off of the stack starts. From that LC point, a sudden increase in scratch depth is observed as the indentation tip abruptly hits the silicon substrate. The sudden increase in scratch load was associated with catastrophic chip-off failure of the whole stack, as well as significant plowing of the tip into the silicon substrate. This plowing action was associated with delamination or chip-off of the whole stack, and it is sustained until the end of the scratch track.

Similar to the nanoindentation tests, in the nanoscratch tests also, the back-grinded stacks exhibit higher LC values (100.73–96.30 mN)

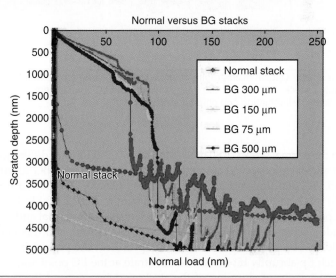

Figure 4.21 Typical nanoscratch depth profiles as a function of normal load for all samples and their comparison.

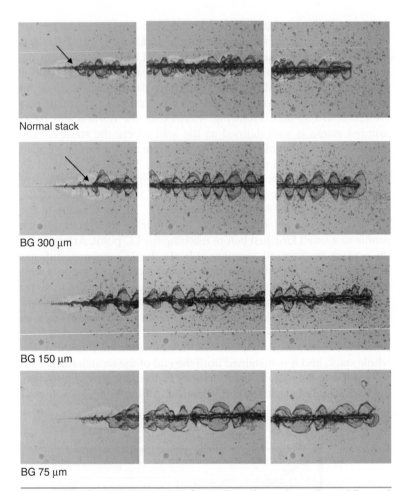

Normal stack

BG 300 μm

BG 150 μm

BG 75 μm

FIGURE 4.22 Optical micrographs of scratch tracks made on the multilayered Cu–low-k stack. Arrows over the image indicate the LC point, where significant delamination and chip-off started.

than the normal stack (72.80 mN). Moreover, no significant difference in LC values was observed among the back-grinded stacks (BG 500 μm = 98.65, BG 300 μm = 100.73, BG 150 μm = 96.00, and BG 75 μm = 96.30 mN), and these results have been summarized in Table 4.6 and compared in Fig. 4.17. In the nanoscratch tests, both normal and back-grinded stacks exhibited significant adhesive failure, and this failure could be observed in the nanoscratch profile at the point where the scratch tip abruptly hit the silicon substrate at the LC point and in the optical images of chipping of the stack as debris around the scratch track, especially after LC. Although the extent of the back-grinded effect over the low-k stack may be different for different back-grinding thicknesses, the nanoscratch test was not able to sense this difference because all the back-grinded stacks showed nearly the same results.

TEM Analysis

As discussed earlier, both nanoindentation and nanoscratch test results indicate that the mechanical integrity of low-k stacks is enhanced after back-grinding. To ascertain the effect of back-grinding on low-k stacks, TEM cross-sectional analysis was conducted on both normal and back-grinded stacks (BG 75 and 150 μm), as shown in Fig. 4.23a–c. From cross-sectional TEM analysis of normal and back-grinded stacks, it is observed that there are significant structural changes at the interfaces of the low-k region, and the interfaces of other regions look almost the same. Low-k interfaces in normal stacks are wavy (see Fig. 4.23a), whereas low-k interfaces in back-grinded stacks of BG 75 and 150 μm (see Figs. 4.23b, c) are smooth. This suggests that the force employed during back-grinding influences the low-k interfaces. Moreover, a significant densification in the low-k layers (BD 1, 2, and 3 μm) was observed in the back-grinded stacks compared with normal stacks, as shown in Fig. 4.23. The densification in the low-k layers of the back-grinded stacks is in the range of 2 to 13 percent compared with normal stacks, and it is expected mainly owing to the weaker physical and mechanical strength of the low-k stack.

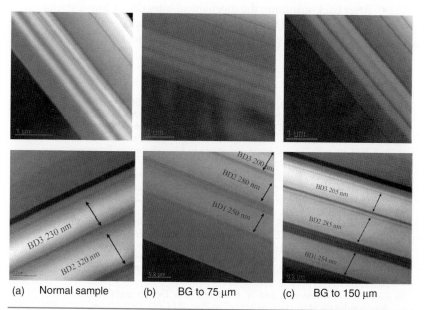

(a)   Normal sample   (b)   BG to 75 μm   (c)   BG to 150 μm

FIGURE 4.23   TEM cross-sectional analysis of normal sample BG 75-μm and BG 150-μm Cu–low-k regions has been zoomed in for densification analysis.

4.3.4 Summary and Recommendations

Some important results have been summarized as follows:

- Normal stacks failed at a load of 456.25 mN and an indentation depth of 2422.41 nm, whereas back-grinded stacks failed at loads in the range of 482.17 to 661.20 mN and indentation depths of 2405.86 to 2979.79 nm.

- One common feature observed among the normal and back-grinded stacks is that the failure of the stack occurs in the low-k region.

- The mechanical properties of normal and back-grinded stacks have been compared at the minimum property region, which represents the low-k region (hardness at 2000 nm and elastic modulus at 500 nm). These properties are $H = 4.27$ GPa and $E = 41.81$ GPa for the normal stack and in the range of $H = 4.94$ to 5.10 GPa and $E = 49.75$ to 53.64 GPa for the back-grinded stacks.

- Similar to the nanoindentation tests, in nanoscratch tests as well, the back-grinded stacks exhibit higher LC values (100.73–96.30 mN) than normal stacks (72.80 mN).

- No significant difference in LC values is observed among the back-grinded stacks in nanoscratch test (BG 300 = 100.73; BG 150 = 96.00; and BG 75 = 96.30 mN).

- After back-grinding, the mechanical integrity of low-k stacks is enhanced, and thus the back-grinded stacks exhibit higher nanomechanical properties than normal stacks.

4.4 References

[1] Miura, H., A. Nishimura, S. Kawai, and G. Murakami, "Structural Effect of IC Plastic Package on Residual Stress in Silicon Chips," *Proc 40th IEEE Electron. Comp. Technol. Conf.*, Vol. 1, Las Vegas, May 1990, pp. 316–321.

[2] Sweet, J. N. "Die Stress Measurement Using Piezoresistive Stress Sensor," in J. H. Lau, ed., *Thermal Stress and Strain in Microelectronics Packaging*, New York: Van Nostrand & Reinhold, 1993, pp. 221–268.

[3] Rahim, M. K., J. C. Suhling, D. S. Copeland, M. S. Islam, R. C. Jaeger, P. Lall, and R. W. Johnson, "Die Stress Characterization in Flip Chip on Laminate Assemblies," *IEEE Trans. Comp. Packag. Technol.*, Vol. 28, No. 3, September 2005, pp. 415–429.

[4] Edwards, D., K. Heinen, S. Groothuis, and J. Martinez, "Shear Stress Evaluation of Plastic Packages," *IEEE Trans. Comp. Hyb. Manuf. Technol.*, Vol. 10, No. 4, December 1987, pp. 618–627.

[5] Lo, T. C. P., and P. C. H. Chan, "Design and Calibration of a 3D Microstrain Gauge for in situ on Chip Stress Measurements," in *Proc. ICSE*, Malaysia, November 1996, pp. 252–255.

[6] Gee, S. A., V. R. Akylas, and W. F. Bogert, "The Design and Calibration of a Semiconductor Strain Gauge Array," in *Proc. IEEE Int. Conf. Microelectron. Test Struct.*, Vol. 1, February 1988, pp. 185–191.

[7] Zhong, Z. W., X. Zhang, B. H. Sim, E. H. Wong, P. S. Teo, and M. K. Iyer, "Calibration of a Piezoresistive Stress Sensor in [100] Silicon Test Chips," in *Proc. 4th Electron. Packag. Technol. Conf.*, December 2002, pp. 323–326.

[8] Beaty, R. E., R. C. Jaeger, J. C. Suhling, R. W. Johnson, and R. D. Butler, "Evaluation of Piezoresistive Coefficient Variation in Silicon Stress Sensors Using a Four-Point Bending Test Fixture," *IEEE Trans. Comp. Hyb. Manuf. Technol.*, Vol. 15, No. 5, October 1992, pp. 904–914.

[9] Suhling, J. C., R. A. Cordes, Y. L. Kang, and R. C. Jaeger, "Wafer-Level Calibration of Stress-Sensing Test Chips," *IEE/ECTC Proceedings*, Washington, DC, May 1994, pp. 1058–1070.

[10] Suhling, J. C., and R. C. Jaeger, "Silicon Piezoresistive Stress Sensors and Their Application in Electronic Packaging," *IEEE Sens. J.*, Vol. 1, No. 1, June 2001, pp. 14–30.

[11] Bittle, D. A., J. C. Suhling, R. E. Beaty, R. C. Jaeger, and R. W. Johnson, "Piezoresistive Stress Sensors for Structural Analysis of Electronic Packages," *J. Electron. Packag.*, Vol. 113, No. 3, September 1991, pp. 203–215.

[12] Kumar, A., X. Zhang, Q. Zhang, M. Jong, G. Huang. V. Lee, V. Kripesh, C. Lee, J. H. Lau, D. Kwong, V. Sundaram, R. Tummula, and G. Meyer-Berg, "Residual Stress Analysis in Thin Device Wafer Using Piezoresistive Stress Sensor", *IEEE Trans. Comp. Packag. Technol.*, Vol. 1, No. 6, June 2011, pp. 841-850.

[13] Ohring, M., *The Materials Science of Thin Films*, Academic Press, San Diego, 1991, pp. 325–552.

[14] Stoney, G. G., "The Tension of Metallic Films Deposited by Electrolysis," *Proc. R. Soc. London A*, Vol. 82, No. 553, May 1909, pp. 172–175.

[15] Sze, S. M., *VLSI Technology*, McGraw-Hill, New Delhi, India, 2003, p. 657.

[16] Takahashi, K., H. Terao, Y. Tomita, Y. Yamaji, M. Hoshino, T. Sato, T. Morifuji, M. Sunohara, and M. Bonkohara, "Current Status of Research and Development for 3D Chip Stack Technology," *Jpn. J. Appl. Phys.*, Vol. 40, No. 4B, 2001, pp. 3032–3037.

[17] Sekhar, V., S. Lu, A. Kumar, T. Chai, V. Lee, S. Wang, X. Zhang, C. Premchandran, V. Kripesh, and J. H. Lau, "Effect of Wafer Back Grinding on the Mechanical Behavior of Multilayered Low-k for 3D-Stack Packaging Applications", *IEEE/ECTC Proceedings*, 2008, pp. 1517-1524..

[18] Reiche, M., and G. Wagner, "Wafer Thinning: Techniques for Ultrathin Wafers," in *Proc. Adv. Packag.*, March 2003.

[19] Yeung, B. H., V. Hause, and T.-Y. Lee, "Assessment of Backside Processes Through Die Strength Evaluation," *IEEE Trans. Adv. Packag.*, Vol. 23, No. 3, August 2000, pp. 582–587.

[20] Chen, S., T.-Y. Kuo, H.-T. Hu, J.-R. Lin, and S.-P. Yu, "The Evaluation of Wafer Thinning and Singulating Processes to Enhance Chip Strength," in *IEEE/ECTC Proceeding*, Vol. 2, May–June 2005, pp. 1526–1530.

[21] Jiun, H. H., I. Ahmad, A. Jalar, and G. Omar, "Effect of Wafer Thinning Methods Toward Fracture Strength and Topography of Silicon Die," *Microelectron. Rel.*, Vol. 46, Nos. 5–6, May–June 2006, pp. 836–845.

[22] Chen, J., and I. De Wolf, "Study of Damage and Stress Induced by Back-Grinding in Si Wafers," *Semicond. Sci. Technol.*, Vol. 18, No. 4, 2003, pp. 261–268.

[23] Blech, A., and D. Dang, "Silicon Wafer Deformation after Backside Grinding," *Solid State Technol.*, Vol. 37, No. 8, 1994, pp. 74–76.

[24] Lee, S., S. Sim, Y. Chung, Y. Jang, and H. Cho, "Fracture Strength Measurement of Silicon Chips," *Jpn. J. Appl. Phys.*, Vol. 36, No. 6A, 1997, pp. 3374–3380.

[25] Gaulhofer, E., and H. Oyer, "Wafer Thinning and Strength Enhancement to Meet Emerging Packaging Requirements," in *Proc. IEMT Eur. Symp.*, Munich, Germany, April 2000, pp. 1–4.

[26] McHatton, C., and C. M. Gumbert, "Eliminating Back-Grind Defects with Wet Chemical Etching," *Solid-State Technol.*, Vol. 41, No. 11, November 1998, pp. 85–90.

[27] DISCO Corporation, Tokyo, Japan [online]. Available at: http://www.disco.co.jp/.

[28] Oliver, W. C., and G. M. Pharr, "An Improved Technique for Determining Hardness and Elastic Modulus using Load and Displacement Sensing Indentation Experiments", *J. Mater. Res.*, Vol. 7, 1992, pp. 1564–1583.

[29] Volinsky, A., D. F. Bahr, M. D. Kriese, N. R. Moody, and W. W. Gerberich, "Nanoindentation Methods in Interfacial Fracture Testing," in *Interfacial and Nanoscale Failure*, Vol. 18, Amsterdam, Elsevier, 2003, Chap. 13.

[30] Campbell, D. S., *Handbook of Thin Film Technology*, McGraw-Hill, New York, 1970, Chap. 12.

[31] Tayebi, N., A. A. Polycarpou, and T. F. Cony, "Effects of Substrate on Determination of Hardness of Thin Films by Nanoscratch and Nanoindentation Techniques," *J. Mater. Res.*, Vol. 19, No. 6, 2004, pp. 1791–1802.

[32] Wang, L., M. Ganor, and S. I. Rokhlin, "Nanoindentation Analysis of Mechanical Properties of Low to Ultralow Dielectric Constant SiOCH Films," *J. Mater. Res.*, Vol. 20, No. 8, 2005, pp. 2080–2093.

CHAPTER 5

Thin-Wafer Handling

5.1 Introduction

In order to have low-profile, lightweight, high-performance, low-power, and wide-bandwidth products with three-dimensional (3D) integrated circuit (IC) integration and 3D Si integration technologies, the thickness of the chip/interposer wafers must be very thin [1, 2]. For memory-chip stacking, the thickness of each chip is no more than 50 μm and eventually is going down to 20 μm. For both active and passive interposers, the thickness is usually 200 μm or less [1, 2]. Thus wafer thinning and thin-wafer handling are the second (right after through-silicon vias [TSVs]) most important key enabling technologies for 3D IC/Si integrations. In this chapter, critical issues of wafer thinning and thin-wafer handling, such as the device/interposer wafer, carrier wafer, temporary bonding, thinning, backside processes, debonding, and assembly, are presented, and their potential solutions are discussed. In addition, the state of the art of material and equipment for thin-wafer handling is examined. Even supporting (carrier) wafer technique is the focus of this chapter, but the carrierless method is mentioned briefly as well.

5.2 Wafer Thinning and Thin-Wafer Handling

Making the wafer thin is not a big problem. Most of the back-grinding machines can do the job and grind wafers to as thin as 5 μm. However, handling thin wafers through all the semiconductor fabrication and packaging assembly processes is difficult. Usually, the chip/interposer wafer is temporarily bonded on a support (carrier) wafer before back-grinding to expose the Cu TSVs (Cu reveal). Then it goes through all the semiconductor fabrication processes, such as metallization, passivation, and underbump metallurgy (UBM), and the packaging processes, such as UBM and solder bumping. After all these are done, removing the thin wafer from the support wafer (debonding) poses another big challenge.

5.3   Adhesive Is the Key

Adhesive is the key-enabling material for thin-wafer handling, and how to select adhesive materials for temporary bonding and debonding of thin-wafer handling is the focus of much research [1–22]. The general requirements of the adhesive materials are as follows: (1) after temporary bonding, the adhesive materials should be able to withstand the process environment and thermal budget; (2) during debonding, the adhesive materials should be able to dissolve and clean up easily; and (3) after debonding, there should be no residue or chipping on the thin wafer.

5.4   Thin-Wafer Handling Issues and Potential Solutions

Thin-wafer handling of 200- and 300-mm wafers has been investigated extensively by Industrial Technology Research Institute (ITRI) and is reported in Refs. 3, 4. It has been found that wafer thinning and plasma-enhanced chemical-vapor deposition (PECVD) of SiO_2 in a vacuum chamber are the most critical steps for thin-wafer handling. Thus it is only necessary to qualify an adhesive with these two conditions: PECVD and wafer thinning.

Also, it has been found that no obvious change or delamination occurred in all the chemical resistance tests in Ref. 3 for different adhesives. Thus, with consultation of the chemical properties of adhesives provided by the vendors, thin-wafer handler can skip this test in selecting adhesives for temporary bonding and debonding. The total-thickness-variation (TTV) performance of composite wafers with thinner adhesive has been found [3] to be much better than that with thicker adhesive (100 μm). Pregrinding the Si carrier could be a short-term solution to compensate for the original poor TTV. Also, it has been found in Ref. 3, because infrared (IR) truly, directly, and precisely "looks" at the thin-wafer layer, that the results of ISIS measurement are more convincing than those by contact-gauge measurement. Also, temporary bonding and debonding of 300-mm-blanket-thin wafers with thicknesses of 50 and 20 μm have been demonstrated successfully in Ref. 3.

Table 5.1 summarizes the thin-wafer handling issues and their potential solutions for the device/interposer wafer, carrier (support) wafer, temporary bonding, thinning, backside processes, debonding, and assembly [4]. It is based on the research work performed in Refs. 3, 4 and new results and the characterization data shown in the sections that follow.

5.4.1   Thin-Wafer Handling of 200-mm Wafers

In the beginning, to assess the temporary bonding adhesive, a 200-mm blanket wafer was temporarily bonded on a supporting wafer, that is,

Check List	Issues	Potential Solutions
Device/ interposer wafer	Edge cracking or chipping after grinding	• Edge trimming. Edge trim depth = Target wafer thickness + 100 µm. Edge trim width = 1 mm • Use a carrier with larger diameter (201 mm or 301 mm)
	Temporary adhesive thickness	• Adhesive thickness > bump height + 20~50 µm
	Temporary adhesive outgassing	• Increase baking temperature and time • Use solvent-free materials
	Voids from the outgassing of the wafer	• Pre-baking at 150°C for 30 min before thin-wafer handling material coating • Add SiN on Oxide to block gas
Carrier (supporting) wafer	Carrier wafer: Si substrate	• Backside polish is needed for backside IR alignment
	Carrier wafer: Glass substrate	• Need to consider the capability of the electrostatic chuck
Temporary bonding	Thin-wafer handling material is too thick and thus overflow after bonding	• Reduce bonding force • Fine-tune edge trim recipe • After coating, add edge bead rinse process
	TTV is too large after bonding	• Improve coating uniformity • Increase bonding force and temperature • Pre-grind the carrier wafer to compensate the original poor TTV
Thinning	Copper contamination after back grinding for TSV Cu protrusion	• Copper protrusion by dry etching • Copper protrusion by Si wet etching
Backside processes	Process temperature limitations on passivation	• Develop higher thermal stability materials • Low temperature process, e.g., polymer passivation instead of PECVD SiO_2
	Missing bumps due to seed layer etching with large undercut	• Reduce seed layer thickness to reduce undercut
	Sn is etched during backside seed layer etching process	• For Cu/Sn microbump structure, typical H_2So_4/H_2O_2 solution may damage Sn, thus a new inhibitor is needed
Debonding	Thin wafer cracks due to film tape shrinkage during chemical cleaning process	• Use good chemical resistance film tape • Protect the film tape during chemical cleaning process
	Delamination between film tape and device	• Use higher adhesion strength film tape during mechanical lift-off debonding
Assembly	Microbump/UBM/RDL peeling during chip or interpose removing from the film tape	• Use UV cured and lower adhesion strength film tape • Increase the adhesion strength of the UBM, microbump, and RDL during backside processes

TABLE 5.1 Thin-Wafer Handling Issues and Potential Solutions

a carrier, with a specific adhesive that was about 20 µm thick. Then backside grinding and a CMP process were conducted to thin down the blanket wafer to 50 µm. Chemical resistance and microbump process flow were tested with step-by-step examinations on the thin composite wafer.

Wafer Thinning after Temporary Bonding

In the very beginning of thin-wafer handling, it was inevitable to examine the result after temporary bonding and following backside grinding. Figure 5.1 shows composite wafers after thinning with adhesive A and adhesive B_v.1, respectively. No cracking or chipping occurred around the entire wafer edge by coating with adhesive A, whereas obvious chipping showed at the wafer edge by coating with adhesive B_v.1. It is important for temporary bonding to use a strong enough adhesion to be supported by the carrier. After strengthening the adhesion of adhesive B, called adhesive B_v.2, the chipping issue was completely overcome, and the wafer looked intact around the edge, as shown in Fig. 5.1. Hence adhesion strengthening of the material is undoubtedly an effective way to ensure that no cracking or chipping occurs on the thin wafer.

Debonding Results

This section discusses the debonding results for 200-mm composite wafers, including a layer of adhesive A at about 20 µm on a thin wafer that is 50 µm thick. This test focuses on various tapes (A, B, C, and D)

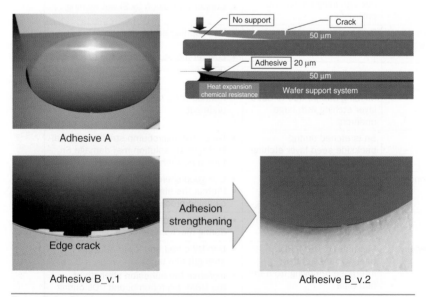

Adhesive A

Adhesion strengthening

Edge crack

Adhesive B_v.1

Adhesive B_v.2

FIGURE 5.1   Blanket-bonded wafers after back-grinding with adhesive A, adhesive B_v.1, and adhesive B_v.2.

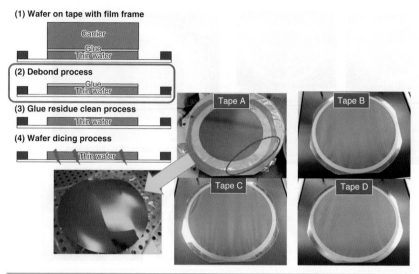

(1) Wafer on tape with film frame

Carrier
Glue
Thin wafer

(2) Debond process
Glue
Thin wafer

(3) Glue residue clean process
Thin wafer

(4) Wafer dicing process
Thin wafer

Tape A
Tape B
Tape C
Tape D

FIGURE 5.2   Schematic flow of debonding with a film frame. For tape A, wafer cracks were seen a few hours after debonding.

for composite wafer mounting on a film frame, and a brief schematic flow is shown in Fig. 5.2. The third step is a cleaning process to remove the remaining glue residue. Care must be taken when selecting the tape; for example, tape A interacted with the cleaning solution and became wavy in places, as shown in Fig. 5.2, which induced nonuniform stress around the whole wafer and automatically caused cracks a few hours later, as shown in Fig. 5.2 (*right*).

In the following, the interaction of four kinds of tape from A to D with the cleaning chemical solution was assessed, as shown in Fig. 5.3. In addition to tape A, mentioned above, the wavy-surface phenomenon also occurred for tapes B, C, and D, but tape D exhibited the least amount of waviness, representing the least interaction with the cleaning solution.

Actually, tapes A, B, and C are the normal dicing tapes, whereas tape D is a back-grinding tape. This type of grinding tape is usually used for dicing despite its higher cost than normal dicing tape.

Therefore, tape D was chosen for debonding use in the following integration process because this type of tape ensures the surface flatness and avoids wafer cracking, as with tape A. Figure 5.3 shows the crack-free experiment with tape D for a shelf life of 96 hours. Furthermore, adhesive A was selected to go through an integrated evaluation on 200-mm wafers at a thickness of 50 µm with active complementary metal-oxide-semiconductor (CMOS) circuits. After successful coating with adhesive A, temporary bonding, thinning, and other packaging processes, the composite wafer proceeded to the final step: debonding.

FIGURE 5.3   Assessment of the interaction with cleaning chemical solution of four kinds of tape. Tape D provides a flat and crack-free surface.

Figure 5.3 is the result just after debonding on tape D with a film frame, and the figure shows a crack-free and chipping-free wafer after glue cleaning, although it is a little wavy of the tape surface. Compared with the reported debonding example, which is a 200-mm wafer around 70 µm thick that was debonded by laser ablation technology [8], this study demonstrated an effective debonding method without high-cost glass handler wafers and will be evaluated further for 300-mm-thin wafers in this study.

Chemical Resistance Test after Temporary Bonding

Thin composite wafers were cut into several quarter pieces for a chemical-resistance test. The test conditions and their corresponding results are listed in Fig. 5.4, including adhesive A and adhesive B_v.2. The purpose of this test is to check whether the chemical solutions usually used in packaging assembly processes would affect the adhesive material or not. The figure shows a photo and the weight of the original state, as well as those after testing by developing, photoresist (PR) stripping, etching, and plating chemicals. For both adhesive A and adhesive B, no obvious change or delamination occurred from all the chemical tests despite a slightly weight gain (<5 percent of the original) after electroplating copper or tin. Therefore, both kinds of adhesives enable common packaging processes. With consultation of the chemical-resistance properties of adhesives provided by the vendors, thin-wafer handler can skip this test procedure in selecting adhesives for temporary bonding and debonding.

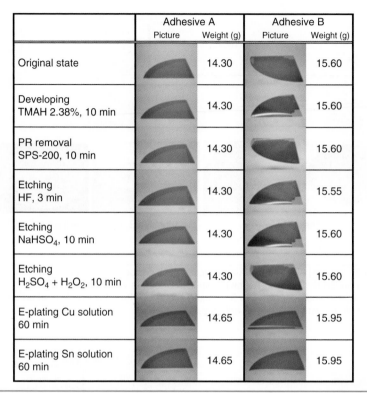

	Adhesive A		Adhesive B	
	Picture	Weight (g)	Picture	Weight (g)
Original state		14.30		15.60
Developing TMAH 2.38%, 10 min		14.30		15.60
PR removal SPS-200, 10 min		14.30		15.60
Etching HF, 3 min		14.30		15.55
Etching NaHSO$_4$, 10 min		14.30		15.60
Etching H$_2$SO$_4$ + H$_2$O$_2$, 10 min		14.30		15.60
E-plating Cu solution 60 min		14.65		15.95
E-plating Sn solution 60 min		14.65		15.95

FIGURE 5.4   Adhesive evaluation of backside microbumping process. There is no delamination for any of the chemical solutions, and the weight change (from the original) is less than 5 percent.

Adhesive Evaluations of Backside Micro Wafer Bumping

In addition to chemical-resistance test, thin composite wafers went through a typical microbump process flow, as shown in Figs. 5.5 to 5.9, sequentially including PECVD, sputtering, patterning, electroplating, PR stripping, and reflow. In general, the dielectric quality and surface conformality of most plasma-deposited insulators increases significantly as the temperature increases from 200 to 300°C [9]. Hence the temperature of PECVD was set at 250°C to deposit 1-μm-thick silicon dioxide, and the process time was around 10 minutes. Titanium at 0.05 μm thick and copper at 0.12 μm thick then were sputtered at around 170°C onto the blanket oxide wafer. After developing the specific mask, copper at 3 μm thick and tin at 4 μm thick were electroplated, and finally, the microbumps at 20-μm diameter and 100-μm pitch formed by reflow at about 240°C. Except for PECVD, sputtering, and reflow at high temperature, all the other processes were conducted at room temperature. Moreover, only PECVD and sputtering should be conducted in vacuum chamber, as indicated in Fig. 5.5.

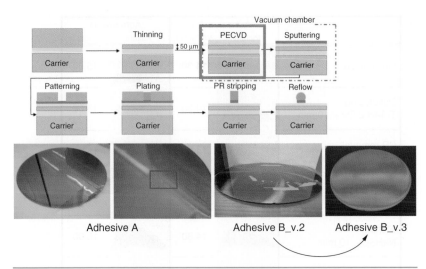

Figure 5.5 (*Above*) A typical backside microbumping process flow for testing 200-mm bonded thin wafers. (*Below*) Post-PECVD results with adhesive A, adhesive B_v.2, and adhesive B_v.3. Dishing occurred with adhesive A (marked by red lines). Wafer was broken with adhesive B_v.2

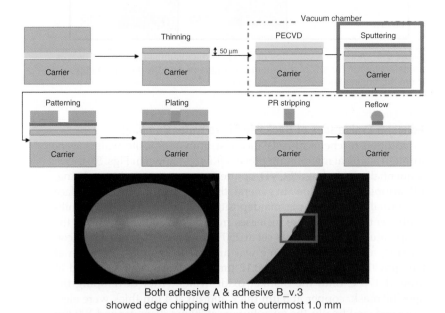

Figure 5.6 Post-sputtering results. Edge chipping occurred (within the outermost 1.0 mm) with both adhesive A and adhesive B_v.3.

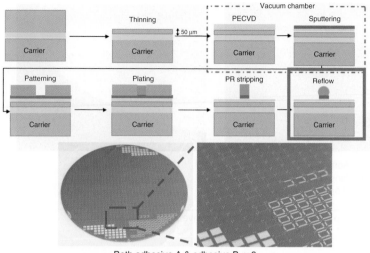

Both adhesive A & adhesive B_v.3
passed even if high-temperature solder was reflowing

FIGURE 5.7   Post-reflow results. High-temperature solder reflow (for microbump) results. Both adhesive A and adhesive B_v.3 passed.

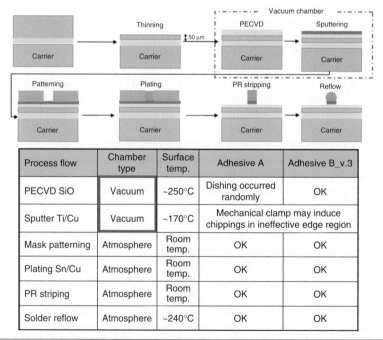

Process flow	Chamber type	Surface temp.	Adhesive A	Adhesive B_v.3
PECVD SiO	Vacuum	~250°C	Dishing occurred randomly	OK
Sputter Ti/Cu	Vacuum	~170°C	Mechanical clamp may induce chippings in ineffective edge region	
Mask patterning	Atmosphere	Room temp.	OK	OK
Plating Sn/Cu	Atmosphere	Room temp.	OK	OK
PR striping	Atmosphere	Room temp.	OK	OK
Solder reflow	Atmosphere	~240°C	OK	OK

FIGURE 5.8   Summary of the backside microbumping process on 200-mm bonded wafers.

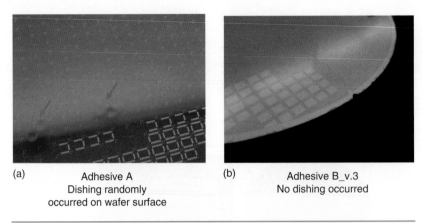

(a)   Adhesive A
Dishing randomly
occurred on wafer surface

(b)   Adhesive B_v.3
No dishing occurred

FIGURE 5.9   Postplating magnification of adhesive A and adhesive B_v.3.

As for adhesive A, no chipping or cracking occurred at the wafer edge after the PECVD process, as shown in Fig. 5.5. However, some dishings as large as chip size (5 × 5 mm) occurred randomly at the wafer surface, as indicated in Figs. 5.5 and 5.9. It is presumed that outgassing from the adhesive layer induced this kind of dishing and that it happened only at high temperatures (250°C) in the vacuum chamber. Based on previous experience, the same adhesive A was able to withstand 180°C in a PECVD process, and no dishing occurred. For adhesive B, the second version provided stronger adhesion and hence no chipping or cracking after thinning, but the wafer was severely broken in the 250°C PECVD process, as shown in Fig. 5.5. The resistance of adhesive B kept improving, and the third version showed a slight decrease in hardness, but it still has enough strength to withstand the stress of backside grinding. After the PECVD process, the composite wafer was much better than the previous one, as shown in Fig. 5.5. No chipping or cracking occurred around the entire edge of the wafer, so adhesive B_v.3 enables PECVD processes at temperatures as high as 250°C in a vacuum chamber.

In the following sputtering process (also in a vacuum chamber), the results for adhesive A and adhesive B_v.3 looked the same, as shown in Fig. 5.6. Both were conducted at around 170°C according to temperature detection by thermal dots pasted on the wafer surface. However, edge chipping appeared within the outermost 1.0 mm for both kinds of adhesives, as shown in Fig. 5.6. The tiny edge chipping may be induced by the mechanical clamp of the tool in the process, but they occurred only in the outermost ineffective region and didn't propagate or enlarge at all. Thus this process step, that is, sputtering, basically can be ignored in selecting a temporary bonding adhesive material.

From the lithography process step to the final reflow for micro-bump formation, all the processes were conducted in atmosphere at room temperature. No matter which was used, adhesive A or adhesive B_v.3, the results were excellent, and no defects appeared in each step. Figure 5.7 shows the post-reflow result, with the micro-bumping process completed with adhesive A. Figure 5.8 summarizes the chamber type (vacuum or atmosphere), wafer surface tempera-ture, and a couple of short comments for each process step of backside microbumping.

The key difference between adhesive A and adhesive B_v.3 is the dishing phenomenon in the PECVD process, as shown and compared in Fig. 5.9. A couple of dishings can be observed randomly on the composite wafer surface with adhesive A, as the arrows indicate in Fig. 5.9a, whereas not at all for adhesive B_v.3, as shown in Fig. 5.9b. The dishing can be seen to be almost as large as a chip in Fig. 5.9a. On the other hand, the vertical gap ranges from less than 1 μm to over 9 μm, the deepest part located at the center of dishing. As for the thin adhesive layer (20 μm), this dishing phenomenon is significant and must influence the uniformity and thin-wafer handling results. With a nondestructive test technique called *scanning acoustic microscopy* (SAM), the dishing phenomenon can be confirmed where imperfect temporary bonding occurred and then revealed in the PECVD pro-cess as shown in Ref. 3.

Brief Summary of 200-mm Wafer Studies

To give a short summary of the 200-mm blanket wafer test, the chem-ical properties of temporary bonding adhesives can be found in mate-rials provided by the vendors. Thus the chemical-resistance test can be skipped. For typical packaging processes, only PECVD conducted in a vacuum chamber is the critical step in selecting adhesives. For all the other processes conducted in atmosphere, even for solder reflow at high temperatures (240°C), the tolerance of adhesives is found to be much wider and thus hardly affects the test results. The adhesives can be made from various polymers, and their thermal stability is a very important consideration in selection, especially after high-temperature processing [9, 11]. The thermal stability of temporary adhesives relates to their ability to resist decomposition and outgas-sing [12]. Some adhesive has been reported to be able to withstand up to 280°C in PECVD [9], but others, such as adhesive A, have not. In addition to the thermal issue mentioned earlier, the vacuum environ-ment is found to be another key factor because the process solder reflow is really okay in this microbumping test, which was in atmo-sphere and at 240°C high temperature.

Therefore, to quickly select a suitable temporary bonding adhe-sive, PECVD is suggested to be the only step to test just after the thin-ning process. In the short term, we suggest a polymer isolation process, for example, polyimide (PI) or poly-benzoxazoles (PBO), as an alternative way to avoid the dishing issue of adhesive A. That's because normal polymer isolation is conducted in atmosphere instead of PECVD in vacuum. Besides, the polyimide base materials are well-known for their high thermal/chemical resistance for thin wafer processing and handling [7]. In the long-term, good thermal stability of temporary adhesives are needed.

5.4.2   Thin-Wafer Handling of 300-mm Wafers

After thin-wafer handling material evaluation with 200-mm wafers, the same two types of adhesives were evaluated on 300-mm wafers. Based on the previous studies on 200-mm wafers, it has been suggested that chemical tests on 300-mm wafers can be skipped, but the thin-wafer handling materials still go through critical processes, for example, PECVD. In this study, three different structures of the 300-mm wafers are considered: (1) blanket wafers, (2) wafers with 80-μm solder bumps, and (3) wafers with TSVs and microbumps. Here, prior to temporary bonding and thinning processes, edge trimming along a width of 500 μm and a depth of 100 μm was conducted in advance because it reportedly could improve the thinning quality and prevent edge cracking or chipping [10].

Material Evaluation of Blanket Wafers

In the 3D IC process, one has to control the total thickness variation (TTV) after the thinning process. The TTV results from deviations in many factors, such as adhesive coating uniformity, the bonding process, and the thinning process. Since the wafer size is increased from 200 to 300 mm, the TTV is supposed to increase as well. For thin-wafer handling material evaluation on blanket wafers, the wafers were thinned down to 50 and 25 μm.

Figure 5.10 shows the TTV data based on contact-gauge measurement before the thinning process. The thicknesses of blanket thin wafer, thin-wafer handling material, and supporting carrier are all included and measured together. For adhesive A, the TTVs for both 25- and 50-μm wafers are less than 3 μm. For adhesive B_v.3, the TTV for 50-μm wafers is around 5 μm and for 25-μm wafers is 4.2 μm. Based on these measurement data, adhesive A obviously has better TTV performance before thinning. This advantage perhaps comes from different material characteristics and the bonding process. Adhesive A was bonded at elevated temperature and pressure, but adhesive B_v.3 was bonded at room temperature with very low bonding force. In addition, Young's modulus of adhesive A is also stronger than that of adhesive B_v.3.

The contact-gauge measurement result includes the thicknesses of the device wafer, adhesive layer, and supporting carrier. For the following stacking process in 3D IC package process integration,

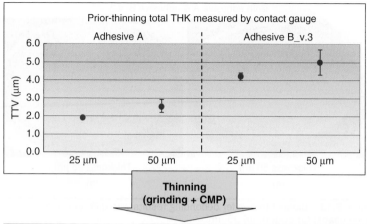

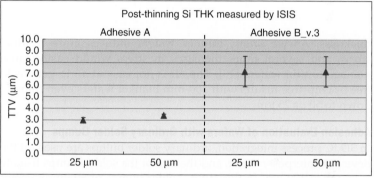

FIGURE 5.10 (*Above*) Total thickness variation (TTV) before the thinning process via contact-gauge measurement. (*Below*) TTV after the thinning process via noncontact infrared (IR) measurement.

measuring only one of these thicknesses of thin device wafers is more important and accurate and, in fact, closely addresses the real consideration. After wafer bonding and thinning, noncontact thickness measurement by infrared light (IR) was conducted, as shown in Fig. 5.10. Based on the IR measurement data, only the thickness of thin blanket wafer was measured. The TTV data show that the TTV of adhesive A is much smaller than that of adhesive B_v.3.

After all the backside processes were finished on the thinned device wafer, the wafer was debonded. The debonding methods are different for the various adhesives. Both adhesive A and adhesive B_v.3 were debonded at room temperature. The debonded blanket wafers showed no chipping or cracking for both the 50- and 25-μm wafer thicknesses. Figure 5.11 illustrates the debonding results for the (1) 50-μm wafer thickness and (2) the 25-μm wafer

Both adhesive A & adhesive B_v.3

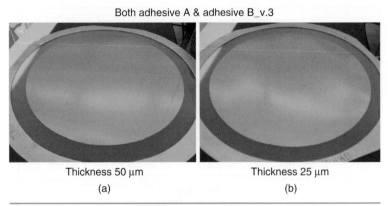

Thickness 50 μm   Thickness 25 μm

(a)   (b)

FIGURE 5.11   Blanket thin-wafer debonding results with different wafer thicknesses: (a) 50 μm; (b) 25 μm.

thickness. As we know, thinner wafers induces more residual stress, and several times increases in stress would easily cause damage to thin wafers [13]. Thus it is not easy at all to get crack-free or chip-free results with 300-mm blanket wafers even thinned down to 50 or 25 μm.

Adhesion Evaluation of Wafers with Ordinary Solder Bumps

In 3D IC process integration, ordinary solder bumps are used on the device or interposer wafer, and the size of the solder bumps usually is less than 100 μm. In this study, solder bumps of 80-μm diameter were fabricated and tested for thin-wafer handling. For the wafers with 80-μm solder bumps, the thickness of the adhesive must be more than 80 μm to fully protect the solder bumps. In order to protect the solder bumps, 100-μm-thick adhesives were adopted to cover solder bump deviations. Based on previous studies, chemical testing is skipped for thin-wafer handling material evaluation. Only the critical process of PECVD of SiO_2 was tested on the wafers with 80-μm solder bumps.

Figure 5.12 demonstrates the results after PECVD of wafers with 80-μm solder bumps with adhesive A. The critical process condition here is PECVD at 180°C. After this process was finished, many dishings were found on wafer surfaces. These dishings may be driven by the outgassing of adhesive A on a scale similar to that seen with 200-mm wafers.

Figure 5.13 shows the inspection result after PECVD under the same process conditions for adhesive B_v.3. Very large voids and delaminations were observed along the outer area of the wafer surface. These phenomena also may result from severe outgassing of adhesive B_v.3. Based on the scanning electron microscope (SEM) image of this bonded wafer, the abnormal brightness of the outer area

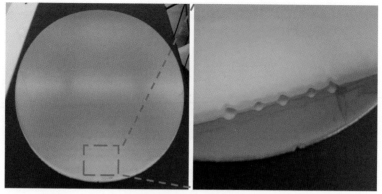

Adhesive A: Dishing still occurred; outgassing is a probable cause
Bump height: 80 μm
Adhesive thickness: 100 μm

FIGURE 5.12 Post-PECVD debonding results for wafers with 80-μm solder bumps for (100-μm-thick) adhesive A. Dishing (the pattern is around the wafer edge) still occurred, and outgassing is a probable cause.

FIGURE 5.13 Post-PECVD inspection of a wafer with 80-μm solder bumps for (100-μm-thick) adhesive B_v.3. Large voids and delaminations were observed along the outer area of the wafer surface.

corresponds to the existence of voids or delaminations. More research is needed to improve thin-wafer handling with ordinary solder bumps.

Both adhesive A and adhesive B are debonded at room temperature. Figure 5.14 shows the debonding result for adhesive B_v.3 on a 300-mm wafer with 80-μm solder bumps. No chipping or cracking was found after the debonding. The figure shows that there is no residue remaining after debonding and cleaning for adhesive B_v.3 on a wafer with 80-μm solder bumps. This also applies to adhesive A.

FIGURE 5.14   Debonding results on a wafer with 80-μm solder bumps for (100-μm-thick) adhesive B_v.3. No residue was found after debonding and cleaning of the wafer with 80-μm solder bumps.

Material/Process Evaluation of Wafers with TSVs and Microbumps

Up to the present, we have established that adhesive A has dishing issues during PECVD, whereas adhesive B_v.3 showed voids and even delaminations for ordinary solder bumps. TTV performance after thinning is obviously better for wafers bonded with adhesive A than adhesive B_v.3. Both kinds of thin-wafer handling adhesives seem to have pros and cons. Because adhesive A performs much better for TTV and the dishing issue scarcely affected the debonding results, it is selected for process evaluation on wafers with TSVs and microbumps (25 μm or less). The TSVs here have diameters of 10 μm, pitches of 40 to 50 μm, and depths of 50 μm. Above the TSVs, there are many microbumps with diameters of 20 μm and pitches of 40 to 50 μm. The adhesive thickness is about 50 μm.

After temporary bonding, the wafer was then thinned down to 50 μm. Figure 5.15 shows the contact-gauge TTV measurement data after thinning with microbumps and TSVs. The TTV can be controlled around 1 μm, which is an improvement over the reported glass-bonded wafer TTV of 2.9 μm [14].

In comparison with the previous TTV measurement result for adhesive A, wafers with 20-μm microbumps were better than those with 80-μm solder bumps. If a large solder bump were used, the adhesive layer needs to be thick enough. Generally speaking, uniformity is very hard to control with very thick films.

Figure 5.15 illustrates the critical process of checking for adhesive A of the wafer with 20-μm microbumps and TSVs. The critical process here is also PECVD at 180°C. Many dishings are observed after the PECVD process. The dishing area is around 1 to 2 and 3 cm from the wafer edge. The dishing mechanism is suspected to be outgassing of the temporary bonding material. This issue may be solved by modifying the coating and baking recipe. If the material is not fully baked after the coating process, it would induce outgassing when the process temperature is raised. This phenomenon will

ID/position	1	2	3	4	5	6	7	8	9	TTV	AVG
1	846.0	846.0	845.9	846.2	846.3	846.2	847.0	847.0	846.9	1.1	846.4
2	847.6	847.8	848.0	847.9	847.5	847.6	848.1	848.3	848.0	0.8	847.9
3	848.9	849.3	849.2	848.3	849.1	849.4	848.7	849.0	849.0	1.1	849.0
4	845.5	845.6	845.7	845.6	845.4	845.4	846.0	846.0	845.8	0.6	845.7

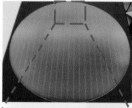

Adhesive A

Microbump size: 20 µm
TSV diameter: 10 µm
Adhesive thk: ~50 µm

Adhesive A

FIGURE 5.15 Post-microbumping results with adhesive A. TSV diameter = 10 µm; microbump size = 20 µm; and adhesive thickness = 50 µm. Good TTV control. Dishing (1–2 cm around wafer edge) still occurred.

become more serious and obvious especially in vacuum chambers at elevated temperature.

Figure 5.15 also shows the debonding results for adhesive A on a wafer with 20-µm microbumps and TSVs. The thickness of the 300-mm wafer is about 50 µm. After debonding, no chipping or cracking was observed. The debonded wafer was handled with a film frame for handling and shipping. There is no residue after debonding and cleaning for adhesive A on wafers with 20-µm microbumps and TSVs [3].

5.5 Effect of Dicing Tape on Thin-Wafer Handling of Wafers with Cu/Au Pads

Figure 5.16 shows the wafer under consideration. The wafer's thickness is 720 µm, and it has many 5- × 5-mm chips. The pads (30 × 30 µm) on the chip are electroplated with Cu and immersed in Au. A very-high-adhesive-strength ultraviolet (UV) tape (film; ~120 µm thick) in a dicing ring is attached on the surface of the wafer. Then the wafer is back-ground to 50 µm. After that, the wafer is diced, and the tape is cut to a depth of 10 to 20 µm. UV exposure equipment is used to dissolve

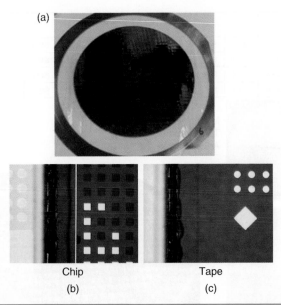

(a)

Chip
(b)

Tape
(c)

FIGURE 5.16   (a) Wafer with 5- × 5-mm chips in a dicing ring with a dicing tape. (b) A chip with missing Au on the Cu pads. (c) The tape with Au immersion.

the tape and reduce the adhesive strength between the chip and the tape. Finally, the tape is peeled off from the chip (see Fig. 5.16) [4].

It can be seen from Fig. 5.16 that given the very high adhesive strength of the dicing tape, some of the immersion Au has been peeled off from the chip pads to the tape. This could be the result of (1) the immersion Au process not being done properly, (2) the adhesive strength of the tape being too great; or (3) not enough UV energy to dissolve the tape. Further experiments involving doubling of the UV time still lead to the same results. Double-checking the immersion Au tank doesn't show anything unusual. Thus the conclusion is that the tape has too great an adhesive strength.

5.6   Effect of Dicing Tape on Thin-Wafer Handling of Wafers with Cu-Ni-Au UBMs

Figure 5.17 shows the wafer under consideration. The chip sizes are 5 mm × 5 mm × 720 μm, and the TSVs have a diameter of 30 μm and a pitch of 60 μm. The processes for making the TSVs are shown in Fig. 5.18, and there are six important steps: (1) via formation by deep reactive ion etch (DRIE), (2) SiO_2 deposition by PECVD, (3) barrier and seed layer deposition by physical vapor deposition (PVD), (4) Cu plating to fill the TSV vias, (5) CMP of Cu plating residues (overburden), and (6) Cu reveal as shown in Fig. 5.18.

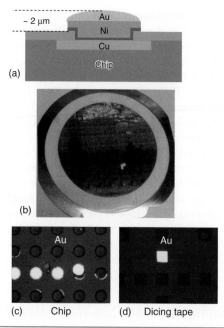

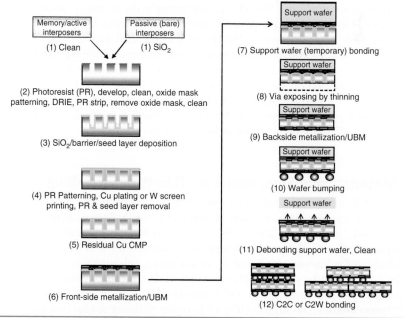

FIGURE 5.17 (*a*) Cu-Ni-Au UBM. (*b*) Wafer with CMOS chip with Cu/Ni/Au UBM and TSVs (30-μm diameter) on 60-μm pitch. (*c*) A chip with missing Au on the Cu-pads and (*d*) The tape with the immersion Au.

FIGURE 5.18 3D IC integration TSV process.

The front-side metallization/UBM can be made on the blind TSV wafer, which is 720 μm thick. Then the TSV wafer is attached to a 720-μm-thick carrier (support) wafer with Brewer Science 9001A adhesive material. As soon as the support wafer is temporarily bonded to the blind TSV wafer, it is thinned down to approximately 50 μm to expose the TSVs (Cu revealing). Then this is followed by backside metallization and UBM (see Fig. 5.18), which is electroplated Cu, electroless Ni, and immersion Au, as shown in Fig. 5.17.

A very-high-adhesive-strength UV (film) tape (~120 μm thick) in a dicing ring is attached on the surface of the wafer with the Cu-Ni-Au UBMs. Then the approximately 50-μm-thick wafer is diced, and the tape is cut to a depth of 10 to 20 μm. UV exposure equipment is used to dissolve the tape and reduce the adhesive strength between the chip and the tape. Finally, the tape is peeled off the chip (see Fig. 5.17). Again, it can be seen that some of the Au immersion is peeled off the chip to the UV (film) tape, whose adhesive strength is just too strong for the immersion Au of the UBMs.

5.7   Effect of Dicing Tape on Thin-Wafer Handling of Interposer with RDLs and Ordinary Solder Bumps

The TSV interposer with redistribution layers (RDLs) on both sides and ordinary solder bumps on the bottom, as shown in Figs. 5.19 and 5.20, is created by the process shown in Fig. 5.18 up to step 11. Before debonding, the TSV/RDL interposer with ordinary solder bumps is attached to a high-adhesive-strength dicing tape with a dicing ring. After debonding (mechanical liftoff) and dicing (step 11 of Fig. 5.18), UV exposure equipment is used to dissolve the tape and reduce the adhesive strength between the chip and the tape. Finally, the tape is peeled off the chip, which is shown in Fig. 5.21. It can be seen that some of the RDL traces and even the ordinary solder bumps have been peeled off the chip. Again, the adhesion strength of this particular dicing tape is too great for thin-wafer handling.

5.8   Materials and Equipments for Thin-Wafer Handling

Table 5.2 shows six thin-wafer handling materials made by five material vendors, namely, Thin Materials (T-MAT), Brewer Science (BSI), 3M, DuPont, and Tokyo Ohka Kogyo (TOK). Based on the constituents of each, the process temperature limitations and available adhesive thicknesses are listed in, respectively, columns 2 and 3 of the table. It can be seen that (1) most materials are able to withstand process temperatures up to 250°C (except HT10.10 at 220°C) and (2) most are thick enough (~100 μm) to cover ordinary solder bumps (except DuPont at <20 μm).

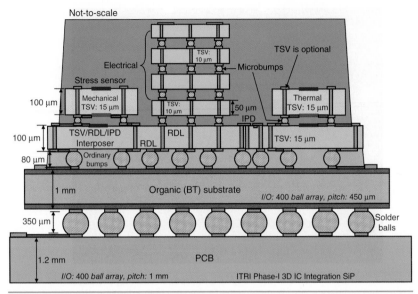

FIGURE 5.19 ITRI's 3D IC integration test vehicle.

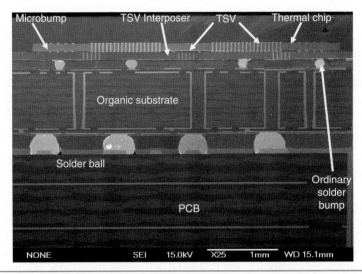

FIGURE 5.20 A cross section of ITRI's 3D IC integration test vehicle.

The carrier (support) wafer plays a very important role in thin-wafer handling. Basically, there are two kinds, namely, Si wafers and glass wafers. In general, a transparent glass wafer is required to serve as the carrier for UV-cured adhesives and light-to-heat-conversion (LTHC) laser release, and it costs more than normal Si carriers.

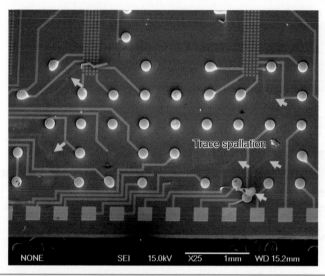

FIGURE 5.21   Bottom side of the TSV/RDL interposer showing the peel-off of traces and ordinary solder bumps owing to the very high adhesion strength of the dicing tape.

The fourth column of Table 5.2 shows the kind of carrier required by each adhesive material.

Equipment plays another key role in thin-wafer handling. Basically, there is equipment for temporary bonding and debonding. The equipment vendors are Electronic Visions Group (EVG), SUSS, Tazmo, and TOK. Usually, TOK equipment is designed only for its adhesives; SUSS equipment can be used for T-MAT, BSI, 3M (without debonder), and DuPont adhesives; EVG equipment can be used for BSI adhesives; and Tazmo equipment can be used for 3M adhesives, as shown in the fifth column of the table.

For temporary bonding of the chip/interposer wafer to the support wafer, all vendors' adhesives have to be applied (via a spin coater) on the chip/interposer wafer, as shown in column 6. For BSI-HT10.10, DuPont, and TOK adhesives, there is no need to apply anything on the carrier wafer. However, for T-MAT adhesive, the carrier requires an elastomer; for BSI9001A adhesive, the carrier requires zone 2 treatment; and for 3M adhesive, the carrier requires an LTHC layer, as shown in column 7 of the table.

For debonding the carrier wafer from the chip/interposer wafer, the operating temperature for all vendors' adhesives is room temperature, except BSI-HT10.10 (~180°C), as shown in column 9 of the table. However, the release methods are different: (1) mechanical (T-MAT), (2) thermal (BSI-HT10.10), (3) solvent (BSI-9001A, TOK), and (4) laser (3M, DuPont), as shown in column 8 of the table.

Materials	Process Temp. Limit (°C)	Adhesive Thickness (µm)	Support Carrier	Equipment (Spin Coating/Bonder/Debonder)	Temporary Bonding		Debonding	
					Device	Carrier	Method	Temp.
T-MAT	~250	20~200	Si/Glass	SUSS	Spin coat Precursor w/plasma	Spin coat Elastomer	(1) Mechanical release (Lift off) (2) Film remain on carrier	Room temp.
B/S (HT10.10)	~220	<100	Si/Glass	EVG SUSS	Spin coat adhesive	—	Thermal release (Slide off)	~180 °C
B/S (9001A) (ZoneBOND)	~250	<120	Si/Glass	EVG SUSS	Spin coat adhesive	Zone 2 treatment	(1) Lift off + solvent release (2) Film remain on device	Room temp.
3M	~250	<125	Glass	Tazmo SUSS w/o debonder	Spin coat adhesive	Spin coat LTHC Bond/UV cure	(1) YAG Laser release (2) Film remain on device	Room temp.
Du-Pont	>350	2–20	Glass	SUSS	Spin coat adhesive	—	Excimer laser release	Room temp.
TOK	~250	<130	Perforated support plate	TOK	Spin coat adhesive	—	Solvent release	Room temp.

TABLE 5.2   Worldwide Thin-Wafer Handling Systems

Different material/equipment vendors provide various temporary bonding and debonding methods that significantly influence material selection, the equipment needed, and choice of silicon/glass carrier. Besides, the thermoplastic/thermosetting properties of the adhesives selected undoubtedly affect the ease of the process, temperature limitations, and final performance of the products' specific features. Each method has its pros and cons, and this is what thin-wafer handling researchers must address so as to select the best adhesive and equipment to meet the products' specific needs.

5.9   Adhesive and Process Guidelines for Thin-Wafer Handling

5.9.1   Some Requirements for Selecting Adhesives

1. It is crucial to know and understand the material properties, characteristics, and limitations of the adhesive during temporary bonding, thinning, back-side metallization, UBM, RDL, and debonding processes.

2. There are two major methods of isolation: PECVD and polymer isolation. Material selection depends on which method is suitable for your products.

3. There must be enough resistance to withstand the chemical process and the thermal budget in vacuum or atmosphere on the thin composite wafers.

4. There must be no Si cracking and chipping after the debonding process.

5. The adhesive materials should be easily removable without any damage and glue remaining on the device wafer and the film tape during the cleaning process after debonding.

5.9.2   Some Process Guideline for Thin-Wafer Handling

Some process guidelines about temporary bonding and debonding in thin-wafer handling are listed in Tables 5.1 and 5.2 that have a significant effect on adhesive selection.

5.10   Summary and Recommendations

- It has been found that wafer thinning and PECVD-SiO_2 in a vacuum chamber are the most critical steps for backside processes in thin-wafer handling. It is only necessary to qualify an adhesive with these two conditions: PECVD and wafer thinning.

- It has been found that no obvious change or delamination occurred in all the chemical resistance tests for different adhesives. Thus, with consultation of the chemical properties of adhesives provided by the vendors, thin-wafer handler can skip this test in selecting adhesives for temporary bonding and debonding.

- The TTV performance of composite wafers with thinner adhesives has been found to be much better than that with thicker adhesives (100 μm). Pregrinding the Si carrier could be a short-term solution to compensate for the originally poor TTV. Good TTV control for thicker adhesives still has to be developed.

- It has been found that because IR truly, directly, and precisely "looks" at the thin-wafer layer, the results of ISIS measurement are more convincing than those by contact-gauge measurement.

- It has been found that both adhesive A and adhesive B successfully demonstrated temporary bonding and debonding on 300-mm blanket-thin wafers with thicknesses of 50 and 25 μm. No cracking or chipping is observed after the debonding process.

- Adhesive B has been shown to be successful in the debonding process on 300-mm wafers with 80-μm solder bumps without chipping and Si cracking, and no adhesive residue remained after the cleaning process.

- By using adhesive A as a temporary bonding material, the 200-mm CMOS thin wafer at 50 μm thick has been successfully debonded without cracking or chipping. Furthermore, the same debonding method and results have been achieved on 300-mm thin wafers with 20-μm microbumps and TSVs (at a depth of 50 μm).

- It has been found that very-high-adhesive-strength UV film tape used on the thin-wafer handling of wafers with Cu-Au pads, Cu-Ni-Au UBMs, and interposer with RDLs and ordinary solder bumps tends to peel off the Au immersion from the pads and UBMs and the traces of the RDL and ordinary solder bumps from the interposer.

- A reasonable-adhesive-strength UV dicing tape is recommended. Also, an increase in UV energy and time helps.

- The critical issues in thin-wafer handling, such as the device/interposer wafer, carrier wafer, temporary bonding, thinning, backside processes, debonding, and assembly, have been presented (in a tabular form) and their potential solutions have been discussed (see Table 5.1).

- The state of the art of materials (Thin Materials, Brewer Science, 3M, DuPont, and TOK) and equipment (SUSS, EVG, Tazmo, and TOK) for thin-wafer handling have been listed (in a tabular form) and examined (see Table 5.2).

- For a specific product, it is recommended that the guidelines for thin-wafer handling are based on a tradeoff (pros and cons) between the critical issues and the materials and equipment shown in Tables 5.1 and 5.2.

5.11   3M Wafer Support System

Figures 5.22 and 5.23 show the new 3M/SUSS wafer support system, which enables conventional back-grinding equipment to be used to produce wafers with a final thickness as low as 20 μm. The key to the 3M system is its ability to provide a rigid, uniform support surface to minimize stress on the wafer as the silicon is removed, resulting in less cracking and chipping. The system includes both the equipment and consumables [3M UV-Curable Liquid Adhesive LC-2201, glass support plate (typically recycled; can be reused many times), and 3M Light-to-Heat Conversion Solution] necessary for mounting, demounting, and removing adhesive from the wafer.

In the 3M system, a glass plate (as shown in Fig. 5.23) is used to support the wafer through the back-grinding process. A UV-curable

Figure 5.22   3M wafer support system.

Figure 5.23   3M wafer support system with a glass supporting wafer.

liquid adhesive is used as the bonding agent between the device wafer and the glass plate (support wafer). After the back-grinding process, as shown in Fig. 5.24, the thinned device wafer is transferred onto a dicing tape, and the support glass is removed by laser debonding of the adhesive-glass interface using a LTHC layer. The adhesive then can be removed from the wafer, leaving behind fewer residues than seen with typical back-grinding tape. This system also works for other semiconductor and packaging processes as long as the thermal-expansion mismatch between the device wafer and the glass plate is within the allowable tolerance.

5.12   EVG's Temporary Bonding and Debonding System

EVG and Brewer Science have developed a solution that enables temporary bonding of a device wafer to a right carrier substrate (support wafer) and allows not only thinning but also a full range of subsequent processes, including high-temperature deposition, etching, lithography, dielectric application and curing, plating, and chemical cleaning, as shown in Figs. 5.25 and 5.26.

5.12.1   Temporary Bonding

The front side of the device wafer and the support wafer are coated with WaferBOND, an HT Brewer Science adhesive material, using the coating chamber (shown in Fig. 5.27), which includes spin- and

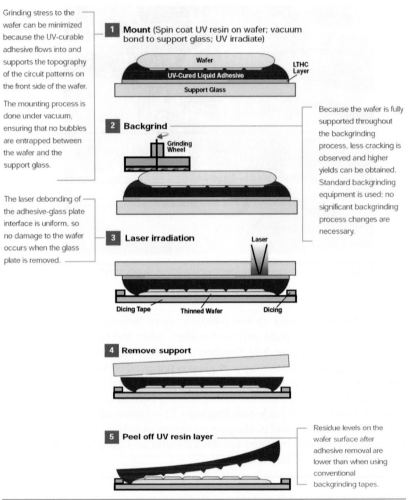

Grinding stress to the wafer can be minimized because the UV-curable adhesive flows into and supports the topography of the circuit patterns on the front side of the wafer.

The mounting process is done under vacuum, ensuring that no bubbles are entrapped between the wafer and the support glass.

The laser debonding of the adhesive-glass plate interface is uniform, so no damage to the wafer occurs when the glass plate is removed.

1 **Mount** (Spin coat UV resin on wafer; vacuum bond to support glass; UV irradiate)

Wafer
LTHC Layer
UV-Cured Liquid Adhesive
Support Glass

2 **Backgrind**

Grinding Wheel

Because the wafer is fully supported throughout the backgrinding process, less cracking is observed and higher yields can be obtained. Standard backgrinding equipment is used; no significant backgrinding process changes are necessary.

3 **Laser irradiation**

Laser

Dicing Tape   Thinned Wafer   Dicing

4 **Remove support**

5 **Peel off UV resin layer**

Residue levels on the wafer surface after adhesive removal are lower than when using conventional backgrinding tapes.

FIGURE 5.24   3M wafer support system process flow (for wafer thinning).

spray-coat capability. Both wafers then are transferred to a bond chamber (EVG850TB), where they are carefully centered and vacuum bonded at elevated temperatures. Once the device wafer is temporarily bonded to the support wafer, it is ready for backside processing, including back-grinding, etching, metallization, TSV formation, etc.

5.12.2   Debonding

During the debonding process, the thin wafer is debonded first via a thermally activated slide lift-off approach from the support

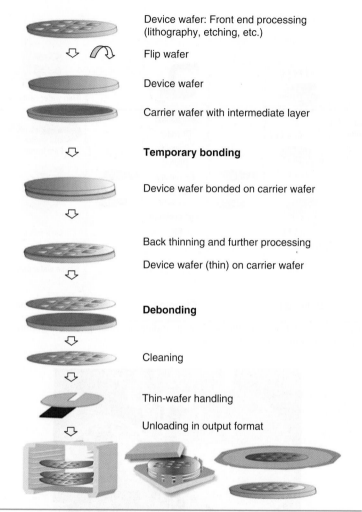

Device wafer: Front end processing
(lithography, etching, etc.)

Flip wafer

Device wafer

Carrier wafer with intermediate layer

Temporary bonding

Device wafer bonded on carrier wafer

Back thinning and further processing

Device wafer (thin) on carrier wafer

Debonding

Cleaning

Thin-wafer handling

Unloading in output format

FIGURE 5.25   EVG's temporary bonding and debonding (of wafers) process flow.

wafer, cleaned in a single-wafer cleaning chamber to remove the remaining adhesive residues, and then transferred to the appropriate output format, such as a film-frame carrier, a dedicated wafer cassette, or a coin-stack packing canister, as shown in Figs. 5.25 and 5.26. The carrier wafer is also cleaned and then can be reused immediately for another debonding process. Recently, EVG introduced their EVG850 "XT Frame" temporary bonding and debonding system with Brewer Science's low temperature ZoneBOND material.

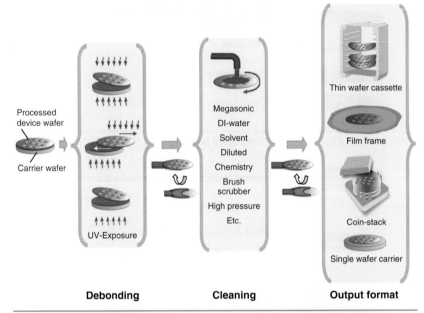

Processed device wafer

Carrier wafer

UV-Exposure

Megasonic
DI-water
Solvent
Diluted
Chemistry
Brush scrubber
High pressure
Etc.

Thin wafer cassette

Film frame

Coin-stack

Single wafer carrier

Debonding **Cleaning** **Output format**

FIGURE 5.26 EVG's debonding process flow, including debonding, cleaning, and outputting the thin wafer in various formats.

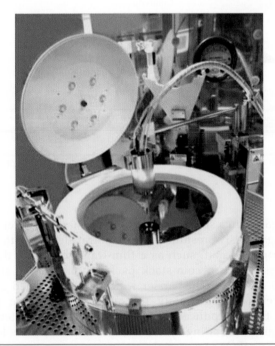

FIGURE 5.27 EVG wafer coating chamber, including spin- and spray-coat capability.

5.13   Thin-Wafer Handling with Carrierless Technology

5.13.1   The Idea

Up to the present, what we have been talking is to use a carrier (support) wafer to attach to the TSV wafer (to increase the bending stiffness and strength) for thin-wafer handling, and the adhesive is the key, with lots of help from advanced equipment. A very innovative method called [6, 15] *carrierless handling* has been developed in which no adhesive is needed (thus it could be very low in cost). With this method, the whole wafer is modified in such a way that the backside of the device/TSV area is thin, whereas the whole wafer is stabilized by an outer rim section. The key idea is to back-grind the inner portion of the wafer and at the same time leave an outer rim section on the wafer. This outer rim section stabilizes the wafer in such a way that no additional support is needed and thus saves process, material, and equipment costs.

5.13.2   The Design and Process

Figure 5.28 shows the process steps, which start from a wafer (~700 μm) that has devices/TSVs on the front side. The wafer is flipped over, and a protective tape or coating is applied to the front side. Then a standard

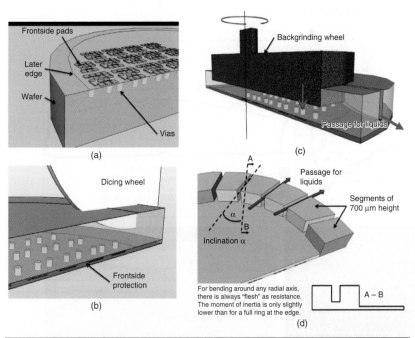

FIGURE 5.28   (a) Initial state of a TSV wafer. (b) Slit generation into the backside.
(c) Wafer during back-grinding. (d) Wafer after slit creation and back-grinding.

dicing saw is used to create a pattern of slits in the back side of the wafer. A key feature of the slit design is the angle of inclination against the radial axis. Following slit generation, back-grinding of the back side of the wafer is accomplished with a special back-grinding tool from the Disco "Taiko" series, which thins only the central portion of the wafer.

Figure 5.29 shows the heart of the design and the finished wafer, and the bending stiffness and strength are demonstrated in Fig. 5.30. It can be seen that sections remain at the outer rim of the wafer. The design of the wafer is such that two conditions are fulfilled [6, 15]:

1. A passage for liquids is created in the outer rim section so that process liquids can leave (be removed from) the wafer in a controlled manner.

2. The geometric moment of inertia of the complete wafer is still intact. This requirement is fulfilled by the radial inclination of the slits against the radial axis.

5.13.3   Summary and Recommendations

Some important results and recommendations are summarized as follows:

- This is a very low-cost thin-wafer handling process because all the costs for the adhesive, temporary bonding, and debonding are gone.

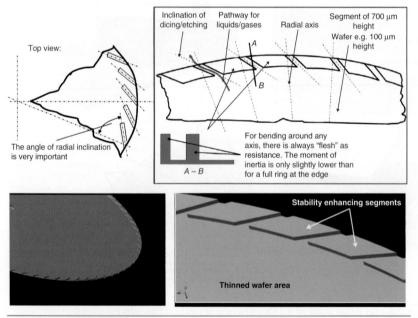

FIGURE 5.29   (*Above*) Angle of inclination for the pathway for liquids/gases. (*Below*) Wafer after back-grinding.

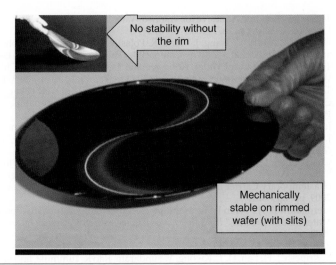

FIGURE 5.30   Mechanical handling of a 200-mm wafer.

- It is a very reliable process because all the nightmares from adhesives such as cleaning residues are gone.

- It is very easy to fully integrate this process into the wet part of the manufacturing line, that is, spinning and related processes.

- More works needed to be done on thinner wafers such as 50 μm or even down to 20 μm for memory-chip stacking and 100 μm for most interposers.

5.14   References

[1] Lau, J. H., *Reliability of RoHS-Compliant 2D and 3D IC Interconnects*, McGraw-Hill, New York, 2011.
[2] Lau, J. H., C. K. Lee, C. S. Premachandran, and A. Yu, *Advanced MEMS Packaging*, McGraw-Hill, New York, 2010.
[3] Tsai, W., H. H. Chang, C. H. Chien, J. H. Lau, H. C. Fu, C. W. Chiang, T. Y. Kuo, Y. H. Chen, R. Lo, and M. J. Kao, "How to Select Adhesive Materials for Temporary Bonding and De-Bonding of Thin-Wafer Handling in 3D IC Integration?" *IEEE/ECTC Proceedings*, Orlando, FL, June 2011, pp. 989–998.
[4] Chang, H., J. H. Lau, W. Tsai, C. Chien, P. Tzeng, C. Zhan, C. Lee, M. Dai, H. Fu, C. Chiang, T. Kuo, Y. Chen, W. Lo, T. Ku, and M. Kao, "Thin Wafer Handling of 300mm Wafer for 3D IC Integration", *44th International Symposium on Microelectronics*, Long Beach, CA, October 2011, pp. 202–207.
[5] Kettner, P., J. Burggraf, and K. Bioh, "Thin Wafer Handling and Processing: Results Achieved and Upcoming Tasks in the Field of 3D and TSV," *11th IEEE Electronic Packaging Technology Conference*, 2009, pp. 787–789.
[6] Bieck, F., S. Spiller, F. Molina, M. Topper, C. Lopper, I. Kuna, T. Seng, and T. Tabuchi, "Carrierless Design for Handling and Processing of Ultrathin Wafers," *60th IEEE/ECTC Proceedings*, 2010, pp. 316–322.
[7] Itabashi, T, and M. P. Zussman, "High Temperature Resistance Bonding Solutions Enabling Thin Wafer Processing (Characterization of Polyimide Base

Temporary Bonding Adhesive for Thinned Wafer Handling)," *60th IEEE/ECTC Proceedings*, 2010, pp. 1877–1880.

[8] Dang, B., P. Andry, C. Tsang, J. Maria, R. Polastre, R. Trzcinski, A. Prabhakar, and J. Knickerbocker, "CMOS Compatible Thin Wafer Processing Using Temporary Mechanical Wafer, Adhesive and Laser Release of Thin Chips/Wafers for 3D Integration," *60th IEEE/ECTC Proceedings*, 2010, pp. 1393–1398.

[9] Tamura, K., K. Nakada, N. Taneichi, P. Andry, J. Knickerbocker, and C. Rosenthal, "Novel Adhesive Development for CMOS-Compatible Thin Wafer Handling," *60th IEEE/ECTC Proceedings*, 2010, pp. 1239–1244.

[10] Justin, W., C. Tai, V. Rao, S. David, D. Fernandez, Y. Li, S. Wen, M. Serene, and J. Lee, "Evaluation of Support Wafer System for Thin Wafer Handling," *12th IEEE/EPTC proceedings*, 2010, pp. 580–584.

[11] Charbonnier, J., S. Cheramy, D. Henry, A. Astier, J. Brun, N. Sillon, A. Jouve, S. Fowler, M. Privett, R. Puligadda, J. Burggraf, and S. Pargfrieder, "Integration of a Temporary Carrier in a TSV Process Flow," *59th IEEE/ECTC Proceedings*, 2009, pp. 865–871.

[12] Hermanowski, J., "Thin Wafer Handling: Study of Temporary Wafer Bonding Materials and Process," *IEEE International Conference on 3D System Integration*, 2009, pp. 1–5.

[13] Zhang, X., A. Kumar, Q. X. Zhang, Y. Y. Ong, S. W. Ho, C. H. Khong, V. Kripesh, J. H. Lau, D.-L. Kwong, V. Sundaram, Rao R. Tummula, and Georg Meyer-Berg, "Application of Piezoresistive Stress Sensors in Ultra Thin Device Handling and Characterization," *Sensors and Actuators A: Physical*, Vol. 156, 2009, pp. 2–7.

[14] Miyazaki, C., H. Shimamoto, T. Uematsu, Y. Abe, K. Kitaichi, T. Morifuji, and S. Yasunaga, "Development of High Accuracy Wafer Thinning and Pickup Technology for Thin Wafer (Die)," *IEEE CPMT Symposium Japan*, August 24–26, 2010, pp. 139–142.

[15] Spiller, S., F. Molina, J. Wolf, J. Grafe, A. Schenke, D. Toennies, M. Hennemeyer, T. Tabuchi, and H. Auer, "Processing of Ultrathin 300-mm Wafers with Carrierless Technology," *IEEE/ECTC Proceedings*, Orlando, FL, 2011, pp. 984–988.

[16] Zhou, S., C. Liu, X. Wang, X. Luo, and S. Liu, "Integrated Process for Silicon Wafer Thinning," *IEEE/ECTC Proceedings*, Orlando, FL, 2011, pp. 1811–1814.

[17] Kwon, W., J. Lee, V. Lee, J. Seetoh, Y. Yeo, Y. Khoo, N. Ranganathan, K. Teo, and S. Gao, "Novel Thinning/Backside Passivation for Substrate Coupling Depression of 3D IC," *IEEE/ECTC Proceedings*, Orlando, FL, 2011, pp. 1395–1399.

[18] Lee, J., V. Lee, J. Seetoh, S. Thew, Y. Yeo, H. Li, K. Teo, and S. Gao, "Advanced Wafer Thinning and Handling for Through Silicon Via Technology," *IEEE/ECTC Proceedings*, Orlando, FL, 2011, pp. 1852–1857.

[19] Halder, S., A. Jourdain, M. Claes, I. Wolf, Y. Travaly, E. Beyne, B. Swinnen, V. Pepper, P. Guittet, G. Savage, and L. Markwort, "Metrology and Inspection for Process Control During Bonding and Thinning of Stacked Wafers for Manufacturing 3D SICs," *IEEE/ECTC Proceedings*, Orlando, FL, 2011, pp. 999–1002.

[20] Jourdain, A., T. Buisson, A. Phommahaxay, A. Redolfi, S. Thangaraju, Y. Travaly, E. Beyne, and B. Swinnen, "Integration of TSVs, Wafer Thinning and Backside Passivation on Full 300-mm CMOS Wafers for 3D Applications," *IEEE/ECTC Proceedings*, Orlando, FL, 2011, pp. 1122–1125.

[21] Knickerbocker, J. U., P. Andry, B. Dang, R. Horton, C. Patel, R. Polastre, K. Sakuma, E. Sprogis, C. Tsang, B. Webb, and S. Wright, "3D Silicon Integration," *58th IEEE/ECTC Proceedings*, 2008, pp. 538–543.

[22] Zhang, X., A. Kumar, Q. X. Zhang, Y. Y. Ong, S. W. Ho, C. H. Khong, V. Kripesh, J. H. Lau, D.-L. Kwong, V. Sundaram, Rao R. Tummula, and Georg Meyer-Berg, "Application of Piezoresistive Stress Sensors in Ultra Thin Device Handling and Characterization," *Sensors and Actuators A: Physical*, Vol. 156, 2009, pp. 2–7.

CHAPTER 6

Microbumping, Assembly, and Reliability

6.1 Introduction

As mentioned in previous chapters, three-dimensional (3D) integrated circuit (IC) integration generally consists of thin chip stacking with through-silicon vias (TSVs) and solder microbumps. Microbump is the third (right after TSV and thin-wafer handling) most important key-enabling technology for 3D IC integration. (It should be noted that 3D Si integration doesn't use solder microbumps because it only integrates all the Si wafers/chips together with TSVs.)

Solders are joining materials for assembling the (1) electronic chips and optoelectronic devices on substrates, (2) electronic and optoelectronic packages/components/modules on printed circuit boards, and (3) 3D IC integration SiPs (systems-in-package). Solders have given us remarkable flexibility and convenience to interconnect electronic and optoelectronic devices/components/modules. The unique properties of solders have facilitated assembly choices that have fueled creative advances in semiconductor assembly developments, for example, 3D IC packaging and 3D IC integration. For these technologies, solder is the electrical and mechanical "glue," and thus solder interconnect assembly and reliability are some of the most critical issues in the development of these technologies.

Green or restriction of the use of certain hazardous substances (RoHS) has been a law in many countries. Among other materials, it bans lead (Pb). In the past 10 years, the electronics industry has spent billions of dollars on compliance with the law. The effort has been directed at eliminating Pb, for example, creating new Pb-free materials, developing high-quality Pb-free soldering processes, performing characterization measurements and reliability assessments of Pb-free solder joints, and so on. In this chapter, the Pb-free solder tin (Sn) is considered.

As mentioned earlier, 3D IC integration is defined [1, 2] as stacking up whatever Moore's law IC chips in the third dimension with

273

TSVs, thin chips or interposers, and solder microbumps to achieve performance, small form factor, low power consumption, wide memory bandwidth, and potentially/eventually low cost. Thus solder microbumps are one of the important enabling technologies for 3D IC integration. The ordinary solder bumps (~100 µm) for flip-chip applications [3, 4] are too big for 3D IC integration SiP applications, which require much smaller solder bumps (≤25 µm) and are called *microbumps*.

The most mature and popular method in wafer bumping of ordinary solder bumps is by electroplating processes [3, 4]. Can we apply the same electroplating method, especially the same process parameters, for wafer bumping of solder microbumps? This is the focus of Part A of this chapter (Secs. 6.2 through 6.7), which involves bumping the 300-mm wafers with Pb-free microbumps. In Part B of this chapter (Secs. 6.8 through 6.11), the focus is on wafer bumping, assembly, and reliability of ultrafine-pitch solder microjoints.

PART A CAN WE APPLY THE WAFER BUMPING METHOD OF ORDINARY SOLDER BUMPS TO SOLDER MICROBUMPS?

6.2 Problem Definition

In this section, the wafer bumping and characterization of fine-pitch lead-free solder microbumps on 300-mm wafers for 3D IC integration are investigated. Emphasis is placed on the Cu-plating solutions (conformal and bottom-up). Also, the amount of Cu and solder (Sn) volumes is examined. Furthermore, characterizations such as the shearing test and aging of the microbumps are provided, and cross sections/scanning electron microscopic (SEM) images of the microbumps before and after testing are discussed. Finally, the process windows of applying the conventional electroplating wafer bumping method of ordinary solder bumps to microbumps are also presented.

6.3 Electroplating Method for Wafer Bumping of Ordinary Solder Bumps

Figure 6.1 shows the electroplating wafer bumping process for ordinary solder bumps such as the controlled collapsed chip interconnection (C4) bumps. Usually the pad size is 100 µm and the target bump height is 100 µm. After redefining the passivation opening (usually it is not required), either Ti or TiW (0.1–0.2 µm) is sputtered over the entire surface of the wafer first, followed by 0.3 to 0.8 µm of Cu. Ti-Cu and TiW-Cu are called *underbump metallurgy* (UBM). In order to obtain a 100-µm bump height, a 40-µm layer of resist then

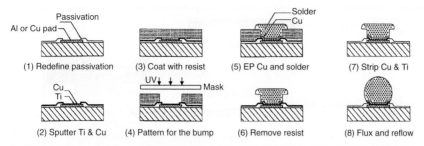

Passivation
Al or Cu pad
(1) Redefine passivation
Cu
Ti
(2) Sputter Ti & Cu
(3) Coat with resist
UV
Mask
(4) Pattern for the bump
Solder
Cu
(5) EP Cu and solder
(6) Remove resist
(7) Strip Cu & Ti
(8) Flux and reflow

FIGURE 6.1   Electroplating processes for wafer bumping of ordinary solder bumps.

is overlaid on the Ti-Cu or TiW-Cu, and a solder bump mask is used to define (ultraviolet [UV] exposure) the bump pattern, as shown in Fig. 6.1. The opening in the resist is 7 to 10 μm wider than the pad opening in the passivation layer. A 6- to 25-μm layer of Cu then is plated over the UBM, followed by electroplating the Pb-free solder. This is done by applying a static or pulsed current through the plating bath with the wafer as the cathode. In order to plate enough solder to achieve the target (100 μm), the solder is plated over the resist coating by about 15 μm to form a mushroom shape. The resist then is removed, and the Ti-Cu or TiW-Cu is stripped off with a hydrogen peroxide etch. The wafer then is reflowed, which creates smooth, truncated, spherical solder bumps owing to the surface tension of the molten solder.

One of the drawbacks of this method is bump height nonuniformity. Usually, owing to the applied electric current at the boundary of the wafer, that is, current density variations across the wafer during the electroplating process, the solder bumps near the edge of the wafer are taller than those near the center of the wafer (e.g., see Refs. 3 and 4).

Can we apply the same wafer bumping processes and parameters to fabricating the mircobumps for 3D IC integration? This is the focus of this investigation.

6.4   Assembly of 3D IC Integration SiPs

There are at least three important steps in assembling a 3D IC integration SiP: (1) microbumping of the Moore's law wafer, (2) fabricating the TSV/redistribution layer (RDL)/integrated passive device (IPD) interposer/chip wafer with UBMs on its top side and either ordinary solder bumps or UBMs on its bottom-side (temporary bonding and debonding the wafer to a supporting wafer usually are required), and (3) Moore's law chip-to-interposer/chip-wafer (C2W) bonding. Figure 6.2 shows some of these processes. The processes for making the TSV/RDL/IPD interposer/chip wafer and wafer bumping have been reported in

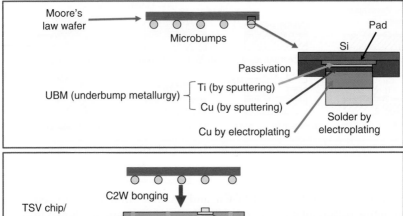

FIGURE 6.2   (*Top*) Solder microbumps on a Moore's law chip wafer.
(*Bottom*) Individual Moore's law chip bonded to the TSV interposer/chip wafer
(C2W) with UBM on its top side and either UBM or C4 bumps on its bottom side.

Refs. 5 through 16 and the C2W bonding process in Refs. 17 through 20.
The focus of this study is on solder wafer microbumping of 300-mm
wafers by the electroplating method with the same UBM (Ti-Cu), solder (Sn), and processes as those (except changing some parameters)
used for ordinary solder bumps [6].

6.5   Electroplating Method for Wafer Bumping of Solder Microbumps

6.5.1   Test Vehicle

In this study, lead-free solder wafer microbumping of 300-mm wafers
with two different chips is investigated. The dimensions and characteristics of these chips are (1) 18×22 mm with more than 7300 pads on
170- and 340-μm pitches (chip 1) and (2) 10×10 mm with more than
2800 pads on 50- and 150-μm pitches (chip 2), as shown in Fig. 6.3.
The passivation opening of both chips is 30 to 35 μm, measured by
the horseradish peroxidase (HRP) and optical microscope (OM)
images, as shown in Fig. 6.4. In order to perform characterization

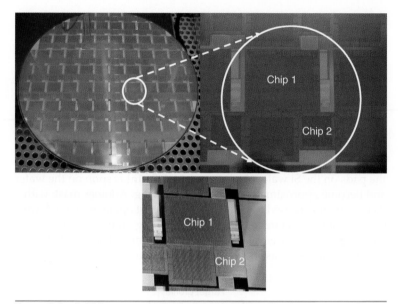

FIGURE **6.3**   A 300-mm test wafer for microbumping. The focus is on chips 1 and 2.

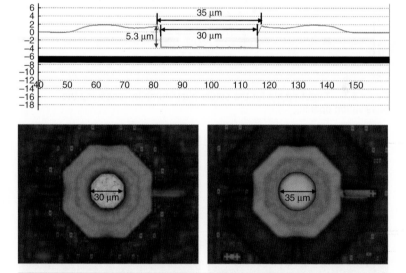

FIGURE **6.4**   Original profile of the passivation opening on the test wafer measured by HRP and OM images.

and reliability assessment of the solder microbumps/joints, both chips have interconnected between pins in an alternating pattern so as to provide a daisy-chain connection when the microbumped chips are soldered to a TSV chip or passive interposer.

6.5.2 Wafer Microbumping of the Test Wafer by Conformal Cu Plating and Electroplating Sn

Figure 6.5 shows the targeted profile of the mircrobumps. It can be seen that the maximum bump size is no larger than 30 μm. Thus the original passivation [first polyimide (PI)] opening has to be modified. In order to do this, a new PI (~4 μm) is coated on top of the original passivation. A mask with a 20-μm opening is used to define the new top passivation opening (~24–25 μm), as shown in Fig. 6.6. A very thin UBM—Ti barrier layer (100 nm) + Cu seed layer (300 nm)—is sputtered by physical vapor deposition (PVD) on top of the new passivation (second PI) and the pads on the Si wafer. A photoresist (~9.5 μm) is spun on the wafer and become approximately 8.5 μm after curing. Another mask with a 25-μm opening is used to define the photoresist opening for conformal Cu plating (11 μm targeted) and electroplating Sn (5 μm targeted), as shown in Fig. 6.7. After plating, photoresist removal, and wet etching, the microbumps are as shown in Figs. 6.8 through 6.10.

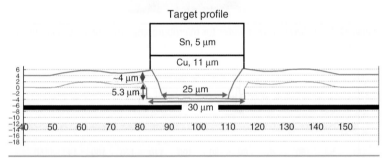

FIGURE 6.5 Target profile for the solder microbump by redefining the passivation opening.

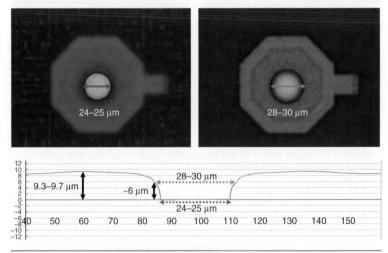

FIGURE 6.6 Passivation opening right after the PI curing.

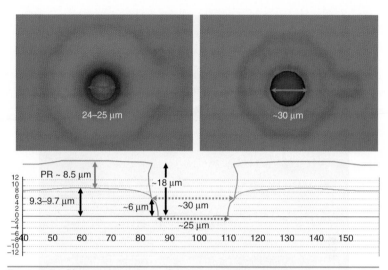

Figure 6.7 Photoresist (on top of the passivation) opening.

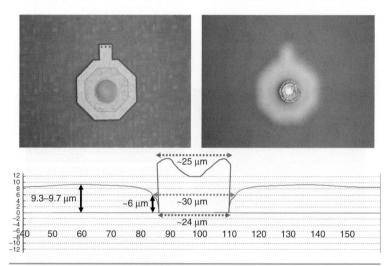

Figure 6.8 Conformal Cu plating and then electroplating of Sn microbumps.

Owing to conformal Cu plating (the Cu is confined by the passivation and photoresist openings), it can be seen from Figs. 6.8 and 6.9 that the Cu has been plated to as high as 18.23 μm (~10.83 μm above the new passivation) along the passivation and photoresist sidewalls, which far exceeds the target (11 μm). Consequently, the electroplated Sn is far below the target (5 μm), as shown in Figs. 6.9 and 6.10.

Figure 6.11 shows the shear test results of the conformal Cu plating and electroplating of Sn mircobumps. It can be seen that the shear

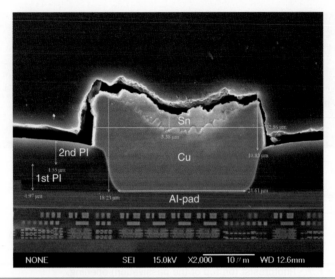

FIGURE 6.9   SEM image of conformal Cu-plating and electroplating of Sn microbumps. (The separation of the microbump is from the potting epoxy.)

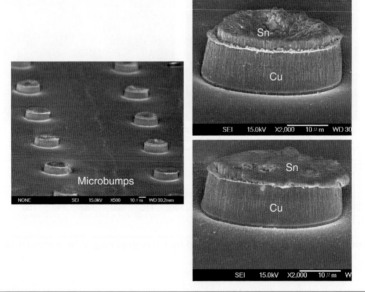

FIGURE 6.10   3D view of conformal Cu plating and electroplating Sn microbumps.

blade is at two different locations: one is 2 μm above the final (second PI) passivation surface, and the other is 6 μm above. The average shear test results at 2 μm (16 g/bump) are much larger than those at 6 μm (4 g/bump). The corresponding failure modes are shown in Fig. 6.12. It can be seen that for the 2-μm standoff, it is shearing the bulk of Cu

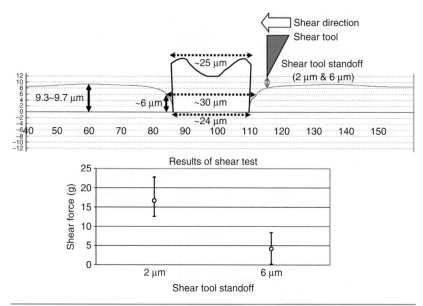

FIGURE 6.11 Shear test results of the conformal Cu plating and electroplating of Sn microbumps.

FIGURE 6.12 Failure models of the shear test of conformal Cu plating and electroplating of Sn.

(no solder). For the 6-μm standoff, it is shearing the Cu and the Sn solder. These microbumps are not acceptable because they may create C2W bonding and solder microjoint reliability problems.

6.5.3   Wafer Microbumping of the Test Wafer by Nonconformal Cu Plating and Electroplating of Sn

Figure 6.13 shows a new target for the lead-free microbumps (9 μm of Cu and 10 μm of Sn). Also, this time nonconformal Cu plating (bottom up) is employed. The masks, new passivation, UBM, photoresist, and wet etching are exactly the same as those in Sec. 6.5.2.

Figures 6.14 and 6.15 show the wafer bumping results of the microbumps by bottom-up Cu plating solution. It can be seen that the

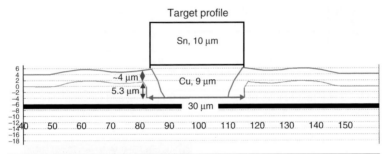

FIGURE 6.13   A new target profile (less Cu and more Sn) of the microbump by nonconformal Cu plating plus electroplating of Sn.

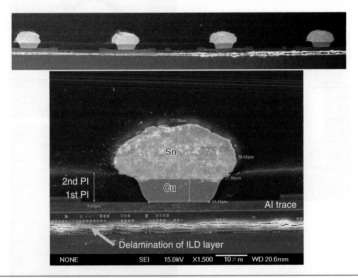

FIGURE 6.14   Cross section and SEM image of the nonconformal Cu-plating and electroplating of Sn microbumps without reflow. [Delamination of the ILD layer is due to sample preparation (grinding).]

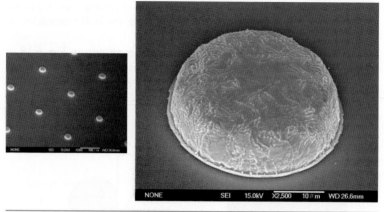

FIGURE 6.15 3D view of the nonconformal Cu plating and electroplating of Sn microbumps without reflow.

solder volume is much more than that by conformal Cu plating. Also, in this case, the solder (Sn) volume is more than that of Cu.

The shear testing results of the microbumps are shown in Fig. 6.16. It can be seen that the shear blade is 2 μm from the final passivation surface, and the average value is about 3.5 g/bump. Of course, this

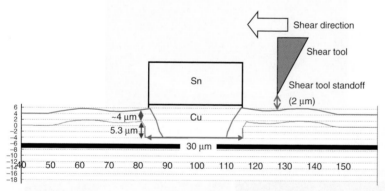

Bump #	Shear force/g	Bump #	Shear force/g	Bump #	Shear force/g	Bump #	Shear force/g	Bump #	Shear force/g
1	2.62	2	3.50	3	3.11	4	3.54	5	3.65
6	3.78	7	3.70	8	3.32	9	3.31	10	4.11
11	2.98	12	3.35	13	3.49	14	3.60	15	3.45
16	3.34	17	3.97	18	3.43	19	2.92	20	3.26
21	3.65	22	3.42	23	3.31	24	3.66	25	3.75
26	3.90	27	3.67	28	3.88	29	3.45	30	2.92

Average of shear force for bumps: 3.47 g/bump

FIGURE 6.16 Shear test results of the nonconformal Cu plating and electroplating of Sn microbumps.

value is smaller than that for the conformal Cu-plating case reported in Sec. 6.5.2 because in this case it is shearing the bulk Sn solder, as shown in the failure mode of Fig. 6.17.

The aging effect of Pb-free microbumps is shown in Fig. 6.18. It can be seen that after aging the microbumps at 150°C for 72 hours, the intermetallic compound (IMC) Cu_5Sn_6 is very obvious. Also, the average value of the shearing test results (2.1 g/bump) is smaller than that without aging (3.5 g/bump).

Failure mode

Solder ball fractured above the surface of the 2nd PI layer in the bulk solder material.

Sn

FIGURE 6.17   Typical failure mode of the nonconformal Cu plating and electroplating of Sn microbumps.

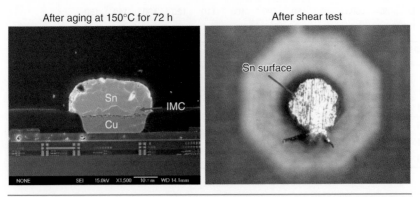

After aging at 150°C for 72 h   After shear test

Sn surface

Sn

IMC

Cu

FIGURE 6.18   Failure mode of the nonconformal Cu plating of microbumps after aging at 150°C for 72 hours.

6.6   Can We Apply the Same Parameters of the Electroplating Method for Ordinary Solder Bumps to Microbumps?

No, we cannot, because microbumps are different from ordinary solder bumps. The biggest difference between them is their solder volume, which is decreased more than 20 times. The effect of IMC, electromigration, and Kirkendall voids (which cannot be observed by x-ray [1]) on the solder microbumps is more critical than for ordinary solder bumps. Thus, unlike ordinary solder bumps, the microbumps usually are not reflowed before joining so as to reduce the growth of IMC, electromigration, and Kirkendall voids. Also, the effect of larger voids (which can be observed by x-ray) on the solder microbumps is more critical than that on ordinary solder bumps. Thus, unlike ordinary solder bump assembly, microbump assembly usually is fluxless to reduce the chance of entrapping flux during solder reflow to form voids. The reliability of the ordinary solder joints should be better than that of solder microjoints. Thus underfill becomes more critical for microbumps. In order to totally fill the gap, underfills with smaller filler sizes should be used.

The size of UBM for ordinary solder bumps and microbumps is different, even though they use the same materials (Ti-Cu). The thickness difference is more than 10 times. Semiconductor equipment such as PVD technology is needed to deposit the Ti-Cu on the wafer. Wet etching to strip off the seed metals is more critical for microbumps than for ordinary solder bumps. The budget for undercuts of the microbumps is much smaller, and thus the process windows for microbumps are smaller.

The photoresist used for ordinary solder bumps is different from that used on microbumps. Not only are the photoresist materials different, but the thickness is also different. Usually, the photoresist thickness of microbumps is four to five times thinner.

Owing to the very large (compared with microbumps) pad size and passivation opening of ordinary solder bumps, Cu plating with either the conformal or the bottom-up approach doesn't make much difference. However, it makes a big difference for microbumps, as shown in this study.

6.7   Summary and Recommendations

Wafer bumping of fine-pitch Pd-free solder microbumps on 300-mm wafers for 3D IC integration has been investigated. Wafer bumping results from two different Cu-plating solutions (conformal and bottom-up) also have been provided and discussed. Furthermore, the effect of Cu and solder (Sn) volumes on microbump quality is examined. Finally, characterizations such as the shearing test and aging of the microbumps and cross sections/SEM images of the microbumps before and after testing have been presented and discussed. Important results and recommendations are summarized as follows [6]:

- The key differences between solder microbumps and ordinary solder bumps by means of the electroplating method have been provided and discussed.

- It has been shown that owing to the tiny geometry of microbump UBM, the conformal Cu-plating solution makes a very big Cu stud with little Sn solder.

- It has been shown that with nonconformal Cu-plating (bottom-up) solution, a reasonable ratio of Cu to Sn solder volume can be achieved.

- Characterization data such as the shearing test of microbumps have been measured, and the average value is about 3.5 g/bump.

- The failure mode of the microbumps after the shearing test has been analyzed by cross section and SEM, and it is found that the cause is the shear fracture of the bulk Sn solder.

- The effect of aging on microbumps is to grow the IMC and reduce its resistance of shear force.

- It is recommended that more characterization data and reliability results are needed and should be performed for the widespread use of the 3D IC integration technology.

PART B WAFER BUMPING, ASSEMBLY, AND RELIABILITY ASSESSMENTS OF ULTRAFINE-PITCH SOLDER MICROBUMPS

Emphasis is placed on the design and fabrication of Cu/Sn lead-free solder microbumps on 10-μm pads with 20 μm of pitch. The chip size is 5×5 mm with thousands of microbumps. A daisy-chain feature is adopted for the characterization and reliability assessment. After pattern trace formation, the microbump is fabricated on the trace by an electroplating technique. A suitable barrier/seed layer thickness is designed and applied to minimize the undercut owing to wet etching but still achieve good plating uniformity. In addition, the shearing test is adopted to characterize bump strength, which exceeds specifications. Also, the Cu-Sn lead-free solder microbumped chip is bonded on a Si wafer [chip-to-wafer (C2W) bonding]. Furthermore, the microgap between the bonded chips is filled with a special underfill. The shear strength of the bonded chips without underfill is measured. The bonding and filling integrity is further evaluated by open/short measurement, scanning acoustic tomographic (SAT) analysis, and cross-sectional with SEM analysis. The stacked ICs are evaluated by reliability (thermal cycling) testing (−55–125°C). Finally, ultrafine-pitch (5-μm pads on 10-μm pitch) lead-free solder microbumping is explored [17].

6.8   Lead-Free Fine-Pitch Solder Microbumping

6.8.1   Test Vehicle

The test vehicles for lead-free fine-pitch solder microbumping and assembly are (1) a 5- × 5-mm chip with 1600 pads with a diameter of 10 μm and a pitch of 20 μm and (2) a 10- × 10-mm chip on a 200-mm wafer. The electroplating process is used for wafer bumping. The bump structure is Cu metal for an RDL and Cu pads with Sn solder for bumps. After electroplating, the Sn layer is reflowed to form microbumps on the Cu pad.

In order to study the influence of microstructural evolution on the bonding strength of microbumps, a Dage 4000 tester was used to shear (rate = 100 μm/s) the microbumps that experienced three episodes of reflow and thermal aging at 100, 125, and 150°C for 15 and 30 minutes and 1, 10, and 30 hours, respectively. Then the samples were cross-sectioned to observe the fracture morphologies and growth of intermetallic compounds (IMCs) by using a field-emission scanning electron microscope (FESEM). The composition of the intermetallic phase was identified by an energy-dispersive spectrometer (EDS).

6.8.2   Microbump Fabrication

Microbump fabrication involves various technical issues. The plator of the Semitool Raider-M is used for the bumping process. The equipment for the aligning requirement used for the entire lithography process is the aligner. Among technical issues of bumping process, etching of the UBM [21–24] is the most critical. For this study, conventional wet etching was used for seed-layer etching.

The electroplating bumping process flow is shown in Fig. 6.19. It can be seen that a thin Ti/Cu seed layer that provides a conducting path for electroplating is evaporated by PVD over the whole test wafer. The seed layer consists of 50 nm of Ti and 120 nm of Cu, for which the Ti is used to improve the adhesion of the copper deposited to the oxide surface. Then a photoresist is spun and developed to form the required pattern. The passivation of oxide is deposited by plasma-enhanced chemical-vapor deposition (PECVD) and patterned by reactive-ion etching (RIE). The bond pads are formed by the combination of seed-layer sputtering, lithography, electroplating, and seed-layer etching. The die consists of 10-μm-diameter pads on 20-μm pitch plated with 3 μm of Cu and 3 μm of Sn.

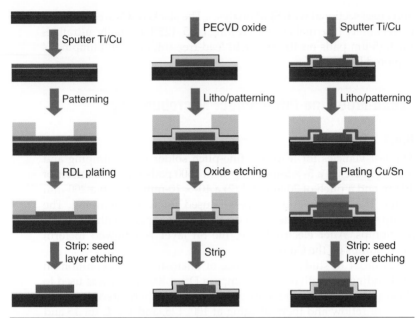

Sputter Ti/Cu

PECVD oxide

Sputter Ti/Cu

Patterning

Litho/patterning

Litho/patterning

RDL plating

Oxide etching

Plating Cu/Sn

Strip: seed
layer etching

Strip

Strip: seed
layer etching

FIGURE 6.19   Process flow for wafer bumping of solder microbumps with RDL.

6.8.3   Characterization of Microbumps

Bump Uniformity

OM and SEM images of the die following CuSn plating are shown in Figs. 6.20 and 6.21, respectively. In this study, we successfully completed process integration of RDLs and microbumps for 10-μm-diameter pads on 20-μm pitch (misalignment is less than 1 μm, as shown in Fig. 6.21). The characteristics of the plated CuSn bump are analyzed after plating. The bump height is measured using Profile Meter system to investigate the uniformity of the plated bumps. A suitable barrier/seed-layer thickness (Ti = 50 nm and Cu = 120 nm) is designed and applied to minimize the undercut owing to wet etching but still achieve good plating uniformity. The variation in bump height including the CuSn bump and RDL is less than 10 percent, as shown in Table 6.1. This bump uniformity is good enough to precede the bonding process.

Figure 6.22 shows the microstructure of the microbumps under various reflow times, in which the scallop-like intermetallic phase within the as-reflowed solder microjoints transformed to a planar one after one time of reflow. The thickness of the Cu_6Sn_5 is defined as the distance from the top of the Cu pillar to the tip of the IMC. (A total of 15 measurements were conducted for each condition.) After one reflow, the average thickness of the Cu_6Sn_5 increased from 2.24 to 2.38 μm. The chemical composition of the IMC was determined to be 53.6Cu-44.4Sn

FIGURE 6.20   OM image of CuSn bumps on 20-μm pitch.

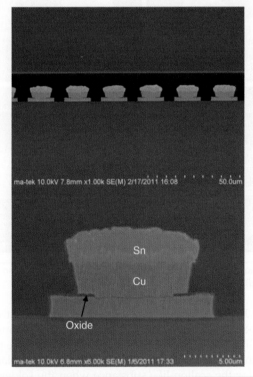

FIGURE 6.21   SEM image of CuSn bumps on 20-μm pitch.

	Bump Height of CuSn (20 μm); 20-μm Pitch
Bottom	8.66
Center	9.44
Left	8.49
Right	8.50
Top	7.76
Uniformity	9.76%

TABLE 6.1   Variation of CuSn Bump Including an RDL for 20-μm Pitch

FIGURE 6.22   Morphologies of microbumps: (a) as reflowed; (b) reflowed once; (c) reflowed twice; (d) reflowed three times.

(in at%) by EDS and was identified as Cu_6Sn_5, the most stable IMC phase in Cu-Sn soldering reaction [25]. Lee et al. [26] demonstrated that the scallop-like Cu_6Sn_5 forms at the interface between the pure solder alloy (Sn) and Cu pillar as reflowed at 240°C, but the planar Cu_6Sn_5 substitutes for the scallop-like substance when the solder alloy contains Cu atoms. Hence the Cu_6Sn_5 formed initially as scallop-like because there were no Cu atoms in the as-plated Sn and then transformed to a planar phase in the following reflow processes.

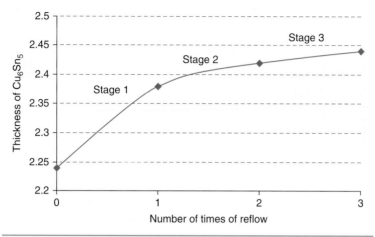

FIGURE **6.23**   Relationship of IMC (Cu_6Sn_5) thicknesses versus numbers of reflows.

The average thicknesses of Cu_6Sn_5 within the microbumps that underwent two and three reflows were increased to 2.42 and 2.44 µm, respectively. The relationship between the thickness of the Cu_6Sn_5 and the number of reflows is given in Fig. 6.23. It can be seen that in stage 1, the growth of Cu_6Sn_5 seems to be reaction-controlled, and then the thickening of Cu_6Sn_5 is retarded in stages 2 and 3 because the flux of the Sn atoms across the Cu_6Sn_5 is significantly less, and the growth of Cu_6Sn_5 becomes a diffusion-controlled reaction [27, 28].

Bump Shear Strength

For each condition, 15 microbumps were sheared to determine the average shear strength. As shown in Fig. 6.24, the shear strength of the as-reflowed microbumps was 174.8 MPa, and then it rose to 217.2 and 228.5 MPa when the microbumps were melted once and twice, respectively, because the Cu_6Sn_5 grows continuously, and its shear modulus is higher than that of Sn [29, 30]. However, the shear strength decreased to 198.5 MPa after the third reflow, which is caused by the voids formed at the interface between the Cu_6Sn_5 and the Cu pillar, as shown in Fig. 6.25a. When a shear force is applied on the microbumps, a crack is initiated and then propagated along the interface until the microbumps separate from the Cu pillar, as shown in Fig. 6.25b. The voids come from the overconsumption of Cu atoms. Hon et al. [31] found the same phenomenon at the interface between the Sn9Zn3.5Ag solder alloy and the Cu pillar after soldering at 350°C for 20 seconds. Although all the samples have a similar failure mode, the voids at the interface speed up the failure of the microbumps that had undergone three reflows, implying that the shear strength of the microbumps might deteriorate after the preconditioning test.

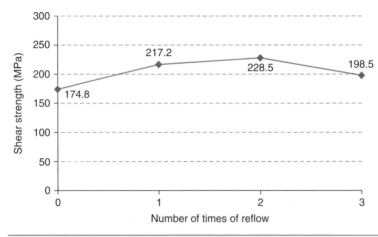

FIGURE 6.24   Relationship of shear strength of microbumps versus number of reflows.

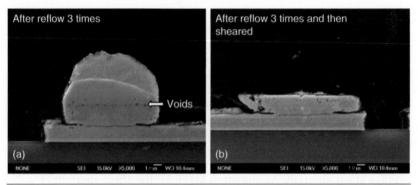

FIGURE 6.25   Morphology of the microbump that underwent (a) three reflows and (b) shear testing.

Aging Effects

The variations in shear strength of the microbumps aged at, respectively, 100, 125, and 150°C for various times are shown in Fig. 6.26, which indicates that the shear strengths of the samples aged for 15 minutes are enhanced, especially for the samples aged at 125 and 150°C, because the Cu_6Sn_5 at the interface between the Cu pillar and the Sn cap keeps growing in this stage, and the values reflect its hardness.

When the aging time was increased to 30 minutes, the shear strengths of the aged samples at 100, 125, and 150°C dropped by 8.8, 89, and 34.7 percent, respectively, which resulted from the defects induced by the volume contraction brought on by overgrowth of Cu_6Sn_5 [32]. This is so because the failure mode of the microbumps

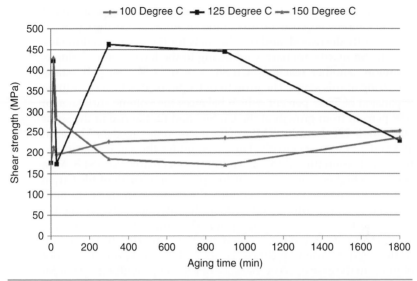

FIGURE 6.26   Shear strength of a microbump aged under various conditions.

after aging is similar to the interfacial fracture rate between the Cu_6Sn_5 and Cu pillar of the reflowed microbumps (Fig. 6.27), meaning that the defects coming from a phase transformation that occurred in the microbumps is more rapid than that in ordinary solder bumps for flip-chip packages and the solder balls for ball-grid-array packages.

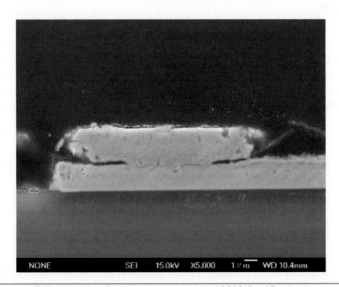

FIGURE 6.27   Failure mode of microbumps aged at 100°C for 15 minutes.

From Fig. 6.23 it can be seen that growth of the Cu_6Sn_5 is inhibited after one reflow because the flux of the Sn atoms across the Cu_6Sn_5 is significantly decreased, and the channels for the Cu atoms diffusing into the Sn matrix are narrowed owing to the lower solder volume. Therefore, another reaction—$Cu_6Sn_5 + 9Cu \rightarrow 5Cu_3Sn$ [32]—starts occurring at this stage. The growth of Cu_3Sn results in (1) a volume expansion of 56.33 percent, which balances the volume shrinkage of 6.39 percent from growth of the Cu_6Sn_5, and (2) the rise in the shear strengths to 226.9 and 461.9 MPa of the microbumps aged for 5 hours at 100 and 125°C, respectively. However, under the same conditions, the shear strength of the microbumps decreases to 184.8 MPa because a higher aging temperature speeds up growth of the Cu_3Sn and the Kirkendall voids. Therefore, the decrease in shear strength resulted from the diffusion defects at the interface between the Cu pillar and the Sn cap. After aging for 30 minutes, the same effect is also found with microbumps aged at 125°C, and the shear strength of the aged microbump drops to 229.7 MPa. However, this is not obvious in the case of 100°C because of a lower growth rate of Cu_3Sn and the Kirkendall voids.

6.9   Lead-Free Fine-Pitch C2C Solder Microbump Assembly

6.9.1   Assembly, Characterization, and Reliability-Assessment Methods

Two methods are used to assemble the test vehicle, namely, thermocompression bonding (TCB) and natural (temperature) reflow. Table 6.2 lists the test items and conditions of reliability assessment. All samples underwent a preconditioning test (JESD22-A113D, LV3) to screen out early failures, and several tens of qualified samples were picked to assess the reliability of the solder microjoints. A temperature cycling test (JESD22-A104B) was performed at −55 to 125°C with a dwell time

Item	Test Condition
Preconditioning	Baking (125°C, 24 h) → Soaking (30°C/60%RH, 192 h) → Reflow (260°C, 3 times)
TCT	−55°C–125°C, 3000 cycles, Dwell time = 5 min, Ramp rate = 15°C/min

TABLE 6.2   Test Conditions for Reliability Assessment

of 5 minutes and a ramp rate of 15°C/min. The resistance was measured regularly, and the test was stopped when the number of cycle reached 3000. The failure criterion in these tests was a variation of 20 percent or more of the initial resistance value. The samples were evaluated by scanning acoustic microscope (SAM) for the detection of voids in the underfill within the microgap after the reliability test.

Finally, the failed samples were mounted in epoxy, ground by SiC paper, and polished using Al_2O_3 paste. The cross sections of the samples were observed by an SEM (JEOL, Japan) to study the morphology of the microjoints, the thickness of IMC, and the interfacial morphologies among chip, underfill, and interposer. Meanwhile, the chemical composition of the intermetallic phase was identified by an EDS (Oxford, UK).

6.9.2 Assembly Process (C2C Natural Reflow)

Lee and Kim [33] developed a fluxless soldering technology using radiofrequency (RF) plasma to remove the oxide layer on the surface of Au20Sn eutectic alloy. In this study, an organic cleaning solvent (MX2302; Kyzen, US) was used first to remove most of the flux residues within the microgaps between the chips and the interposer, and then a plasma etcher (Creating Nano Technologies; Taiwan) with a reactant of N_2 was enforced to clean the other residues that did not dissolve in the chemical. Finally, a SAM (Hitachi; Japan) with a 75-MHz probe was used to determine the voids in the underfill.

6.9.3 Characterization of C2C Reflow Assembly Results

Figure 6.28 shows the solder microjoints formed by conventional reflow, in which the interconnection of the microbumps is precise, and the intermetallic phase formed at the interface is identified to be scallop-shaped Cu_6Sn_5 with an average thickness of 2.23 μm, like that found in Fig. 6.22a.

To study the shear strength of the solder microjoints, the Dage 4000 tester was used to shear the top chip at a rate of 100 μm/s, and the results indicate that the average shear strength of the solder microjoint is 138.1 MPa, which is 21 percent lower than that of the as-reflowed microbump because some microbumps with a lower bump height did not interconnect during reflow, as shown in Fig. 6.29.

The test vehicles with well-interconnected solder microjoints were sieved out for reliability assessment. Before that, the flux residues within the microgap between the top and bottom chips were removed by a compound cleaning method [34]. After underfill dispensing, the microgap was inspected by SAT, and the results are displayed in Fig. 6.30. It can be seen that all the flux residues have been cleaned using wet etching by organic solvent plus dry etching by N_2 plasma, and no gas voids were found within the underfill.

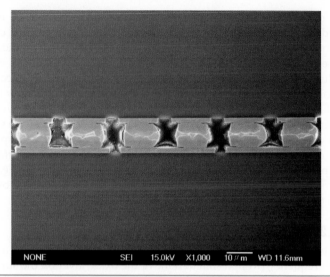

FIGURE 6.28   Solder microjoints formed by conventional reflow.

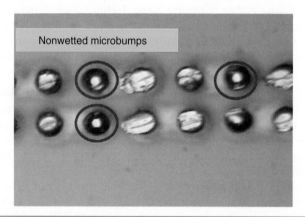

FIGURE 6.29   Nonwetted microbumps by conventional reflow.

6.9.4   Assembly Process (C2C Thermocompression Bonding)

The microbumps were pretreated by plasma to remove the oxides and the contamination on their surface. Then the top chip was interconnected to the bottom chip by a SÜSS FC-150 thermocompression bonder. The heating profile of the TCB is shown in Fig. 6.31. Subsequently, an underfill material from Namics was used to seal the microgaps between the chips and the interposer (the bottom chip), and then the unit was cured at 150°C for 30 minutes. The characteristics of those two capillary underfill materials for microgap filling are listed in Table 6.3.

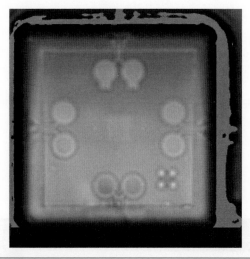

FIGURE 6.30   SAT image of the microgap cleaned by solvent plus plasma and then sealed by underfill.

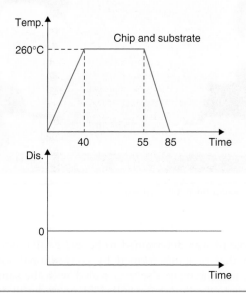

FIGURE 6.31   Heating profile for TCB.

6.9.5   Characterization of C2C TCB Assembly Results

The interfacial morphology of the solder microjoints formed by TCB is exhibited in Fig. 6.32, which shows that the Sn bumps are completely reacted with the Cu pillar to form Cu_6Sn_5 (besides the squeezing of Sn) at the interface, and the average shear strength of a single

Item	Underfill
T_g (°C) by	135
Viscosity at	9 pa/s
CTE (ppm/°C)	42
Filler content	50
Filler size	0.3
Modulus	6.5

TABLE 6.3   Material Properties of the Underfill Materials

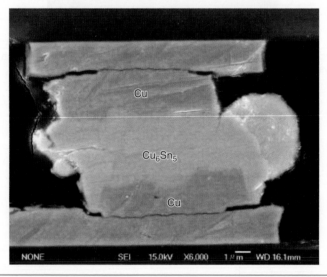

FIGURE 6.32   Solder microjoint formed by TCB.

solder microjoint was determined to be 292.4 MPa, which is much higher than that of joints formed by conventional solder reflow. Afterwards, the microgap also was sealed with the same underfill, and according to the reported results [35], no underfill delamination was found after preconditioning.

6.9.6   Reliability Assessments of Assemblies

Thermal Cycling Test Conditions and Results
The reliability assessment of CuSn solder microjoints (by TCB) was investigated by the thermal cycling test method. The test conditions are shown in Table 6.2, and they were from −55 to 125°C and back; the

dwell time was 5 minutes, and the ramp rate was 15°C/min. The sample size was 34. The test was stopped at 3000 cycles, and there were 16 failures. Figure 6.33 and Table 6.4 show a Weibull plot of the test results, and it can be seen that for the median (50 percent) rank, the Weibull slope is 3.31, and the sample characteristic life

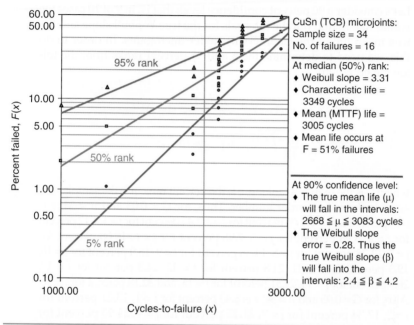

CuSn (TCB) microjoints:
Sample size = 34
No. of failures = 16

At median (50%) rank:
♦ Weibull slope = 3.31
♦ Characteristic life =
   3349 cycles
♦ Mean (MTTF) life =
   3005 cycles
♦ Mean life occurs at
   F = 51% failures

At 90% confidence level:
♦ The true mean life (μ)
   will fall in the intervals:
   2668 ≦ μ ≦ 3083 cycles
♦ The Weibull slope
   error = 0.28. Thus the
   true Weibull slope (β)
   will fall into the
   intervals: 2.4 ≦ β ≦ 4.2

FIGURE 6.33 Weibull plots of CuSn (TCB) solder microjoints under thermal cycling testing (−55 ↔ 125°C, dwell time = 5 minutes, and ramp rate = 15°C/minute.) The failure criterion was more than 20 percent resistance changes.) The required confidence level was 90 percent.

Solder Microjoints	No. of Failures/ Sample Size	Weibull Slope	Characteristic Life (63.2% Failed) (Cycles)	Mean Life (Cycles)	% Failed at Mean Life
CuSn (TCB)	16/34	3.31	3,349	3,005	51
Sn2.5Ag (Reflow)	45/47	1.02	980	972	63
Sn2.5Ag (TCB)	27/72	1.15	5,006	4,765	61

TABLE 6.4 Sample Size, Number of Failures, Characteristic Life, Mean Life, and Percent Failed at Mean Life of CuSn (TCB), Sn2.5Ag (Reflow), and Sn2.5Ag (TCB) Solder Microjoints Based on the Thermal Cycling Test (−55 ↔ 125°C, Dwell Time = 5 Minutes, and Ramp Rate = 15°C/Minute; Failure Criterion Was More than 20 Percent Resistance Changes)

(63.2 percent failures) is 3349 cycles. The sample mean life of the CuSn (TCB) solder microjoints is defined as the mean time to failure (MTTF) = $3349\Gamma(1 + 1/3.31) = 3005$ cycles, where Γ is the Grammar function. This mean life occurs at $F(3005) = 1 - \exp[-(3005/3349)^{3.31}] = 0.51$, that is, 51 percent failures.

Test Results at 90 Percent Confidence

Let's consider a 90 percent confidence level; that is, in 9 of 10 cases we would like to find out the intervals for the true Weibull slope and the true mean life of the CuSn (TCB) solder microjoints. In order to do this, we need to determine the 5 and 95 percent ranks from the following equation:

$$1-(1-z)^n - nz(1-z)^{n-1} - \frac{n(n-1)}{2!}z^2(1-z)^{n-2} - \cdots$$

$$-\frac{n(n-1)\cdots(n-j+1)}{(j-1)!}z^{j-1}(1-z)^{n-j+1} = G$$

where G is the required ranking, z is the percent rank, n is the sample size, and j is the failure order number. Thus, for $G = 0.05$ and $n = 34$, z is 0.15 percent for $j=1$, 1.06 percent for $j=2$, 2.45 percent for $j=3$, 4.12 percent for $j = 4$, 5.98 percent for $j = 5$, 7.98 percent for $j = 6$, 10.08 percent for $j=7$, 12.28 percent for $j=8$, 14.56 percent for $j=9$, 16.9 percent for $j = 10$, 19.3 percent for $j = 11$, 21.8 percent for $j = 12$, 24.3 percent for $j = 13$, 26.9 percent for $j = 14$, 29.51 percent for $j = 15$, and 32.18 percent for $j = 16$. Also, for $G = 0.95$ and $n = 34$, z is 8.43 percent for $j = 1$, 13.21 percent for $j = 2$, 17.34 percent for $j = 3$, 21.25 percent for $j = 4$, 24.93 percent for $j = 5$, 28.46 percent for $j = 6$, 31.89 percent for $j = 7$, 35.22 percent for $j=8$, 38.47 percent for $j=9$, 41.65 percent for $j=10$, 44.76 percent for $j = 11$, 47.82 percent for $j = 12$, 50.82 percent for $j = 13$, 53.78 percent for $j = 14$, 56.68 percent for $j = 15$, and 59.5 percent for $j = 16$. The 5 and 95 percent ranks are plotted in Fig. 6.33. Thus, for the 90 percent confidence level, the CuSn (TCB) solder microjoint life will fall into the intervals bounded by the 5 and 95 percent. For example, the true mean life (μ) of the CuSn (TCB) solder microjoint will fall into the intervals $2668 \leq \mu \leq 3083$ cycles. This means in 90 of 100 cases (the other 10 cases, no one knows), the true MTTF of the CuSn (TCB) solder microjoints will be no less than 2668 cycles but no larger than 3083 cycles. This is very useful because the lower bound (2668 cycles) of CuSn (TCB) solder microjoints has been determined and can be used to compare against a specification (depending on the test condition and application). In this case, the risk (uncertainty) is 10 percent. In order to reduce the risk by using higher confidence levels, the MTTF intervals must be wider and the lower bound smaller.

True Weibull Slope and True Characterization Life

Because there are only 16 failures of the test samples, the error of the Weibull slope is important and can be determined by searching from the following equation:

$$\frac{1}{\sqrt{2\pi}} \int_{-\infty}^{E\sqrt{2N}} e^{-t^2/2} dt = \frac{(1+C)}{2}$$

where E is the Weibull slope error, N is the number of failures, and C is the required confidence level. For the present case, $C = 90$ percent and $N = 16$; then, after some search efforts, we have $E = 28$ percent. Thus the true Weibull slope β of the CuSn (TCB) solder microjoint falls into the intervals $2.4 \leq \beta \leq 4.2$, which is shown in Fig. 6.33 and Table 6.5.

Figure 6.34 shows the Weibull median-rank plots of the CuSn solder microjoints by TCB along with those of the Sn2.5Ag solder microjoints fabricated by the conventional (natural) reflow and TCB methods. The thermal cycling test conditions and failure criterion were the same for all three cases. The sample size for Sn2.5Ag (reflow) was 47 and for Sn2.5Ag (TCB) was 72. The number of failures for Sn2.5Ag (reflow) was 45 and for Sn2.5Ag (TCB) was 27. The sample Weibull slope, characteristic life, and mean life for Sn2.5Ag (reflow), respectively, were 1.02, 980 cycles, and 972 cycles and for Sn2.5Ag (TCB), respectively, were 1.15, 5006 cycles, and 4765 cycles.

Comparison of MTTF of Different Solder Microjoints

In many situations, the qualities and uniformities of two products are compared based on knowledge of limited test data. One of the difficult tasks in life testing is to draw conclusions about a population from a small sample size. It is even more difficult to compare the populations of two products based on knowledge of their limited test

Items		CuSn Solder Microjoints
Mean (MTTF) life	Sample	3005 cycles
	Percent failed at Mean	51%
	True mean (μ) at 90% confidence	2668 % $\mu \leq$ 3083 cycles
Weibull slope	Sample	3.31
	Weibull slope error at 90% confidence	28%
	True Weibull slope at 90% confidence	$2.4 \leq \mu \leq 4.2$

TABLE 6.5 True Weibull Slope and True Mean Life of CuSn (TCB) Solder Microjoints at the 90 Percent Confidence Level

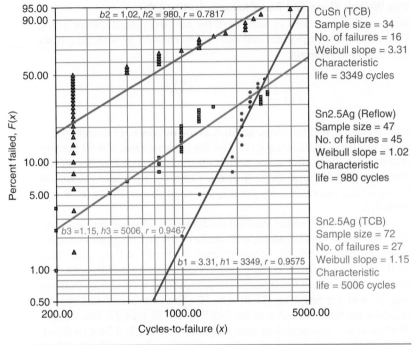

FIGURE 6.34   Median plots of CuSn (TCB), Sn2.5Ag (reflow), and Sn2.5Ag (TCB) solder microjoints under thermal cycling test (−55 ↔ 125°C, dwell time = 5 minutes, and ramp rate = 15°C/minute; failure criterion was more than 20 percent resistance changes).

data. If one product is found to be superior to another, how confident *P* (this is different from the confidence *C* discussed earlier) can one be that the same is true of their populations? Here, a simple approach is used to determine whether one product's quality (mean life, or MTTF) is better than another without inquiring as to what the actual differences are [1]. Therefore,

$$P = \frac{1}{1+\dfrac{\log 1/q}{\log 1/(1-q)}}$$

where

$$q = 1 - \frac{1}{\left[1+\left(\dfrac{t+4.05}{6.12}\right)^5\right]^{40/7}}$$

$$t = \frac{\sqrt{1+\sqrt{T}}\,(\rho-1)}{\rho\Omega_2 + \Omega_1}$$

$$T = (r_A - 1)(r_B - 1)$$

$$\Omega_1 = \sqrt{\frac{\Gamma(1 + 2/\beta_B)}{\Gamma^2(1 + 1/\beta_B)} - 1}$$

$$\Omega_2 = \sqrt{\frac{\Gamma(1 + 2/\beta_A)}{\Gamma^2(1 + 1/\beta_A)} - 1}$$

$$\rho = \frac{M_A}{M_B}$$

and M_A is the mean life of sample A (S_A), M_B is the mean life of sample B (S_B), r_A is the number of failures in S_A, r_B is the number of failures in S_B, β_A is the Weibull slope of S_A, and β_B is the Weibull slope of S_B. It can be seen that this confidence P (which needs to be determined) is different from the confidence C (which is a given value).

Table 6.6 shows comparisons of the mean life between the CuSn (TCB), Sn2.5Ag (reflow), and Sn2.5Ag (TCB) solder microjoints and the determined confidence level P. It can be seen that the solder joint quality (mean life) of CuSn (TCB) is superior to that of Sn2.5Ag (reflow) with 96 percent confidence. This means that in 96 of 100 cases, the expected (mean) life of the CuSn (TCB) solder microjoints will be better than that of the Sn2.5Ag (reflow) solder microjoints.

It also can be seen from the table that the solder joint quality of Sn2.5Ag (TCB) is "better" than that of CuSn (TCB), but with only 51 percent confidence (i.e., in 51 of 100 cases). Since the confidence is so low, the last statement is not statistically significant. Essentially, the mean lives of the SnCu (TCB) and Sn2.5Ag (TCB) solder joints are statistically the same!

Finally, by comparing the mean life of the Sn2.5Ag (TCB) with Sn2.5Ag (reflow) solder microjoints, as shown in Table 6.6, it can be seen that Sn2.5Ag (TCB) is superior to Sn2.5Ag (reflow) with 92 percent confidence. This means that in 92 of 100 cases the quality of the solder microjoints of Sn2.5Ag (TCB) assembly will be better than that of the Sn2.5Ag (reflow) assembly.

Comparing the Mean Life between	MTTF (Mean Life) Ratio	Determined Confidence Level (P)
Sn2.5Ag (TCB) and Sn2.5Ag (reflow)	4,765/972	92%
Sn2.5Ag (TCB) and CuSn	4,765/3005	51%
CuSn (TCB) and Sn2.5Ag (reflow)	3,005/972	96%

TABLE 6.6 Comparing the Quality (Mean Life) of CuSn (TCB), Sn2.5Ag (Reflow), and Sn2.5Ag (TCB) Solder Microjoints and the Calculated Confidence Level

6.10   Wafer Bumping of Lead-Free Ultrafine-Pitch Solder Microbumps

6.10.1   Test Vehicle

The test vehicle consists of 3200 5-μm-diameter pads on 10-μm pitch. The sizes for the chip and substrate are 5×5 mm and 10×10 mm, respectively, on a 200-mm wafer. The bump structure of the 5-μm-diameter pads on 10-μm pitch is the same as that of the 10-μm-diameter pads on 20-μm pitch.

6.10.2   Microbump Fabrication

The bumping process for the 10-μm-pitch microbumps is the same as that for the 20-μm-pitch microbumps. Some preliminary results are shown below.

6.10.3   Characterization of Ultrafine-Pitch Microbumps

Figure 6.35 is an OM image of 5-μm-pad/10-μm-pitch solder micro-bumps. SEM images of the bumps are shown in Fig. 6.36, and it can be seen that the grain size of the plated Sn for the 10-μm-pitch microbumps is larger than for the 20-μm-pitch microbumps. The uniformity of Sn volume of the 10-μm-pitch microbumps is not as good as that for the 20-μm-pitch microbumps. However, there is a nice result, which is the barrier (Ti = 50 nm)/seed layer (Cu = 120 nm) thickness design to reduce the undercut issue; that is, the undercut of Ti is controlled to only 0.7 μm (Fig. 6.37).

10-μm pitch and 5-μm opening

FIGURE 6.35   OM image of microbumps on 10-μm pitch.

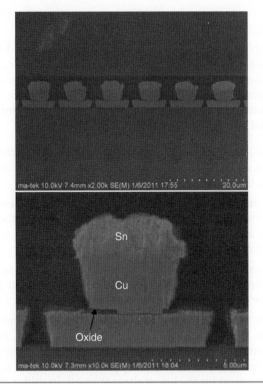

FIGURE 6.36   SEM image of microbumps on 10-μm pitch.

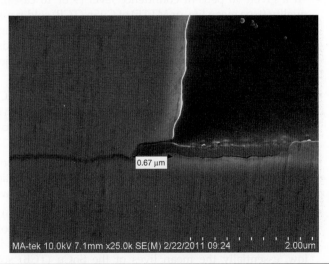

FIGURE 6.37   The undercut (0.63 μm) of Ti on a 10-μm-pitch solder microbump.

6.11   Conclusions and Recommendations

Wafer bumping, assembly, and reliability of fine-pitch lead-free solder microjoints for 3D IC integration have been studied. Characterization data and some reliability results also have been presented. Furthermore, ultrafine-pitch (5-μm pads on 10-μm pitch) microbumping has been explored. Some important results and recommendations are summarized as follows [17]:

- For lead-free solder microbumping, a good process integration of RDLs and microbumps for 10-μm pads on 20-μm pitch has been developed.

- In the as-reflowed solder microjoints, the intermetallic phase at the interface between the Sn microbump and the Cu pillar has been identified to be scallop-shaped Cu_6Sn_5, and it takes a planar shape in subsequent reflows.

- After three reflows, some microvoids have been found at the interface between the planar Cu_6Sn_5 and the Cu pillar, and they degrade the shear strength of the solder microjoints.

- The shear strength of solder microjoints formed by TCB has been found to be higher than that for microjoints formed by conventional reflow. This is so because the whole Sn bump reacts with the Cu pillar to form the Cu_6Sn_5. Also, the reason for the lower shear strength in solder microjoints formed by conventional reflow is the nonuniformity of the microbumps.

- At a given 90 percent confidence level (9 of 10 cases), the quality (true mean life, or MTTF) of CuSn (TCB) solder microjoints will fall into the interval [2668, 3083]. By comparing the lower bound (2668 cycles) with a specification (depends on the application), the reliability of the CuSn (TCB) solder joints can be roughly estimated.

- At 96 percent confidence (in 96 of 100 cases), the mean life of CuSn (TCB) solder microjoints is better than that of Sn2.5Ag (with conventional reflow) solder microjoints. The other 4 percent, no one knows.

- At 92 percent confidence, the mean life of Sn2.5Ag (TCB) solder microjoints is better than that of Sn2.5Ag (with conventional reflow) solder microjoints. The other 8 percent are unknown.

- At 51 percent confidence, the mean life of Sn2.5Ag (TCB) solder microjoints is better than that of CuSn (TCB) solder microjoints. Since the confidence level (51 of 100 cases) is so low, essentially the mean lives of the SnCu (TCB) solder microjoints and that of Sn2.5Ag (TCB) solder microjoints are statistically the same.

- It has been demonstrated that a new design and process are feasible to reduce the undercut (to 0.7 μm) for 5-μm-pad on 10-μm-pitch solder microbumps. Of course, more characterization data and reliability results are needed to support this.

6.12 References

[1] Lau, J. H., *Reliability of RoHS-Compliant 2D and 3D IC Interconnects*, McGraw-Hill, New York, 2011.

[2] Lau, J. H., C. K. Lee, C. S. Premachandran, and A. Yu, *Advanced MEMS Packaging*, McGraw-Hill, New York, 2010.

[3] Lau, J. H., *Flip-Chip Technology*, McGraw-Hill, New York, 1995.

[4] Lau, J. H., *Low-Cost Flip-Chip Technology*, McGraw-Hill, New York, 2000.

[5] Hsin, Y. C., C. Chen, J. H. Lau, P. Tzeng, S. Shen, Y. Hsu, S. Chen, C. Wn, J. Chen, T. Ku, and M. Kao, "Effects of Etch Rate on Scallop of Through-Silicon Vias (TSVs) in 200-mm and 300-mm Wafers," *IEEE/ECTC Proceedings*, Orlando, FL, June 2011, pp. 1130–1135.

[6] Lau, J. H., P-J Tzeng, C-K Lee, C-J Zhan, M-J Dai, Li Li, C-T Ko, S-W Chen, H. Fu, Y. Lee, Z. Hsiao, J. Huang, W. Tsai, P. Chang, S. Chung, Y. Hsu, S-C Chen, Y-H Chen, T-H Chen, W-C Lo, T-K Ku, and M-J Kao, J. Xue, and M. Brillhart,, "Wafer Bumping and Characterizations of Fine-Pitch Lead-Free Solder Microbumps on 12″ (300mm) wafer for 3D IC Integration", *Proceedings of IMAPS International Conference*, Long Beach, CA, October 2011, pp. 650-656.

[7] Chen, J. C., P. J. Tzeng, S. C. Chen, C. Y. Wu, J. H. Lau, C. C. Chen, C. H. Lin, Y. C. Hsin, T. K. Ku, and M. J. Kao, "Impact of Slurry in Cu CMP (Chemical Mechanical Polishing) on Cu Topography of Through-Silicon Vias (TSVs), Re-distributed Layers, and Cu Exposure," *IEEE/ECTC Proceedings*, Orlando, FL, June 2011, pp. 1389–1394.

[8] Tsai, W., H. H. Chang, C. H. Chien, J. H. Lau, H. C. Fu, C. W. Chiang, T. Y. Kuo, Y. H. Chen, R. Lo, and M. J. Kao, "How to Select Adhesive Materials for Temporary Bonding and De-Bonding of Thin-Wafer Handling in 3D IC Integration?" *IEEE/ECTC Proceedings*, Orlando, FL, June 2011, pp. 989–998.

[9] Dorsey, P., "Xilinx Stacked Silicon Interconnect Technology Delivers Breakthrough FPGA Capacity, Bandwidth, and Power Efficiency," Xilinx white paper: Virtex-7 FPGAs, WP380, October 27, 2010, pp. 1–10.

[10] Banijamali, B., S. Ramalingam, K. Nagarajan, and R. Chaware, "Advanced Reliability Study of TSV Interposers and Interconnects for the 28nm Technology FPGA," *IEEE/ECTC Proceedings*, Orlando, FL, June 2011, pp. 285–290.

[11] Banijamali, B., S. Ramalingam, N. Kim, and R. Wyland, "Ceramics vs. Low CTE Organic Packaging of TSV Silicon Interposers," *IEEE/ECTC Proceedings*, Orlando, FL, June 2011, pp. 573–576.

[12] Kim, N., D. Wu, D. Kim, A. Rahman, and P. Wu, "Interposer Design Optimization for High Frequency Signal Transmission in Passive and Active Interposer using Through Silicon Via (TSV)," *IEEE/ECTC Proceedings*, Orlando, FL, June 2011, pp. 1160–1167.

[13] Chai, T., X. Zhang, J. H. Lau, C. Selvanayagam, K. Biswas, S. Liu, D. Pinjala, G. Tang, Y. Ong, S. Vempati, E. Wai, H. Li, B. Liao, N. Ranganathan, V. Kripesh, J. Sun, J. Doricko, and C. Vath, "Development of Large Die Fine-Pitch Cu/Low-*k* FCBGA Package with Through Silicon Via (TSV) Interposer," *IEEE Transactions on Components, Packaging and Manufacturing Technology*, Vol. 1, No. 5, 2011, pp. 660–672.

[14] Selvanayagam, C., J. H. Lau, X. Zhang, S. Seah, K. Vaidyanathan, and T. Chai, "Nonlinear Thermal Stress/Strain Analysis of Copper Filled TSV (Through Silicon Via) and Their Flip-Chip Microbumps," *IEEE/ECTC Proceedings*, Orlando, FL, May 27–30, 2008, pp. 1073–1081. Also, *IEEE Transactions on Advanced Packaging*, Vol. 32, No. 4, November 2009, pp. 720–728.

[15] Yu, A., N. Khan, G. Archit, D. Pinjalal, K. Toh, V. Kripesh, S. Yoon, and J. H. Lau, "Development of Silicon Carriers with Embedded Thermal Solutions for High-Power 3D Package," *IEEE Transactions on Components, Packaging and Manufacturing Technology*, Vol. 32, No. 3, September 2009, pp. 566–571.

[16] Zhang, X., A. Kumar, Q. X. Zhang, Y. Y. Ong, S. W. Ho, C. H. Khong, V. Kripesh, J. H. Lau, D.-L. Kwong, V. Sundaram, Rao R. Tummula, and Georg Meyer-Berg, "Application of Piezoresistive Stress Sensors in Ultrathin Device Handling and Characterization," *Journal of Sensors & Actuators: A. Physical*, Vol. 156, November 2009, pp. 2–7.

[17] Lee, C. K., T. C. Chang, Y. Huang, H. Fu, J. H. Huang, Z. Hsiao, J. H. Lau, C. T. Ko, R. Cheng, K. Kao, Y. Lu, R. Lo, and M. J. Kao, "Characterization and Reliability Assessment of Solder Microbumps and Assembly for 3D IC Integration," *IEEE/ECTC Proceedings*, Orlando, FL, June 2011, pp. 1468–1474.

[18] Zhan, C., J. Juang, Y. Lin, Y. Huang, K. Kao, T. Yang, S. Lu, J. H. Lau, T. Chen, R. Lo, and M. J. Kao, "Development of Fluxless Chip-on-Wafer Bonding Process for 3D Chip Stacking with 30-μm Pitch Lead-Free Solder Micro Bump Interconnection and Reliability Characterization," *IEEE/ECTC Proceedings*, Orlando, FL, June 2011, pp. 14–21.

[19] Huang, S., T. Chang, R. Cheng, J. Chang, C. Fan, C. Zhan, J. H. Lau, T. Chen, R. Lo, and M. Kao, "Failure Mechanism of 20-μm-Pitch Micro Joint Within a Chip Stacking Architecture," *IEEE/ECTC Proceedings*, Orlando, FL, June 2011, pp. 886–892.

[20] Lin, Y., C. Zhan, J. Juang, J. H. Lau, T. Chen, R. Lo, M. Kao, T. Tian, and K. N. Tu, "Electromigration in Ni/Sn Intermetallic Micro Bump Joint for 3D IC Chip Stacking," *IEEE/ECTC Proceedings*, Orlando, FL, June 2011, pp. 351–357.

[21] Yu, A., A. Kumar, S. W. Ho, H. W. Yin, J. H. Lau, K. C. Houel, Sharon L. P. Siang, X. Zhang, D.-Q. Yu, N. Su, M. C. Bi-Rong, J. M. Ching, T. T. Chun, V. Kripesh, C. Lee, J. P. Huang, J. Chiang, S. Chen, C.-H. Chiu, C.-Y. Chan, C.-H. Chang, C.-M. Huang, and C.-H. Hsiao, "Development of Fine Pitch Solder Microbumps for 3D Chip Stacking," *IEEE/ECTC Proceedings*, December 2008, pp. 387–392.

[22] Yu, A., J. H. Lau, S. W. Ho, A. Kumar, W. Y. Hnin, D.-Q. Yu, M. C. Jong, V. Kripesh, D. Pinjala, and D.-L. Kwong, "Study of 15-μm-Pitch Solder Microbumps for 3D IC Integration," *IEEE/ECTC Proceedings*, Orlando, FL, June 2009, pp. 6–10.

[23] Reed, J. D., M. Lueck, C. Gregory, A. Huffman, and J. M. Lannon, Jr., "High Density Interconnect at 10-μm Pitch with Mechanically Keyed Cu/Sn-Cu and Cu-Cu Bonding for 3D Integration," *IEEE/ECTC Proceedings*, Orlando, FL, June 2010, pp. 846–852.

[24] Zhang, W., P. Limaye, Y. Civale, R. Labie, and P. Soussan, "Fine Pitch Cu/Sn Solid State Diffusion Bonding for Making High Yield Bump Interconnections and Its Application in 3D Integration," *IEEE Proceedings of Electronics System Integration Technology Conference*, Berlin, Germany, 2010, pp. 1–4.

[25] Tu, K. N., F. Ku, and T.Y. Lee, "Morphological Stability of Solder Reaction Products in Flip Chip Technology," *J. Electronic Materials*, Vol. 30, No. 9, 2001, pp. 1129–1132.

[26] Lee, T. Y., W. J. Choi, K. N. Tu, J. W. Jang, S. M. Kuo, J. K. Lin, D. R. Frear, K. Zeng, and J. K. Kivilahti, "Morphology, Kinetics, and Thermodynamics of Solid-State Aging of Eutectic SnPb and Pb-Free Solders (Sn3.5Ag, Sn3.8Ag0.7Cu, and Sn0.7Cu) on Cu," *Journal of Materials Research*, Vol. 17, No. 2, 2002, pp. 291–301.

[27] Shen, J., Y. C. Chan, and S. Y. Liu, "Growth Mechanism of Ni_3Sn_4 in a Sn/Ni Liquid/Solid Interfacial Reaction," *Acta Materialia*, Vol. 57, 2009, pp. 5196–5206.

[28] Dybkov, V. I., "Reaction Diffusion and Solid State Chemical Kinetics," Chapter 1, IPMS Publications, New York, 2002.

[29] Frear, D. R., S. N. Burchett, H. S. Morgan, and J. H. Lau, eds., *The Mechanics of Solder Alloy Interconnects*, Van Nostrand Reinhold, New York, 1994, p. 60.

[30] Schwartz, Mel, "Tin and Alloys: Properties," in *Encyclopedia of Materials, Parts and Finishes*, 2nd ed., CRC Press, Boca Raton, FL, 2002.

[31] Hon, M. H., T. C. Chang, and M. C. Wang, "Phase Transformation and Morphology of the Intermetallic Compounds Formed at the Sn9Zn3.5Ag/Cu Interface in Aging," *Journal of Alloys and Compounds*, Vol. 458, 2008, pp. 189–199.

[32] Bader, S., W. Gust, and H. Hieber, "Rapid Formation of Intermetallic Compounds by Interdiffusion in the Cu-Sn and Ni-Sn Systems," *Acta Metall. Mater.*, Vol. 43, No. 1, 1995, pp. 329–337.

[33] Kim, J., and C. C. Lee, "Fluxless Wafer Bonding with Sn-Rich SnAu Dual-Layer Structure," *Materials Science and Engineering A*, Vol. 417, Nos. 1–2, 2006, pp. 143–148.

[34] Chang, T. C., R. Cheng, P. Chang, Y. Hung, J. Chang, T. Yang, and S. Huang., "Reliability Characterization of 20-μm-Pitch Microjoints Assembled by a Conventional Reflow Technique," *IEEE Proceedings of International Conference on Electronics Packaging*, Nara, Japan, 2011, pp. 221–226.

[35] Chang, J. Y., S. Huang, R. Cheng, F. Leu, C. Zhan, and T. Chang, "High Throughput Chip on Wafer Assembly Technology and Metallurgical Reactions of Pb-Free Micro-Joints Within a 3D IC Package," *IEEE Proceedings of International Conference on Electronics Packaging*, Hokkaido, Japan, 2010, pp. 159–164.

CHAPTER 7

Microbump Electromigration

7.1 Introduction

As mentioned in previous chapters, ordinary solder bumps (~80–100 μm) are too big for three-dimensional (3D) integrated circuit (IC) integrations, which require much smaller bumps (25 μm or less). The smaller bump size will result in a substantial increase in current flow in each solder microjoint compared with larger bump sizes. For instance, when applying a current of 0.05 A to a solder bump with a diameter of 20 μm, the current density will reach a value on the order of $10^4 \, A/cm^2$, which is much larger than the value in ordinary solder joints with a diameter of 100 μm under the same applying current.

Second, the effect of current crowding induced by electromigration is expected to become more prominent with the trend in reducing bump size. Third, experimental data from the literature show that microstructural change accompanying large numbers of Kirkendall voids in fine-pitch Cu/Sn solder microbump joints is occurred during electromigration. These results mean that electromigration would enhance the phase transformation happening within solder microjoints and cause a conspicuous microstructure degradation. This circumstance is more noticeable in fine-pitch solder microjoints than in to solder joints of larger size. Fourth, intermetallic compounds (IMCs) may play a key role in the electromigration of solder microjoints because of the small volume of solder. The current crowding-induced electromigration failure of ordinary flip-chip solder joints has been studied and published extensively in, for example, Refs. 1 through 42. The current stress-induced failure of multiphase solder microjoints is examined in this chapter. First, larger-pitch and larger-solder-volume microjoints [43] are examined, followed by smaller and finer-pitch microjoints [44].

7.2   Solder Microjoints with Larger Solder Volumes and Pitch

7.2.1   Test Vehicles and Methods

Solder Bumps

Figure 7.1 shows the solder bump used for the test vehicles, namely, Sn2.5wt%Ag solder on the Cu stud. Right after plating, the dimensions of the SnAg solder on the Cu-stud bump are 20-μm SnAg solder and 20-μm Cu stud. The pad diameter is approximately 25 μm. The chip size is 10 × 8 mm with 13,413 bumps on either 50- or 100-μm pitch. The underbump metallurgy (UBM) on the substrate carrier chip is electronless Ni (5 μm) and immersion Au (0.1 μm) and is approximately 20 μm in diameter. For both chips, the Al trace is 1.5 μm thick and 25 μm wide.

Test Methods

Figure 7.2 shows the assembled test vehicle, and Fig. 7.3 shows a schematic of a Kelvin structure for continuous resistance measurements of a single solder bump. It can be seen that (1) there are four point measurements of the bump through the Kelvin circuit, (2) there are two abundant bumps adjacent to the bump under test, (3) the test vehicle is flipped and attached onto the (20- × 20-mm) carrier substrate, (4) the carrier is die-attached and Au wire-bonded onto a printed circuit board (PCB) as shown in Fig. 7.2, and (5) the directions

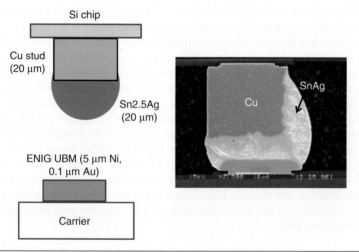

FIGURE 7.1   Test vehicle with Cu stud + SnAg solder and the electronless nickel and immersion gold (ENIG) UBM.

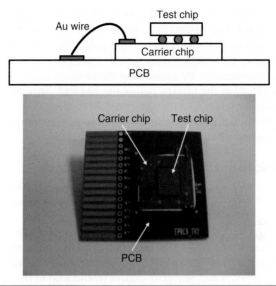

FIGURE 7.2   The assembled test vehicle.

of current flow and voltage are measured with two probes for applying current and the other two probes for sensing the voltage drop. In order to avoid the current-induced failure at the wire-bonding interface, four Au wires are used for one pair of pads. The current stress is at the solder joints near the edges of the chip.

7.2.2   Test Procedures

Electromigration (EM) tests are conducted with a modular integrated reliability analyzer (MIRA, QualiTau) system at 140°C with

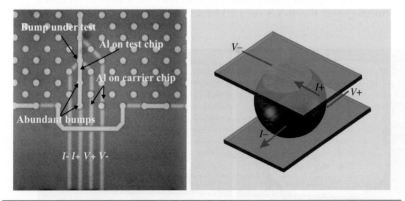

FIGURE 7.3   The four-point Kelvin structure and the current flow direction and voltage measurement.

two current densities, 2.04 and 4.08×10^4 A/cm^2, with respect to bump diameter. The direction of the electron flow in the bump is from electronless nickel and immersion gold (ENIG) to Cu post. For each condition, six samples were tested. Joule heating of the bumps was taken into account by measuring the temperature coefficient of resistance (TCR) of the bumps before the EM test. This method uses the bumps themselves as a thermometer. According to measurement results of the two different current densities, the real temperatures of the test chips are about 148 and 165°C, respectively.

7.2.3   Microstructures of Samples Before Tests

Figure 7.4 shows two bump interconnects from the assembled chip with a Cu stud + SnAg + ENIG UBM before tests [43]. According to energy-dispersive x-ray spectrometry (EDX) results, $(Cu,Ni)_6Sn_5$ compounds are formed on both the Cu side and the ENIG side. From lectures, it is known that when SnAg solder is wetted with Ni, NiSn intermetallic compounds (IMCs) will form [36]. Also, it is reported that the change in Cu content in SnCu and SnAgCu solders would lead to the IMC transformation, that is, from $(Cu,Ni)_3Sn_4$ to $(Cu,Ni)_6Sn_5$. Furthermore, a trace amount of Ni would decrease the Cu solubility in liquid Sn and could accelerate the growth of Cu_6Sn_5 on the Cu UBM [36]. Considering the short distance between the Cu UBM and the ENIG pad (10–20 µm) in the present microbumps, Cu is expected to diffuse to the ENIG pad in a shorter time, and the Cu contents can reach the threshold for the formation of $(Cu,Ni)_6Sn_5$ compounds on ENIG. The typical Ag_3Sn IMCs for SnAg solder joints can be seen in Fig. 7.4.

The resistance of the Cu-stud and solder-bump interconnect is in the range of 10 to 40 mΩ. The difference in resistance is caused mainly by the different initial microstructures after reflow. For the small-bump interconnects, the initial interfacial microstructure may be slightly different after reflow, even though they are neighbor bumps on the same chip. For example, in Fig. 7.4a, the IMC thickness on ENIG is thinner, and the IMCs on UBM and ENIG are separated by the

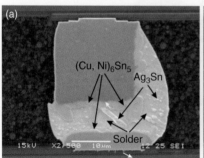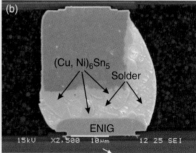

FIGURE 7.4 Assembled solder joints with Cu stud and SnAg solder.

remaining solder. By way of contrast, in Fig. 7.4b, thick IMC layers formed on UBM and ENIG and meet together at some regions. Since IMCs have larger resistance values than solder, the resistance of the solder joint in Fig. 7.4b should be larger than that of the solder joint in Fig. 7.4a.

7.2.4 Samples Tested at 140°C with Low Current Density

Samples Tested at 140°C with a Current Density of 2.04 × 10⁴ A/cm²

The resistance evolution of three solder-joint interconnects subjected to a current density of 2.04×10^4 A/cm^2 is shown in Fig. 7.5a. It can be seen that there is no bump failure, and the resistance value increases gradually up to 1000 hours. Figure 7.5b shows the plot of the relative degradation versus stress time, which is easier to observe than the resistance changing rate. For the so-called relative degradation, the measured resistance was normalized to its initial value. The three curves in Fig. 7.5b show that the resistance of the solder-joint interconnects increased to 125 to 150 percent of their original values at 140°C with a current density of 2.04×10^4 A/cm^2. Cross-sectional observation of the microbump interconnections reveals the microstructure evolution under current stressing.

Microstructures of Samples Tested at Low Current Density

Figure 7.6 shows the typical microstructure of the Cu-stud solder joint under a low current density of 2.04×10^4 A/cm^2 for 1000 hours. Obviously, $(Cu,Ni)_6Sn_5$ becomes thicker, and less solder material remains at the interface. Cu_3Sn compounds also form on the Cu side. In summary, the following reactions occur under present test conditions:

$$Cu + Sn + Ni \rightarrow (Cu, Ni)_6Sn_5 \qquad (7.1)$$
$$Cu + Cu_6Sn_5 \rightarrow Cu_3Sn \qquad (7.2)$$

For Eq. (7.1), solder is consumed and resistance increases because IMCs have a high resistivity. For Eq. (7.2), the formation and growth of Cu_3Sn will decrease "the increasing resistance" because it has a low resistivity compared with Cu_6Sn_5. Therefore, the resistance change is closely related to the diffusion of materials and the growth of the IMCs. The resistivity for Cu, SnAg solder, Cu_3Sn, and Cu_6Sn_5 are 1.7, 11.5, 8.3, and 17.5 μΩ-cm, respectively [45]. Thus the consumption of Cu and solder and the growth of IMCs would lead to the increasing resistance of the solder joint.

Few Kirkendall voids are detected in the thin Cu_3Sn. Kirkendall voids are formed because the diffusion rate of Cu from Cu_3Sn to $(Cu, Ni)_6Sn_5$ is faster than that of Sn from $(Cu,Ni)_6Sn_5$ to Cu_3Sn [38]. With the growth of Cu_3Sn, more Kirkendall voids would form as a result.

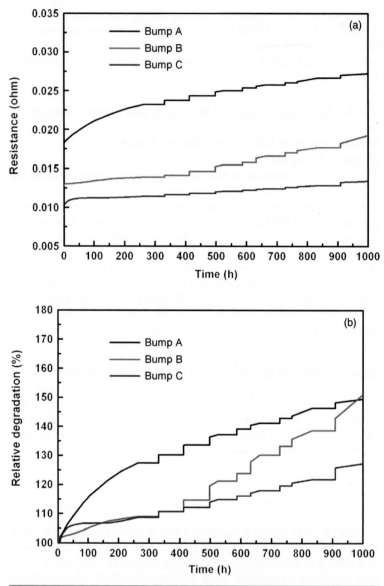

Figure 7.5 (a) Resistance versus current stress time and (b) relative degradation versus stress time (under low current density, 2.04×10^4 A/cm^2).

Xu et al. [31] have studied the electromigration of 90-μm SnAgCu solder on a Cu column at 135°C with a current density of 1.6×10^4 A/cm^2. Since the volume ratio between solder and Cu is larger, the main IMCs are $(Cu,Ni)_6Sn_5$ after 1290 hours of current stressing. Kirkendall voids and crack formation could not be found [31].

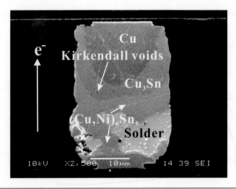

FIGURE 7.6   Typical microstructure of solder joint under low current density
$(2.04 \times 10^4 \text{ A/cm}^2)$.

7.2.5   Samples Tested at 140°C with High Current Density

Samples Tested at 140°C with a Current Density of $4.08 \times 10^4 \text{ A/cm}^2$

When samples are tested with a higher current density $(4.08 \times 10^4 \text{ A/cm}^2)$, there are failures. Figure 7.7a shows the resistance degradation curves of three samples from the chips with Cu stud + SnAg + ENIG UBM solder joint. After the resistance reaches a certain value, it remains nearly constant until an abrupt increase, which means the solder-joint interconnect is opened at this time. In the three curves shown in Fig. 7.7a, two of them show a continuous increase in resistance before reaching the constant values. However, the resistance of one sample increases first and then drops after 100 hours before it reaches the plateau stage. Apparently, it is hard to attribute the abnormal change in resistance to a restructuring effect of the solder matrix that is found in the larger solder bump interconnects because the spreading of resistance occurred at the beginning stage [5]. Also, after 100 hours of testing under high temperature and high current density, the solder is nearly consumed, as can be seen in Fig. 7.8a.

Microstructures of Samples Tested at High Current Density

According to the discussion in Sec. 7.2.4, we know that the evolution of IMCs plays an important role in the evolution of resistance. As shown in Fig. 7.7a, the initial resistance of bump D is about 0.01 Ω, which is smaller than those of the others. As mentioned in Sec. 7. 2.3, it can be deduced that fewer IMCs are formed at the beginning in solder joint D. Thus, during the EM test, with the growth of $(Cu,Ni)_6Sn_5$ IMCs, the resistance would increase faster, which can be seen in Fig. 7.7b, and the resistance is three times larger than its initial value. When the growth of Cu_3Sn becomes dominant, the resistance decreases until an equilibrium state is reached. In solder joints E and F, thicker

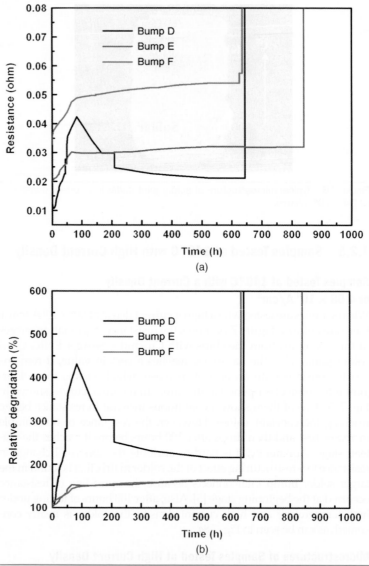

FIGURE 7.7   (a) Resistance versus current stress time and (b) relative degradation versus stress time under high current density (4.08 × 10⁴ A/cm²).

(Cu,Ni)$_6$Sn$_5$ IMCs are formed after reflow. When under EM test, the growth rate of (Cu,Ni)$_6$Sn$_5$ is slower, and Cu$_3$Sn would grow up earlier than in solder joint D. Therefore, there is no backward effect of Cu$_3$Sn formation. The diminishing of resistance owing to Cu$_3$Sn formation is compensated for by the growth of (Cu,Ni)$_6$Sn$_5$.

Figure 7.8a shows the solder-joint interconnect under high current density (4.08 × 10⁴ A/cm²) for 132 hours. It can be seen that

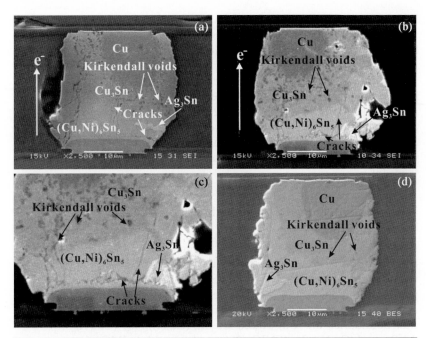

FIGURE 7.8 SnAg solder joint under high current density (4.08×10^4 A/cm^2): (a) 132 hours; (b) 640 hours; (c) 640-hour close-up. In part d, the solder joint is not under any current density.

(1) owing to the effect of electron force and Joule heating, the diffusion rate of materials is accelerated, (2) there is no residual solder anymore, (3) $(Cu,Ni)_6Sn_5$ and Cu_3Sn compounds become thicker, and (4) a large number of Kirkendall voids form in the Cu_3Sn compounds. When the low-temperature solder is consumed and interdiffusion between IMCs reaches equilibrium, the resistance would not change. This explains why a plateau of the resistance curve exists, as shown in Fig. 7.7. It also can be seen from Fig. 7.8a that vertical (Kirkendall voids) cracks are found, and a small horizontal crack is initiated adjacent to the ENIG pad.

Figure 7.8b shows the failed solder joint subjected to high-current stressing for 640 hours. Vertical cracks linking with Kirkendall voids penetrate into $(Cu,Ni)_6Sn_5$ compounds. Also, a continuous thin crack is formed in $(Cu,Ni)_6Sn_5$ compounds along the ENIG pad, which causes the solder joint to open. More detailed microstructure of the ENIG side is revealed in Fig. 7.8c. Such a thin crack is quite different from that (wider) formed at the normal flip-chip solder joint under current stressing [26, 27]. The reason is that in the present case, the crack is formed by mechanical stress, and it propagates along the grain or phase boundary of high-temperature components. It is interesting to note that after testing, some of the failed solder joints can resist slightly again, but when they are heated up to the test temperature, the openings reappear.

Figure 7.8*d* shows the microstructure of the neighboring solder joints without current stressing. It can be seen that (1) there are fewer Kirkendall voids and cracks, (2) no apparent cracks are found at the interface, and (3) the thickness of Cu_3Sn compounds is thinner and the size of the Kirkendall voids is also smaller. This shows that current stressing could enhance/accelerate the formation of Kirkendall voids [5].

7.2.6   Failure Mechanism of the Multiphase Solder-Joint Interconnect

It is known that electromigration triggers failures at zero-flux boundaries, that is, a heterogeneous boundary or an interface under high-current-density stressing [42]. For microbump solder joints under high-current-density stressing, it is found that (1) when all the solders convert to high-temperature IMCs, the potential zero-flux boundaries are Cu/Cu_3Sn, $Cu_3Sn/(Cu,Ni)_6Sn_5$, $(Cu,Ni)_6Sn_5/Ag_3Sn$, and $(Cu,Ni)_6Sn_5/ENIG$; (2) in such a multiple-IMC joint, current stressing enhances the formation of Kirkendall voids at Cu/Cu_3Sn and their growth at $(Cu,Ni)_6Sn_5/ENIG$; and (3) vertical and horizontal cracks are formed and related to the Kirkendall voids.

It is well known that fractures along interfacial boundaries are accelerated by interfacial stresses generated by differences in physical properties. The difference in coefficients of thermal expansion (CTEs) of Cu_3Sn and Cu_6Sn_5 (16.3 and $19 \times 10^{-6}/°C$, respectively) would produce stress in the solder joint. In addition, since the solder joint under current stressing is near the edge of the chip, the CTE mismatch between the chips and the carrier also would generate stress in the bilateral solder joints at the test temperature. Moreover, the fracture toughnesses of Cu_3Sn and Cu_6Sn_5 are 5.72 and 2.80 $MPa/m^{1/2}$, respectively [39]. Therefore, we can extrapolate that cracks propagate more easily in the $(Cu,Ni)_6Sn_5$ solder when the joints are under high-current-density stressing and mechanical stress, compared with Cu_3Sn. This explains why the opening of the solder joint interconnect is near the $(Cu,Ni)_6Sn_5/ENIG$ interface instead of Cu_3Sn where a huge number of Kirkendall voids are formed. On the other hand, the fact that the microbump joints survive being subjected to the lower-current-density stressing is due to the following: (1) The electron wind force is smaller, (2) the joule heating effect is not significant, which also results in a smaller mechanical stress in the joint, and (3) the residual solder could release the mechanical stress in the solder joint.

7.2.7   Summary and Recommendations

The electromigration test for microjoints of the Cu-stud + SnAg + ENIG UBM was studied. It was found when the microjoint was subjected to a current density of 2.04×10^4 A/cm^2 at 140°C, no failure was found at up to 1000 hours. When the joints were tested under a current density of 4.08×10^4 A/cm^2 at the same temperature, owing to the effect of joule heating and electron force, the initial low-temperature solder was quickly consumed, and multiple IMCs, that is, Cu$_3$Sn and (Cu,Ni)$_6$Sn$_5$, were more prevalent within the joint. At the bump interconnect, besides the huge number of Kirkendall voids formed inside the Cu$_3$Sn compound, voids also were formed at the interface between (Cu,Ni)$_6$Sn$_5$ and the ENIG, where crack formation and propagation triggered sudden failure of the joint. The present study revealed that the formation of multiple IMCs within a microjoint would be a great concern for electromigration failure when subjected to high-temperature and high-current-density stressing. Some of the important results and recommendations are summarized as follows:

- The increase in electrical resistance of the microbump solder joints under high-temperature (140°C) and high-current stressing (2.04×10^4 and 4.08×10^4 A/cm^2) is due mainly to the formation of high-temperature IMCs.

- For the Cu stud with SnAg + ENIG UBM solder joints tested at 140°C with a higher current density of 4.08×10^4 A/cm^2, the resistance would reach the plateau region after a certain time (50–200 hours), where the diffusion between different materials, that is, Cu, Ni, and Sn, reaches equilibrium and the IMCs become quite stable. A large number of Kirkendall voids in Cu$_3$Sn compounds are found under high-temperature current densities. Owing to the brittle nature, large cracks along (Cu,Ni)$_6$Sn$_5$ grain boundaries are also found, which would cause sudden failure of the solder joints.

- For the Cu stud with SnAg + ENIG UBM solder joints tested at 140°C with a lower current density of 2.04×10^4 A/cm^2, the resistances do not reach the plateau region after 1000 hours of testing owing to the decreased joule heating effect. The electron flow direction has an effect on the diffusion bonds of materials. The resistance increases faster when electrons flow from Cu UBM to ENIG.

- Test results indicate that the EM failure mode and mechanism of microbump solder joints are totally different from those for ordinary solder joints because the resistance increase in microbump solder joints is not due to the traditional pancake crack formation but rather the IMC growth and phase transformation.

7.3   Solder Microjoints with Smaller Volumes and Pitches

In this section, electromigration tests and results of SnAg solder microjoints with a pitch of 30 μm are presented [44]. The bump structure, which consists of SnAg solder with 5 μm Cu/3 μm Ni UBM, is used to compare with the Cu UBM system. Kelvin structure is implemented to detect the slight resistance change of solder microbumps because the resistance change in the microjoints is very small, whereas daisy-chain structure is applied to evaluate the joule heating effect on the fine-pitch solder microbump interconnection during electromigration. The experimental results reveal that solder microjoints show a superior electromigration resistance when compared with flip-chip solder bump joints. This superior property could be attributed to the microstructure evolution that the most residual Sn is transformed to be the IMC within solder microjoints during the electromigration test.

7.3.1   Experimental Setup and Procedure

Test Vehicles

Chip-on-chip (COC) structures are used as test vehicles. The COC test vehicle includes a top chip and a bottom chip with outside dimensions 5.1 × 5.1 mm and 15 × 15 mm, respectively. Both have the same chip thickness, bumping structure, and more than 3000 microbumps with a 30-μm pitch layout. The size of the solder microbump is 8 μm and the height is 3 μm. The solder microbump structure is schematized in Fig. 7.9. The UBM above the Al pad is comprised 5 μm Cu,

Specifications of test vehicles		
Item	Top chip	Bottom chip
Chip size (X, Y)	(5.1 mm, 5.1 mm)	(15 mm, 15 mm)
Chip thickness	750 μm	
Bump pitch	30 μm	
Bump size	Diameter: 8 μm	
Bump height (Sn2.5Ag/Ni/Cu)	5 μm/3 μm/5 μm	

Appearance photo

Schematic structure

Figure 7.9   Schematic structure of the chip-on-chip test vehicle.

3 μm Ni, and 5 μm Sn2.5Ag. The Sn2.5Ag solder material is con-structed by electroplating and then reflow. After that, an Sn2.5Ag layer is formed on top of the UBM. Detailed specifications of test vehicles are listed in Fig. 7.9.

Test Vehicle Assembly

The assembly process is shown in Fig. 7.10. Plasma treatment was applied to clean both surfaces of the top and bottom chips. The con-tamination and oxidation layers on solder microbumps could be removed via plasma treatment. After the treatment process, alignment and bonding between top and bottom chips were carried out. Toray FC3000WS was used as the bonder in this study, and gap-control thermobonding was executed. The gap-control thermobonding pro-cess could control the gap between the top and bottom chips during bonding. The top and bottom chips were flip-chip bonded at 300°C for 15 seconds. A reasonably good microbump (before annealing) is shown in Fig. 7.11a.

Thermal Annealing

For study of the electromigration behaviors of SnAgCuNi solder microjoints experimentally, two kinds of solder microjoints were used: type I original solder microjoints and type II thermal annealing

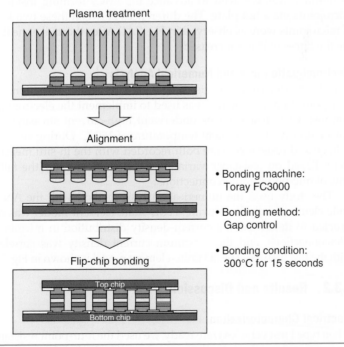

FIGURE 7.10   Chip-on-chip flip-chip bonding process.

(a)   (b)

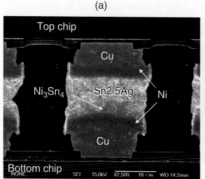 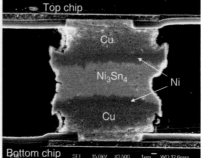

Type I microjoint (before thermal aging)   Type II microjoint (after thermal aging)

• Two kinds of solder microjoints were used: type I original solder microjoints and type II thermal aging ones.

• Thermal aging of 300°C for 15 minutes was enforced to advance the IMC forming inside the type II microjoints by hot pate.

FIGURE 7.11   (a) Cross-sectional image of type I microjoint (before thermal annealing). (b) Cross-sectional image of type II microjoint (after thermal annealing).

ones. Compared with type I samples, thermal annealing at 300°C for 15 minutes was enforced to advance the IMCs forming inside the microjoints via a hot plate. The differences between these two kinds of microjoints were as observed in Fig. 7.11. After thermal annealing, the thickness of IMCs increased.

Electromigration Test and Numerical Analysis

In order to determine precise single-joint resistance variations, the four-point Kelvin structure was used to implement the electromigration test. COC test vehicles underwent high-current stressing by a power supply at an ambient temperature of 150°C. During stressing, voltage and resistance were both recorded with the in-situ measurement. Based on resistance variations, we could monitor the failure time of microbump interconnections.

The study used the numerical-analysis method with the ANSYS finite-element analysis software. A 3D finite element model was constructed to determine the current-density distribution in microbump interconnections, and the maximum current density was correlated with the EM lifetime. The 3D finite-element model is shown in Fig. 7.12.

7.3.2   Results and Discussion

Electrical Characterization

When type I test vehicles were ready, we used the four-point resistance-measurement method to determine precise single-joint resistance.

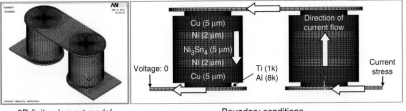

3D finite-element model   Boundary conditions

Material property

	Cu	Ni	Al trace	Ni₃Sn₄	Ti
Resistivity (mΩ-cm)	1.7	6.8	2.7	28.5	43.1

FIGURE 7.12   3D finite-element model.

We used a four-point Kelvin structure to measure the voltage value for current stressing between –10 and 10 mA. And single-joint resistance could be evaluated based on voltage and current. The results are illustrated in Fig. 7.13. When the current value increased higher than 5 mA, resistance was close to 15 mΩ. As the current value decreased lower than ±5 mA, the divergence of resistance had a tendency to increase rapidly.

The electrical characteristics of thermal-annealed COC test vehicles (type II) also were measured. For comparison of resistance

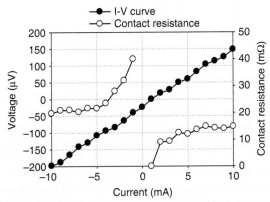

- Four-point Kelvin structure to measure the voltage value for current stressing between –10 and 10 mA.
- Single-joint resistance could be evaluated from voltage and current.
- When current ≧5 mA, resistance was closed to 15 mΩ.
- As the current <±5 mA, the divergence of resistance tends to increase rapidly.

FIGURE 7.13   Results of four-point resistance measurement (type I).

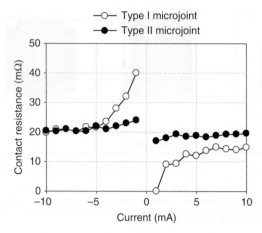

- As applied current was lower than ±5 mA, the average of microjoints resistance was more stable than type I.
- Because of IMC transforming fully in type II micro joint, it resulted in higher and more stable resistance, especially in ultralow-current stressing.

FIGURE 7.14   Comparison of Type I and Type II microjoints on four-point resistance measurements.

between the two types of microjoints, their resistance results were combined (the graphic representation in Fig. 7.14). When applied current was higher than ±5 mA, resistance of type II microjoints tended to converge and close to 20 mΩ. When applied current was lower than ±5 mA, the average microjoint resistance was more stable than that for type I microjoints. Owing to IMCs transforming fully into type II microjoints, the result was higher and more stable resistance, especially in ultralow-current stressing.

Numerical-Analysis Results

Figure 7.15*a* shows the current-density distribution in the microjoint interconnections under 0.13 A of current stress. The analysis results indicate that the maximum current density was generated in the corner of the aluminum trace near copper UBM, especially when the electron flow entered the corner and left. By the finite-element analysis (FEA), maximum current density in each material under different current stresses was evaluated. When 0.13 A of current was applied, maximum current density values in the trace, UBM, and bump, were 1.25×10^6, 6.94×10^5, and 5.96×10^4 A/cm², respectively. Also, Fig. 7.15*b* shows the maximum current density (J_{max}) versus the applied current. It can be seem that (1) the larger the current, the larger is J_{max}, (2) J_{max} at Al is larger than at the UBM and bump, (3) J_{max} at the bump is the smallest, and (4) J_{max} is a linear function of the applied currents.

(a)   (b)

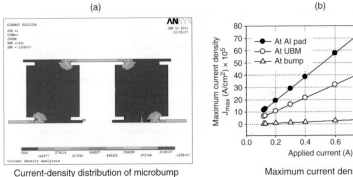

Current-density distribution of microbump joint   Maximum current density vs. applied current

(1) The larger the current the larger the J_{max}.
(2) The J_{max} at Al is larger than at UBM and bump.
(3) The J_{max} at the bump is the smallest.
(4) J_{max} is a linear function of the applied currents.
(5) The J_{max} at Al could be larger than 10^6 A/cm^2 even though the J_{max} at the bump is still less than 10^4 A/cm^2.

FIGURE 7.15 Numerical-analysis results: (a) Current-density distribution of microbump interconnections; (b) maximum current density (J_{max}) versus applied current.

Experimental Results

The experimental results of EM tests in a four-point Kelvin structure are summarized in Figs. 7.16 and 17. We showed two kinds of lifetimes. One was the timing when the samples were open circuit completely. The other was when samples with more than 20 percent resistance variation were reached by in-situ resistance measurement.

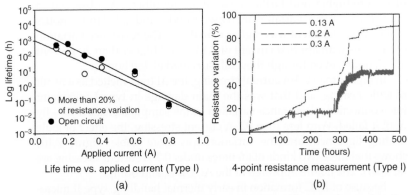

Life time vs. applied current (Type I)   4-point resistance measurement (Type I)

(a)   (b)

• The lifetime of more than 20% resistance variation was lower and very close to the lifetime of open circuit in high-current stressing.

• As applied current decreased, the lifetimes of open circuit samples were higher and farther away from the others.

FIGURE 7.16 Type I samples. (a) Lifetime versus applied current. (b) Four-point resistant measurement.

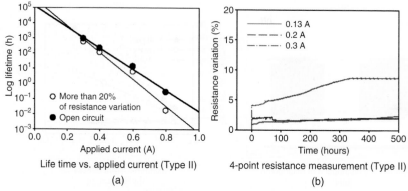

Life time vs. applied current (Type II)   4-point resistance measurement (Type II)

(a)   (b)

- Under the same current density, lifetimes of type I samples without thermal annealing (Fig. 7.16a) were shorter than type II samples obviously.

- In high applied current, the lifetime of more than 20% variation was lower and farther away from the life of open circuit samples.

FIGURE 7.17   Type II samples. (a) Lifetime versus applied current. (b) Four point resistant measurement.

The applied currents were 0.13, 0.2, 0.3, 0.4, 0.6, and 0.8 A. All the type I samples were open circuit inside 600 hours. But their resistance variations were higher than 20 percent inside half the open time. Figure 7.16 plots the lifetimes of type I samples for each applied current. The Y axis is explained by a common logarithm. We found that the lifetimes of more than 20 percent resistance variation were lower and very close to the lifetimes of open-circuit samples in high-current stressing. As applied current decreased, the lifetimes of open-circuit samples were higher and farther away than the others. As observed in Fig. 7.16, resistance variation increased rapidly and failed in 100 hours when applied current was higher than 0.3 A. The resistance variation of samples with applied current of 0.13 A had higher than 20 percent at 292 hours, even though they still survived at 480 hours.

In higher than 0.3-A current stressing, the Al trace melted as a result of joule heating, and that was the main damage in type I microjoint interconnections. Resistance variation in the sample increased rapidly until it failed. With less than 0.2-A current stressing, the effect of joule heating lightened, and the variation was more stable. Figure 7.16a shows that these two kinds of lifetimes under 0.2-A current stressing are far apart. The failure appeared at the Al, UBM, and bump.

Because of IMC formation in early thermal handling, type II microjoints have strong EM resistance. The results for type II joints are shown in Fig. 7.17. Under the same current densities, the lifetimes of type I samples without thermal annealing were shorter than those of type II samples obviously. In high applied current, the lifetimes of more than 20 percent variation were lower and farther away from the lifetimes of open-circuit samples. In low applied current, the lifetimes of open-circuit samples were still longer than the others, but their differences were close. But 0.2- and 0.13-A stressed samples still survived and had less than 10 percent resistance variation, so they weren't plotted in Fig. 7.17.

As observed in Fig. 7.17, 0.3-A stressed type II microjoints failed in 926 hours, and 0.2-A stressed type II microjoints survived to 1470 hours and were still going. They had low resistance variation of less than 10 percent. The 0.13-A stressed samples had been stressed to 276 hours and were still going. Variation in the 0.13-A stressed samples also was lower than 10 percent. We expect that they could survive over 1000 hours. Because of strong EM resistance, type II microjoints had low resistance variation for a long time.

Failure Modes

After 0.13-A current stressing, cross-sectional inspection of type I microjoints was performed by SEM and is shown in Figs. 7.18 and 7.19. The direction of electron flow is marked by arrows. In high current density, residual solder in the microjoint had transformed to IMC gradually because of long thermal aging caused by current crowding. Based on these two SEM photos, we found three kinds of cracks, voids inside solder microjoint, voids in UBM, and melting of the Al material, after the electromigration test. In Fig. 7.18a, the bump joint was powered by 0.13-A electron flow from the bottom to the top. Voids appeared at middle side of the microjoint and UBM on cathode side of the joint.

In Fig. 7.18b, the joint with the downward electron flow showed UBM cracks both at the cathode and the anode of the joint. Melting of the Al material occurred on the anode side of the joint. Experimental

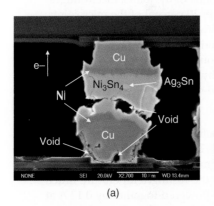
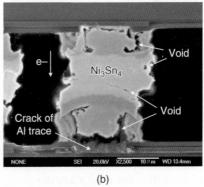

(a)   (b)

- This sample is failed. Failures appeared at the Al, UBM, and bump.
- In left figure, void appeared at middle side of microjoint and UBM in cathode side of the joint.
- In right figure, UBM cracks appeared both at cathode and anode of the joint. Melting of Al material occurred in anode side of joint.

FIGURE 7.18   Failure mode of type I microjoint under current stress of 0.13 A at 484 hours.

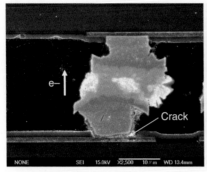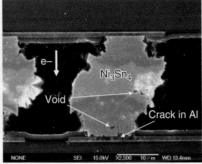

The failure appeared at the Al, UBM, and bump when more than 20% variation of resistance was reached.

FIGURE 7.19   Failure mode of type I microjoint (current stress = 0.13 A and resistance variation >20%).

data from the literature [41] indicated that electromigration damage would appear in both the cathode and the anode of Al interconnects when applied current density in the Al interconnects reached 10^6 A/cm^2. The applied current density in the Al trace was 1.25×10^6 A/cm^2. The joule heating induced by current crowding increased the temperature of the Al trace and then enhanced its electromigration. At the same time, the diffusion of Ni and Cu also was accelerated.

SEM images of type I microjoints that show over 20 percent variable resistance under 0.13-A applied current are shown in Fig. 7.19. We also found voids in the cathode and anode of the joints. Electromigration damage in the Al trace happened in 20 percent of the microjoints. The failures of type II microjoints with more than 0.3 A of current stressing were similar to these. This indicates that current density of the Al trace still could be too great in our COC structures. Compared with bigger solder joints, voids caused by electromigration were almost at the cathode. But the microbumps did not conform to this situation.

Testing on Daisy-Chain

Besides testing the electromigration behavior, we also did some preliminary electromigration tests on this daisy-chain structure. Different from an EM test on a single pair of microbumps, the joule heating generated by the Al trace can elevate the testing temperature much more. When the four daisy chains are powered together by 0.12 A at the furnace temperature of 56°C, the surface temperature of the single test vehicle can reach to 150°C. Therefore, at the same EM testing conditions, the lifetime of daisy-chain structures would be much shorter than that of a single pair of microjoints. And the failure mechanism of the daisy chain would be different as well. To study the polarity effect of Ni consumption induced by electron flow, only one daisy chain of a type I microjoint was powered by 0.12 A at 150°C for 1 day and then was polished to compare with its neighboring

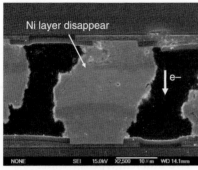
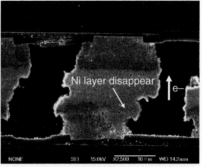

Bump A   **Bump B**

- One daisy chain of type I microjoint was powered by 0.12 A at 150°C for 1 day and then was polished.
- At the cathode of the powered bumps, the Ni layers are always thinner and shorter, even tend to be disappeared.
  - → Phase transformation–induced failure in solder bumps, which will tend to happen at high temperature.

FIGURE 7.20   Daisy chain with current stressing: (a) bump A; (b) bump B.

daisy chains. Since an Si chip has good thermal conductivity, it is reasonable to assume that the neighboring daisy chains suffer the same elevated temperature generated by joule heating from the powered daisy chain.

Figures 7.20 and 21 show four bumps. Bump A and bump B come from the powered daisy chain, whereas bump C and bump D come from the a neighboring unpowered daisy chain. Obviously, compared with the unpowered bumps C and D, at the cathode of the powered bumps, the Ni layers are always thinner and shorter, even though they tend to disappear.

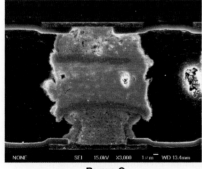
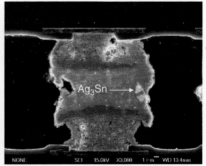

Bump C   **Bump D**

Phase transformation–induced failure isn't found in these solder bumps
without current stressing.

FIGURE 7.21   Daisy chain without current stressing: (a) bump C; (b) bump D.

As we know, in this COC test vehicle, the current crowding effect in microjoints may not be as critical as in flip-chip solder bumps. Then there are two rest reliability problems that probably are critical in the EM test. One is electromigration in the Al trace, which will happen with high current density; the other is phase transformation–induced failure in the solder bumps, which will tend to happen at high temperatures.

7.3.3   Summary and Recommendations

In this study, the electromigration investigation of 30-μm-pitch COC solder microjoints has been carried out. Two different types of solder microjoints have been constructed via the thermal annealing process. Some important results and recommendations are as follows [44]:

- It has been found that annealed solder microjoints have more stable and higher EM resistance than the nonannealing ones.

- During the EM test, it has been found that the resistance of type I solder microjoints (with current density of more than 10^4 A/cm^2) increased rapidly and then failed at 600 hours. On the other hand, type II solder microjoints have been found to have strong EM resistivity and survived longer than type I joints.

- The EM failures have occurred in the Al trace and the UBM, which implies that there is high current density in the Al trace of the microbump interconnection. These failures have been caused by joule heating induced from current crowding. High current density applied in the Al trace could be one of the serious reliability issues in microbump interconnection.

- It has been found that the situation of Ni layer–induced failure is following the direction of electron flow in the daisy chain. This phase transformation–induced failure is another reliability problem in fine-pitch microbump interconnection.

- For further EM research, applied current density in the Al trace must be less than 10^6 A/cm^2. It is expected that the EM test will last for a long time, and more stressing channels are helpful to raise efficiency.

7.4   References

[1] Tu, K. U., "Recent Advances on Electromigration in Very Large-Scale-Integration of Interconnects," *J. Appl. Phys.*, Vol. 94, No. 9, 2003, pp. 5451–5473.
[2] Lin, K. L., and G. P. Lin, "The Electromigration Investigation on the Newly Developed Pb-Free Sn-Zn-Ag-Al-Ga Solder Ball Interconnect," *IEEE/ECTC Proceedings*, Reno, NV, May 29–June 1, 2007, pp. 1467–1472.
[3] Lin, K. L., and S. M. Kuo, "The Electromigration and Thermomigration Behaviors of Pb-Free, Flip-Chip Sn-3Ag-0.5Cu Solder Bumps." In *IEEE/ECTC Proceedings*, San Diego, CA, May 30–June 2, 2006, pp. 667–672.

[4] Nah, J. W., J. O. Suh, K. N. Tu, S. W. Yoon, C. T. Chong, V. Kripesh, B. R. Su, and C. Chen, "Electromigration in Pb-Free Solder Bumps with Cu Column as Flip-Chip Joints," *IEEE/ECTC Proceedings*, San Diego, CA, May 30–June 2, 2006, pp. 657–662.

[5] Ebersberger, B., R. Bauer, and L. Alexa, "Reliability of Lead-Free SnAg Solder Bumps: Influence of Electromigration and Temperature," *IEEE/ECTC Proceedings*, Lake Buena Vista, FL, May 31–June 3, 2005, pp. 1407–1415.

[6] Rinne, G., "Electromigration in SnPb and Pb-Free Solder Bumps," *IEEE/ECTC Proceedings*, Las Vegas, NV, June 1–4, 2004, pp. 974–978.

[7] Jang, S. Y., J. Wolf, W. S. Kwon, and K. W. Paik, UBM (Under Bump Metallization) Study for Pb-Free Electroplating Bumping: Interface Reaction and Electromigration," *IEEE/ECTC Proceedings*, San Diego, CA, May 28–31, 2002, pp. 1213–1220.

[8] Gan, H., and K. Tu, "Effect of Electromigration on Intermetallic Compound Formation in Pb-free Solder-Cu Interfaces." In *IEEE Proceedings of Electronic & Components Technology Conference*, San Diego, May 28–31, 2002, pp. 1206–1212.

[9] Huang, Z., Chatterjee, R., Justison, P., Hernandez, R., Pozder, S., Jain, A., Acosta, E., Gajewski, D. A., Mathew, V., and Jones, R. E. "Electromigration of Cu-Sn-Cu Micropads in 3D Interconnect," *IEEE/ECTC Proceedings*, Lake Buena Vista, FL, May 27–30, 2008, pp. 12–17.

[10] Rinne, G., "Emerging Issues in the Physics and Control of Electromigration in SnPb and Pb-Free Solder Bumps," *IEEE/ECTC Proceedings*, Singapore, December 10–12, 2003, pp. 72–76.

[11] Su, P., T. Uehling, D. Wontor, M. Ding, and P. S. Ho, "An Evaluation of Electromigration Performance of SnPb and Pb-free Flip Chip Solder Joints," *IEEE/ECTC Proceedings*, Lake Buena Vista, FL, May 31–June 3, 2005, pp. 1431–1436.

[12] Lu, M., D. Y. Shih, P. Lauro, S. Kang, C. Goldsmith, and S. K. Seo, "The Effects of Ag, Cu Compositions and Zn Doping on the Electromigration Performance of Pb-Free Solders," *IEEE/ECTC Proceedings*, San Diego, CA, May 25–29, 2009, pp. 922–929.

[13] Nicholls, L., R. Darveaux, A. Syed, S. Loo, T. Y. Tee, T. A. Wassick, and B. Batchelor, "Comparative Electromigration Performance of Pb-Free Flip Chip Joints with Varying Board Surface Condition," *IEEE/ECTC Proceedings*, San Diego, CA, May 25–29, 2009, pp. 914–921.

[14] Su, P., L. Li, Y. S. Lai, Y. T. Chiu, and C. L. Kao, "A Comparison Study of Electromigration Performance of Pb-Free, Flip-Chip Solder Bumps," *IEEE/ECTC Proceedings*, San Diego, CA, May 25–29, 2009, pp. 903–908.

[15] Lee, J. H., G. T. Lim, Y. B. Park, S. T. Yang, M. S. Suh, Q. H. Chung, and K. Y. Byun, "Size Effect on Electromigration Reliability of Pb-Free, Flip-Chip Solder Bump," *IEEE/ECTC Proceedings*, Lake Buena Vista, FL, May 27–30, 2008, pp. 2030–2034.

[16] Lu, M., P. Lauro, D. Y. Shih, R. Polastre, C. Goldsmith, D. W. Henderson, H. Zhang, and M. G. Cho, "Comparison of Electromigration Performance for Pb-Free Solders and Surface Finishes with Ni UBM," *IEEE/ECTC Proceedings*, Lake Buena Vista, FL, May 27–30, 2008, pp. 360–365.

[17] Chae, S. H., B. Chao, X. Zhang, J. Im, and P. S. Ho, "Investigation of Intermetallic Compound Growth Enhanced by Electromigration in Pb-Free Solder Joints," *IEEE/ECTC Proceedings*, Reno, NV, May 29–June 1, 2007, pp. 1442–1449.

[18] Chae, S. H., J. Im, P. S. Ho, and T. Uehling, "Effects of UBM Thickness, Contact Trace Structure and Solder Joint Scaling on Electromigration Reliability of Pb-Free Solder Joints," *IEEE/ECTC Proceedings*, Lake Buena Vista, FL, May 27–30, 2008, pp. 354–359.

[19] Chae, S. H., X. Zhang, H. L. Chao, K. H. Lu, P. S. Ho, M. Ding, P. Su, T. Uehling, and L. N. Ramanathan, "Electromigration Lifetime Statistics for Pb-Free Solder Joints with Cu and Ni UBM in Plastic Flip-Chip Packages," *IEEE/ECTC Proceedings*, San Diego, CA, May 30–June 2, 2006, pp. 650–656.

[20] Kwon, Y. M., and K. W. Paik, "Electromigration of Pb-Free Solder Flip-Chip Using Electroless Ni-P/Au UBM," *IEEE/ECTC Proceedings*, Reno, NV, May 29–June 1, 2007, pp. 1472–1477.

[21] Ebersberger, B., R. Bauer, and L. Alexa, "Reliability of Lead-Free SnAg Solder Bumps: Influence of Electromigration and Temperature," *IEEE/ECTC Proceedings*, Lake Buena Vista, FL, May 31–June 3, 2005, pp. 1407–1415.

[22] Iwasaki, T., M. Watanabe, S. Baba, Y. Hatanaka, S. Idaka, Y. Yoloyama, and M, Kimura, "Development of 30-Micron Pitch Bump Interconnections for COC-FCBGA," *Proceedings of the 56th Electronic Components and Technology Conference*, 2006, pp. 1216–1222.

[23] Gan, H., S. L. Wright, R. Polastre, L. P. Buchwalter, R. Horton, R. Horton, P. S. Andry, C. Patel, C. Tsang, J. Knickerbocker, E. Sprogis, A. Pavlova, S. K. Kang, and K. W. Lee, "Pb-Free Micro-Joints (50-µm Pitch) for the Next Generation Micro-Systems: The Fabrication, Assembly and Characterization," *Proceedings of the 56th Electronic Components and Technology Conference*, 2006, pp. 1210–1215.

[24] Sunohara, M., T. Tokayunaga, T. Kurihara, and M. Higashi, "Slicon Interposer with TSVs (Through-Silicon Vias) and Fine Multilayer Wiring," *IEEE/ECTC Proceedings*, Lake Buena Vista, FL, May 27–30, 2008, pp. 847–852.

[25] Ren, F., X. Zhang, J. W. Nah, K. N. Tu, L. Xu, and J. H. L. Pang, "In-Situ Study of the Effect of Electromigration on Strain Evolution and Mechanical Property Change in Lead-Free Solder Joints," *IEEE/ECTC Proceedings* San Diego, CA, May 30–June 2, 2006, pp. 1160–1163.

[26] Yeh, E., W. J. Choi, K. N. Tu, P. Elenius, and H. Balkan, "Current-Crowding-Induced Electromigation Failure in Flip Chip Solder Joints," *Appl. Phys. Lett.*, Vol. 80, 2002, pp. 580–582.

[27] Zhang, L., S. Ou, J. Huang, K. N. Tu, S. Gee, and L. Nguyen, "Effect of Current Crowding on Void Propagation at the Interface Between Intermetallic Compound and Solder in Flip Chip Solder Joints," *Appl. Phys. Lett.*, Vol. 88, 2006, pp. 012106–1.

[28] Ding, M., G. Wang, B. Chao, P. S. Ho, P. Su, and T. Uehling, "Effect of Contact Metallization on Electromigration Reliability of Pb-Free Solder Joints," *J. Appl. Phys.*, Vol. 99, 2006, pp. 094906–094906-6.

[29] Nah, J., J. O. Suh, K. N. Tu, S. W. Yoon, V. S. Rao, V. Kripesh, and F. Hua, "Electromigration in Flip Chip Solder Joints Having a Thick Cu Column Bump *and a Shallow* Solder Interconnect," *J. Appl. Phys.*, Vol. 100, 2006, pp. 123513–123513-5.

[30] Lai, Y., Y. T. Chia, C. W. Lee, Y. H. Shao, and J. Chen, "Electromigration Reliability and Morphologies of Cu Pillar Flip-Chip Solder Joints," *IEEE/ECTC Proceedings*, Lake Buena Vista, FL, May 27–30, 2008, pp. 330–335.

[31] Xu, L., J. K. Han, J. J. Liang, K. N. Tu, and Y. S. Lai, "Electromigration Induced High Fraction of Compound Formation in SnAgCu Flip Chip Joints with Copper Column," *Appl. Phys. Lett.*, Vol. 92, 2008, pp. 262104–262104-3.

[32] Chang, Y., T. H. Chiang, and C. Chen, "Effect of Voids Propagation on Bump Resistance Due to Electromigration in Flip-Chip Solder Joints Using Kelvin Structure," *Appl. Phys. Lett.*, Vol. 91, 2007, pp. 132113–132113-3.

[33] Yeo, A., B. Ebersberger, and C. Lee, " Consideration of Temperature and Current Stress Testing on Flip Chip Solder Interconnects," *Miroelectron Reliab.*, Vol. 48, 2008, pp. 1847–1856.

[34] Lu, M., D. Y. Shih, P. Lauro, C. Goldsmith, and D. W. Henderson, "Effect of Sn Grain Orientation on Electromigration Mechanism in High Sn-Based Pb-Free Solders," *Appl. Phys. Lett.*, Vol. 92, 2008, pp. 211909–211909-3.

[35] Yang, P., C. C. Kuo, and C. Chen, "The Effect of Pre-Aging on the Electromigration of Flip-Chip Sn-Ag Solder Joints," *J Mater.*, 2008, pp. 77–80.

[36] Laurila, T., V. Vuorinen, and J. K. Kivilahti, "Interfacial Reactions Between Lead-Free Solders and Common Base Materials," *Mater. Sci. Eng.*, Vol. R49, 2005, pp. 1–60.

[37] Wei, C., C. F. Chen, P. C. Liu, and C. Chen, "Electromigration in Sn-Cu Intermetallic Compounds," *J. Appl. Phys.*, Vol. 105, 2009, pp. 023715–023715-41.

[38] Zeng, K., R. Stierman, T. Chiu, D. Edwards, K. Ano, K. N. Tu, "Kirkendall Voids Formation in Eutectic SnPb Solder Joints on Bare Cu and Its Effects on Joint Reliability," *J. Appl. Phys.*, Vol. 97, 2005, pp. 024508–024508-8.

[39] Lee, C., P. Wang, and J. Kim, "Are Intermetallics in Solder Joints Really Brittle?" *IEEE/ECTC Proceedings*, Reno, NV, May 29–June 1, 2007, pp. 648–652.

[40] Labie, R., P. Limaye, K. W. Lee, C. J. Berry, E. Beyne, and I. D. Wolf, "Reliability Testing of Cu-Sn Intermetallic Micro-Bump Interconnections for 3D-Device Stacking," *International Interconnect Technology Conference*, 2008, pp. 19–21.

[41] Ouyang, F.-Y., K. N. Tu, C.-L. Kao, and Y.-S. Lai, "Effect of Electromigration in the Anodic Al Interconnect on Melting of Flip Chip Solder Joints," *Applied Physics Letters*, Vol. 90, No. 2, 2007, pp. 294–296.

[42] Liu, C., J. Chen, Y. Chuang, L. Ke, and S. Wang, "Electromigration-induced Kirkendall voids at the Cu/Cu3Sn Interface in Flip-Chip Cu/Sn/Cu Joints," *Appl. Phys. Lett.*, Vol. 90, Issue 11, 2007, pp. 112114–112114-3.

[43] Yu, D., T. Chai, M. Thew, Y. Ong, V. Rao, L. Wai, and J. H. Lau, "Electromigration Study of 50 μm Pitch Micro Solder Bumps Using Four-Point Kelvin Structure," *IEEE/ECTC Proceedings*, San Diego, CA, May 25–29, 2009, pp. 930–935.

[44] Lin, Y., C. Zhan, J. Juang, J. H. Lau, T. Chen, R. Lo, M. Kao, T. Tian, and K. N. Tu, "Electromigration in Ni/Sn Intermetallic Micro Bump Joint for 3D IC Chip Stacking", *IEEE/ECTC Proceedings*, Orlando, FL, May 2011, pp. 351–357.

[45] Frear, D., H. Morgan, S. Burchett, and J. H. Lau, *The Mechanics of Solder Alloy Interconnects*, Van Nostrand Reinhold, New York, 1994.

CHAPTER 8

Transient Liquid-Phase Bonding: Chip-to-Chip, Chip-to-Wafer, and Wafer-to-Wafer

8.1 Introduction

For three-dimensional (3D) integrated circuit (IC) integration (not 3D Si integration), the device chip is bonded with the chip/wafer by either chip-to-chip (C2C) or chip-to-wafer (C2W) bonding methods. Owing to chip yields and chip size differences, the wafer-to-wafer (W2W) bonding method is seldom used except micro-electro-mechanical systems (MEMS) applications. Most current bonding methods use temperatures higher than 300°C [1–3], and the throughput is very slow owing to cooling. Thus low-temperature bonding [4–39] is desired to reduce process time. Also, it reduces the damage of the 3D IC integration systems-in-package (SiPs) owing to the thermal-expansion mismatch of the bonding structure (less bow).

The bonding temperature for silicon C2C, C2W, and W2W bonding can be as low as room temperature [13] (e.g., surface-active bonding). However, for this kind of bonding, the bonding surfaces must be very flat and clean, a requirement that does not permit high-volume production and is beyond the scope of this book. Low-melting solders such as InAg, InCu, InSn, InNi, and InSnCu, with bonding temperatures of less than 200°C, will be considered and presented in this chapter.

8.2 How Does Low-Temperature Bonding with Solder Work?

The basics of low-temperature bonding are shown in Fig. 8.1. In this book, low-temperature bonding is the same as such well-defined processes as transient liquid-phase (TLP) bonding, solid-liquid interdiffusion

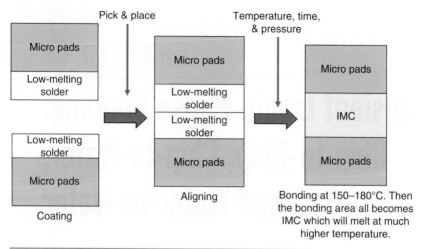

FIGURE 8.1   Fundamental of transient liquid-phase bonding.

(SLID) bonding, and intermetallic compound (IMC) bonding. The low-temperature solder (e.g., InAg, InCu, InSn, InNi, InAu, InSnAu, InCuNi, and InSnCu) is coated on the specially designed and fabricated underbump metallurgy (UBM; e.g., Cu, Cu/Au, Ti/Au, Ni/Au, Ti/Cu, and Cu/Ti/Au). After aligning and bonding at a temperature of less than 200°C, all the solders will react with the UBM on the chips/wafers and become intermetallic compounds (IMCs) with melting points that are a couple of hundred degrees higher than that of the solders.

This feature is welcome by 3D IC chip stacking and 3D MEMS packaging. For example, after bonding the first two chips at a low temperature, all the bonding interconnects are converted into IMCs with a much higher remelting temperature. When the third chip is bonded on top of the first two chips with low temperature, the interconnects (IMCs) between those two chips will not be reflowed and moved. For another example, after the bonding of a MEMS device to an application-specific IC (ASIC) wafer with a low-melting solder, all the bonding (solder-interconnect) areas become IMCs with a very high remelting temperature. When a cap wafer is bonded to the ASIC wafer (which is already bonded with the MEMS device), the interconnect between the ASIC and the MEMS devices will not be reflowed and moved. Furthermore, when the whole 3D IC chip-stacking and/or 3D MEMS packaging is attached to the printed circuit board (PCB) with surface-mount technology (SMT) lead-free soldering (with a maximum reflow temperature as high as 260°C), the solder interconnects (IMCs) between the chips for the 3D chip stacking and between the ASIC and MEMS (for 3D MEMS) will not be reflowed and moved.

8.3   Low-Temperature C2C [(SiO$_2$/Si$_3$N$_4$/Ti/Cu) to (SiO$_2$/Si$_3$N$_4$/Ti/Cu/In/Sn/Au)] Bonding

8.3.1   Test Vehicle

Figures 8.2 and 8.3 show schematic cross sections of hermetic C2C and face-to-face low-temperature bonding. The C2W bonder (Fig. 8.4) is used for the C2C bonding. The size of the bonding ring on the silicon base chip is 8×8 mm, and the width of the bonding ring is 300 μm.

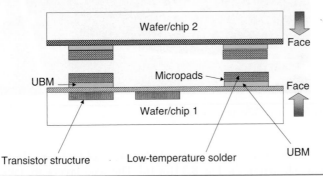

FIGURE 8.2   Face-to-face low-temperature bonding.

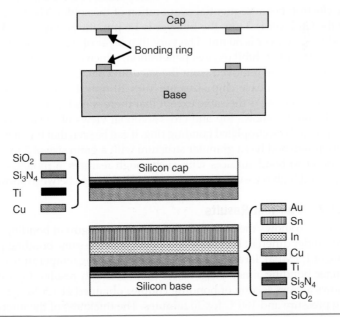

FIGURE 8.3   Schematic of cap chip and base chip with bonding ring (*above*) and their multilayer structures on the bonding ring.

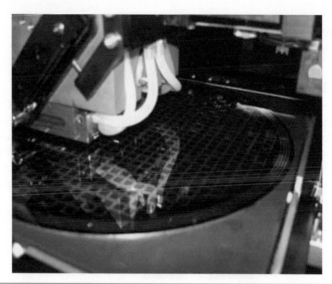

FIGURE 8.4   The C2W bonder (SUSS).

The seed layers (SiO_2, Si_3N_4, Ti, and Cu) and solder layers (In, Sn, and Au) are shown in the Fig. 8.3 [15, 16]. The seed layers are fabricated by a sequential evaporation process along the bonding ring from a silicon wafer. This is followed by electroplating of the Cu and, finally, by electron-beam sequential evaporation of the solder. The thickness of the Cu layer is 3 μm, the In layer is 1.6 μm, the Sn layer is 1.4 μm, and the Au layer is 50 nm. The very thin layer of Au is deposited on the Sn layer to inhibit oxygen penetration, which could happen when the samples are removed from the vacuum chamber and exposed to air. The silicon-base chip sizes are 10×10 mm. For the silicon cap chip, everything is the same except that there is no In, Sn, or Au.

Figure 8.5 shows top and cross-sectional views of the as-received evaporated/electroplated bonding ring. It can be seen that the surface is very rough and has a granular structure with a grain size of 2 to 3 μm. In order to bond surfaces with such rough features, a molten-solder layer and high pressure are necessary to fill the gaps between surfaces.

8.3.2   Pull-Test Results

Figure 8.6 shows the pull-test results (bond strength) of bonding couples under different bonding conditions (with the same bonding pressure of 1.5 MPa). It can be seen that higher bonding temperatures with longer bonding times yield better bond-strength results. Figure 8.7 shows a cross section of bonding couples obtained at 1.5-MPa bonding pressure and 180°C for 20 minutes. The thickness of the interconnect is uniform throughout the whole contact area, with a typical value of 6.2 ± 0.5 μm. The bond couples are in good contact, and no

FIGURE 8.5   Scanning electron microscopic (SEM) images of as-received evaporated composite coating surface (*left*) and its cross section (*right*).

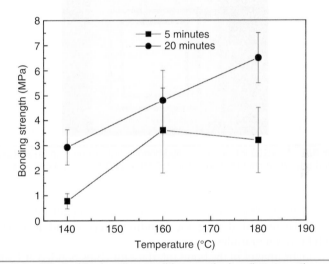

FIGURE 8.6   Effect of the bonding conditions on the bonding strength.

FIGURE 8.7   Cross section of a bonding ring bonded at 1.5 MPa, 180°C, and 20 minutes.

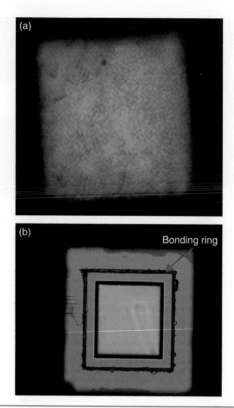

FIGURE 8.8   Acoustic images of a good bonded flat silicon chip (a) and good bonded patterned chip (b) bonded at 1.5 MPa, 180°C, and 20 minutes.

interface line is observed, which means that the solder intermediate layers have melted during the bonding process and leveled the trough between granules.

Acoustic images of the bond interface for chips bonded at 1.5 MPa of bonding pressure and 180°C for 20 minutes are shown in Fig. 8.8. The whole bonding ring is a uniform gray color, indicating that the bonded couples have intimate contact and that no voids or delaminations occurred. The helium-leak-rate test results show that the bonding couples are hermetically sealed with an average leak rate of 5.8×10^{-9} (atm · mL)/s (the average of five samples).

8.3.3   X-Ray Diffraction and Transmission Electron Microscope Observations

The fracture surfaces after pull tests are studied by x-ray diffraction (XRD) to examine the IMCs evolved in the bonding reaction zone. The $AuIn_2$, Cu_6Sn_5, and $Cu_{11}In_9$ phases are found in the XRD analyses shown in Fig. 8.9. No peak of Sn or In is detected.

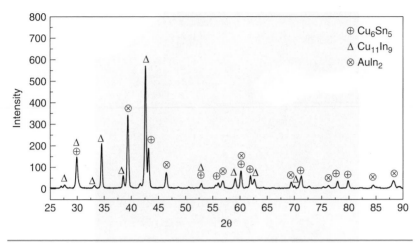

FIGURE 8.9   XRD results of interface after pull tests.

Since the IMC layers are relatively thin, the transmission electron microscope (TEM) technique is used because it provides local chemical analysis. Figures 8.10 and 8.11, respectively, show TEM images of the region close to the base material silicon and interfacial reaction region between the Cu and solder intermediate layers. Figure 8.10 shows that on top of the silicon, the first layer is silicon oxide, followed by silicon nitride, titanium, and copper. The interfaces between each layer are void-free and uniform. The silicon oxide and nitride

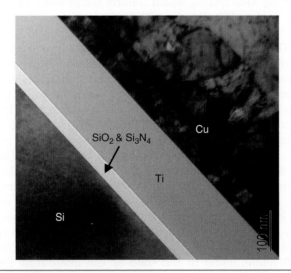

FIGURE 8.10   TEM image at silicon region.

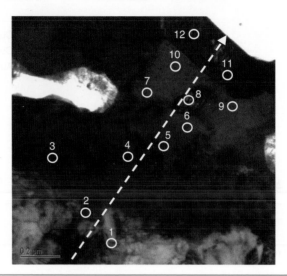

FIGURE 8.11   TEM image at interface region.

layers provide good adhesion between the subsequent metal layer and silicon base.

Figure 8.11 shows the reaction zone between the Cu and SnIn solder interlayer. Small grains are clearly visible in the reaction zone. The elemental composition results of the selected points are summarized in Table 8.1. Two IMCs are identified along the main reaction zone, that is, $Cu_6(Sn, In)_5$ and $Cu_{11}(In, Sn)_9$. This result is very consistent with the XRD results, which confirm the presence of these two IMCs in the bonding interface. The composition of Cu gradually reduces from 56 to 48 at% from the near-Cu region to the near-In region. The energy dispersive x-ray (EDX) analysis suggests that there is still unreacted Cu that is not used for joint formation. At the edge of the bonding interface, Cu composition drops dramatically, and only In and Au dominate, with the composition shown as points 11 and 12 in Table 8.1, that is, $Cu_{5.9}Au_{34}In_{59.0}Sn_{1.1}$ and $Cu_{5.1}Au_{33}In_{60.2}Sn_{1.7}$, which correspond to an equilibrium-phase $AuIn_2$.

It is interesting to note that the overall composite solder is molten at 180°C, and the melting temperature of Sn is 232°C. This is so because the molten In layer is mixed with a solid-state Sn layer and forms a thin, low-melting In-Sn layer at the interface. From the binary diagram, the eutectic In-Sn alloy has a lower melting point of 118°C, and at room temperature, the extent of Sn content in β phase ranges from 15 to 28 wt%, and the extent of In in γ phase can reach approximately 28 wt%. This means that In and Sn can easily diffuse into each other. Therefore, liquid In and solid Sn can interdiffuse into each other, and as a result, the eutectic In-Sn layer grows at the expense of

Point	Composition (at%)				Phase
	Cu	In	Sn	Au	
1	99.5	0.3	0.2	0	Cu
2	99	0.3	0.2	0.5	Cu
3	56.0	28.5	15.5	0	$Cu_{11}(In,Sn)_9$
4	54.6	31.3	13.7	0.4	$Cu_{11}(In,Sn)_9$
5	52.1	30.4	17.0	0.5	$Cu_{11}(In,Sn)_9$
6	51.4	28.3	19.6	0.7	$Cu_{11}(In,Sn)_9$
7	50.2	31.2	18.3	0.3	$Cu_{11}(In,Sn)_9$
8	48.1	21.0	30.4	0.5	$\eta-Cu_6(SnIn)_5$
9	49.8	23.4	26.6	0.2	$\eta-Cu_6(SnIn)_5$
10	50.3	20.7	29.0	0	$\eta-Cu_6(SnIn)_5$
11	5.9	59.0	1.1	34.0	$Au(In,Sn)_2$
12	5.1	60.2	1.7	33.0	$Au(In,Sn)_2$

TABLE 8.1 Summary of Point EDX Analysis Along the Interface Shown in the TEM Image (Fig. 8.11)

the single In layer and Sn layer until both solder layers are completely consumed.

Notice that the initial overall composition of In and Sn present in the as-received base chip is the eutectic composition, that is, 48 wt% Sn. Although the In also reacts with Au, it consumes only a small portion of In, which leaves the overall In and Sn compositions slightly shifted to the off-eutectic point, that is, with 50 wt% Sn. Thus, with such a near-eutectic liquid layer, the intrinsic roughness of the evaporated surface can be overcome, and a good joint interface is observed, as described earlier. During the soldering process, the subsequent interdiffusion of Cu and Sn or In takes place in the interface region and forms various IMCs. Because of the fully molten status of the intermediate layers at 180°C, this may explain why the In atoms can interact with Au and form $AuIn_2$ in the case where the Sn layer initially lies between the In layer and the Au.

The composition analysis and XRD results also suggest that no unreacted In or Sn is present after bonding. The solder materials have completely transformed into IMCs by reacting with copper. This finding indicates that the bonding joint produced at low temperature can withstand high service temperatures owing to the presence of high-melting IMCs. From a consideration of the phase diagrams, the melting points of Cu_6Sn_5, $Cu_{11}In_9$, and $AuIn_2$ are 430, 320, and 580°C, respectively. Therefore, the thermal stability of the joint is expected to be higher than 300°C.

8.4   Low-Temperature C2C [(SiO₂/Ti/Cu/Au/Sn/In/Sn/Au) to (SiO₂/Ti/Cu/Sn/In/Sn/Au)] Bonding

8.4.1   Test Vehicle

Figure 8.12 shows a slightly different solder system [10, 11], where Cu/Au is used for the UBM, and the formation of mixed In-Cu and In-Au IMC phases is expected at the joint after bonding. The IMC phases and their melting temperatures could be checked in the phase diagrams (Fig. 8.13). The structure and dimensions of various solder elements are shown in Fig. 8.12, where the bonding conditions, such as pressure (2.5 MPa), temperature (180°C), and time (20 minutes) are given. Figure 8.14 shows a void-free cross section and the differential scanning calorimetry (DSC) curve indicating that the remelting temperature is higher than 400°C.

8.4.2   Qualification Test Results

Figure 8.15 shows the very good helium-leakage test results and shear-strength test results. The low-temperature interconnects also pass some reliability assessments, such as the pressure-cook test (PCT), the high-temperature-storage test (HTS), and the thermal-cycling test (TCT).

8.5   Low-Temperature C2W [(SiO₂/Ti/Au/Sn/In/Au) to (SiO₂/Ti/Au)] Bonding

For 3D IC chip stacking, as mentioned in Chap. 1, the chip is bonded with the base chip or other chips by either C2C or C2W bonding methods. Owing to IC chip yield and chip size differences, the W2W bonding

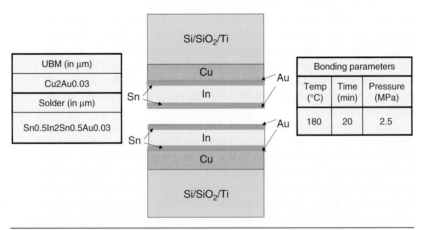

FIGURE 8.12   Schematic of the cap chip and base chip and their multilayer structures on the bonding ring.

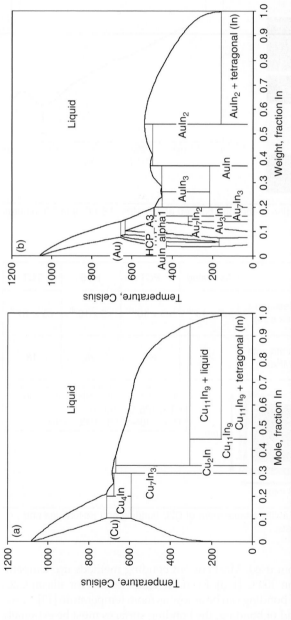

FIGURE 8.13 (a) Phase diagram of Cu-In. (b) Phase diagram of Au-In.

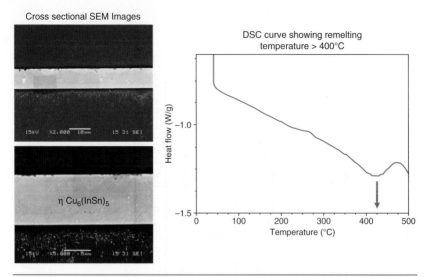

FIGURE. 8.14   Cross section of the joint bonded at 1.5 MPa, 180°C, and 20 minutes (*left*) and its DSC curve.

	As sealed	PCT	HTS	TCT
He leakage (atm × cc/s)	$<5 \times 10^{-8}$	$<5 \times 10^{-8}$	$<5 \times 10^{-8}$	$<5 \times 10^{-8}$
Shear strength (MPa)	27	34	28	18

PCT: 121°C, 100% RH, 2 atm, 300 h
HTS: 125°C, 1000 h
TCT: −40/+125°C 1000 cycles

Seal ring

8 mm

8 mm

FIGURE 8.15   Reliability assessment of C2C bonding and He-leakage and shear-strength tests.

method is seldom used. Most current bonding methods use temperatures higher than 300°C [1–3]. Even the temperatures for silicon C2C, C2W, and W2W bonding can be as low as room temperature [13]. However, for this kind of bonding, the bonding surfaces must be extremely flat and clean, a requirement that prevents it from being used for high-volume production and thus is beyond the scope of this book.

Low-melting solders such as InAg, InCu, InSn, InNi, InAu, InSnAu, InCuNi, and InSnCu with a bonding temperature of less than 200°C [4–39] are desired to reduce damage (warpage) to the microstructures resulting from the thermal-expansion mismatch of the bonding chips [one of them with copper-filled through-silicon vias (TSVs)]. Also, the solder joints form robust IMC interconnects with high remelting temperatures, which allows stable 3D IC chip stacking.

It should be emphasized that with the In-base solder (melting point ≈156°C), the choice of the right composition of solder and UBM to form nothing but IMCs with a high remelting temperature is extremely important. Since the In atoms with additional elements tend to form a ternary eutectic phase, which has an even lower melting temperature than the In binary eutectic point [9], it is necessary to avoid forming the ternary eutectic phase by properly designing the suitable layer combination.

8.5.1 Solder Design

The use of indium with Ni, Cu, or Au has been studied for the low-temperature bonding in 3D integrations [4–7]. Usually, a barrier layer such as Ti is inserted between the In layer and the other metal layers to prevent the excessive interdiffusion [8].

In this section, the In-based solder layer is chosen for the low-temperature solder and bonded to a thin Au layer to form AuIn-based IMCs after bonding. Since the diffusion of Au into the In layer is fast enough to form AuIn intermetallics even at room temperature, a thin layer of Sn is inserted between the In and Au layers to minimize the interdiffusion before bonding; that is, the Sn reacts with the Au and temporarily forms a diffusion barrier layer before bonding so that the In layer is able to be preserved. On the other hand, this barrier layer should not interrupt the interfacial reaction between the In and Au layers during and after bonding to form the IMCs completely.

8.5.2 Test Vehicle

The test vehicle for the 3D IC chip stacking is shown in Figs. 8.16 through 8.18 and Table 8.2. The silicon test chip (8×8 mm^2) has approximately 1700 solder bumps, and each consists of 100-μm-diameter Ti/Au (0.1/1 μm) UBM and 80-μm-diameter Sn/In (0.5/2 μm) solder with a flesh of Au (0.05 μm). The other side of the chip has 120×120 μm of Ti/Au (0.1/1 μm) UBM only.

The test vehicle was fabricated on a 200-mm Si wafer with SiO$_2$ coating. On the front side (face) of the wafer, 0.1 μm of Ti and 1 μm of Au were sputtered on the SiO$_2$ and were patterned to form the 120- × 120-μm UBM. Then the backside (back) of the wafer was sequentially (1) background to 200 μm thick, (2) coated with SiO$_2$ for the passivation layer, (3) deposited with 0.1 μm of Ti and 1 μm of Au (in a sputter chamber), (4) wet etched to pattern the 100-μm-diameter (Ti/Au) UBM pads,

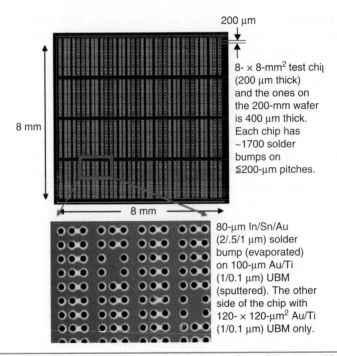

200 μm

8- × 8-mm² test chip (200 μm thick) and the ones on the 200-mm wafer is 400 μm thick. Each chip has ~1700 solder bumps on ≦200-μm pitches.

8 mm

8 mm

80-μm In/Sn/Au (2/.5/1 μm) solder bump (evaporated) on 100-μm Au/Ti (1/0.1 μm) UBM (sputtered). The other side of the chip with 120- × 120-μm² Au/Ti (1/0.1 μm) UBM only.

FIGURE 8.16   Solder bumps and pads on the top surface of the test vehicle.

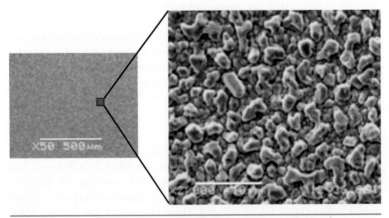

FIGURE 8.17   SEM images of as-evaporated composite coating surface morphology.

(5) laminated with a 20-μm dry film, (6) the openings for the solder bumps are developed, (7) the Sn (0.5 μm) is deposited on the Au (of the UBM) and then the In (2 μm) on the Sn in an evaporation chamber, and (8) the dry film is stripped (a lift-off process). Figure 8.17 shows the top view of the as-evaporated composite coating surface morphology.

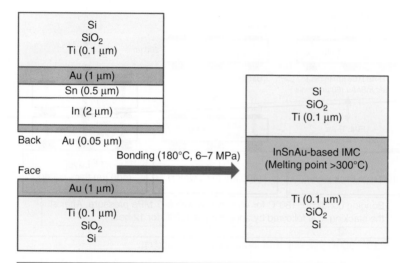

FIGURE 8.18   Schematic of (*left*) the chip with UBM + solder bump and the chip with UBM only and (*right*) the IMCs after bonding.

The surface is rough and has a granular structure with a grain size of 2 to 3 µm. In order to bond a surface with such rough features, a molten-solder layer and high pressure are necessary to fill the gaps between surfaces.

8.5.3   3D IC Chip Stacking with InSnAu Low-Temperature Bonding

Figure 8.19 shows the assembly process of 3D IC chip stacking with low-temperature bonding. The first layer is the 200-mm wafer (400 µm thick) with the 120- × 120-µm UBM (face). The second chip with the lead-free InSnAu solder bump (back) is bonded on the face of the wafer

Items	Values
Wafer size	200 mm
Chip size	8 mm
Stacking chip thickness	200 mm
Base chip thickness	400 mm
Solder bump diameter	80 µm
Solder bump pitch	≦200 mm
Number of bumps	~1700
UBM (Ti/Au)	0.1/1 µm
Solder (In) bump	2.5 µm
Total bump height	3.6 µm

TABLE 8.2   Test Vehicle Dimensions

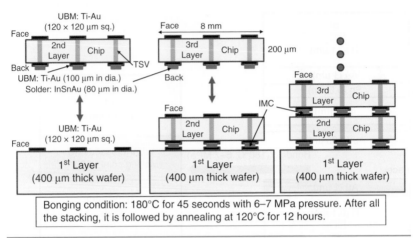

Bonging condition: 180°C for 45 seconds with 6–7 MPa pressure. After all the stacking, it is followed by annealing at 120°C for 12 hours.

Figure 8.19 3D IC stacking with low-temperature bonding.

at 180°C and 6 to 7 MPa for 45 seconds. After bonding, the lead-free solder interconnects become InSnAu IMCs, and the remelt temperature is a couple hundred degrees higher than the melting point of In. Then the back of the third chip with solder bumps is bonded to the face of the second chip under the same conditions. The stacking is followed by annealing at 120°C for 12 hours. Figure 8.20 shows the assembled three-chip stacking by the present low-temperature bonding process.

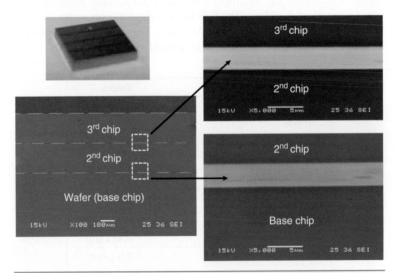

Figure 8.20 The assembled module and SEM cross-sectional views of the three-layer stacked chips consisting of 1700 solder microjoints on each interconnect layer.

The cross sections show that the bonding is in very good condition, and there are no visible voids. The design of experiments leads to the present optimal bonding condition given in Ref. 10.

8.5.4   SEM, TEM, XDR, and DSC of the InSnAu IMCs

Since the IMC layers are relatively thin, the TEM technique is used because it provides local chemical analysis. Figure 8.21 shows TEM images of the region close to the base-material silicon and interfacial reaction region between the AuTi and solder intermediate layers. The interfaces between each layer are void-free, uniform, and fully converted to IMCs. The elemental composition results (at%) of the selected points are labeled in Fig. 8.21. Two IMCs are identified along the main reaction zone, that is, InAu and InSnAu. This result is consistent with the XRD results, which confirmed the presence of these two IMCs in the bonding interface. Figure 8.22 shows the DSC curve indicating that the remelting temperature is higher than 365°C. The SEM image (Fig. 8.21) shows that uniform bonding has occurred without any defects.

8.5.5   Young's Modulus and Hardness of the InSnAu IMCs

The elastic modulus and hardness of the AuSnIn-based IMC interconnects are determined by a nanoindentation method. Since the size

FIGURE 8.21   SEM (above) and TEM (below) images at the interface region.

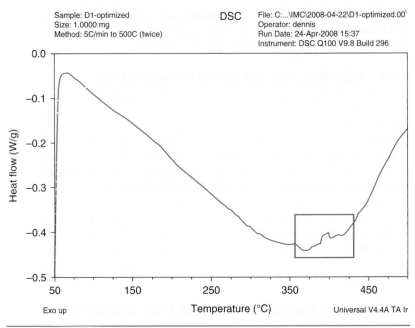

Sample: D1-optimized
Size: 1.0000 mg
Method: 5C/min to 500C (twice)

DSC

File: C:...\IMC\2008-04-22\D1-optimized.00¹
Operator: dennis
Run Date: 24-Apr-2008 15:37
Instrument: DSC Q100 V9.8 Build 296

Exo up

Temperature (°C)

Universal V4.4A TA Ir

FIGURE 8.22   The DSC curve clearly shows the melting point at approximately 360°C.

of the IMC layers from the cross section of a sample is too thin for the tip of an optical microscope, the measurements are carried out along the fractured surfaces of the chips after shear tests. It is found that the Young's modulus of InSnAu is about 81 GPa, which is a little higher than that of Au (78 GPa) [40] and much lower than that of Cu_6Sn_5 (112.6 GPa) and Cu_3Sn (132.7 GPa) [41]. The measured hardness is approximately 1.5 GPa.

8.5.6   Three Reflows of the InSnAu IMCs

There are at least two different kinds of low-temperature assemblies—namely, one is for chip-level interconnects in a package and the other is for board-level interconnects. For board-level interconnects, because it is the final assembly, low-melting solders can be used to attach the packages on the board as long as the solder joints are reliable. However, for chip-level interconnects in a package, because the package has to be assembled on the board with lead-free solders, the package must be qualified before it is mounted on the printed circuit board (PCB). Some of the qualifying tests, among others, include

- *Temperature cycling test* (−40 to 125°C for 1000 cycles) (The package should be preconditioned with humidity exposure and at least three lead-free SMT reflow cycles.)

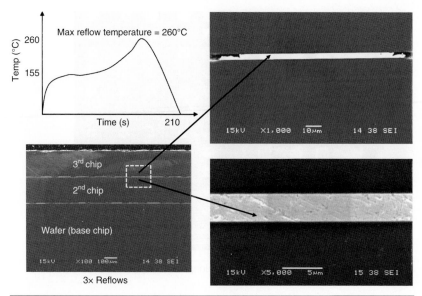

FIGURE 8.23   Cross sections of the 3D IC stacking after three reflows with a maximum temperature of 260°C.

- *Biased HAST* (130°C/85% relative humidity, 1.8 V for 100 hours)
- *Pressure-cook (steam) test* (121°C, 100% relative humidity, at 2 atm for 168 hours)

In general, the quality and reliability requirements of the chip-level interconnects are always tougher than those of board-level interconnects; that is, the low-temperature solders that work for board-level interconnects may not work for chip-level interconnects. Figure 8.23 shows a typical reflow profile for qualifying packages. It also shows the cross section at different magnifications of the IMC interconnects. After three reflows, there is no visible change in the InSnAu IMC interconnects of the 3D IC chip stacking.

8.5.7   Shear Strength of the InSnAu IMCs

Since the InSnAu IMC interconnects are very strong and can resist shear, many of the 200-μm chips are broken during shear (horizontal push) tests. Thus, to determine the shear strength of the InSnAu IMC interconnects, 400-μm chips are used for the tests. Figure 8.24 shows the typical failure mode of the InSnAu IMC interconnects under shear tests. The fracture surfaces occur along the IMC joint and the interface between the TiAu UBM and the InSnAu IMC joint.

Figure 8.25 shows the shear-strength test results of the InSnAu IMC interconnects as fabricated and after three reflows. It can be seen that (1) the shear strength of the first IMC joint (between the base chip

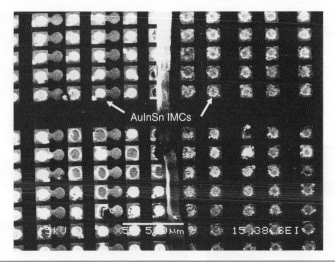

FIGURE 8.24   The fracture surfaces occur along the IMC joints and the interface between the UBM and IMC joints.

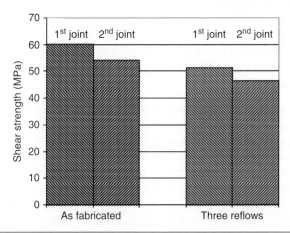

FIGURE 8.25   Bonding shearing strength of the three-layer stack: as fabricated and with three reflows.

and the second chip) and the second IMC joint (between the second chip and the third chip) is reduced by approximately 15 percent because of the three reflows; (2) even with the reduction, however, the shear strength is still much higher than the required 20 MPa; and (3) for both cases, the shear strength of the first IMC joint is slightly higher (~8 percent) than that of the second IMC joint. This could be due to the shear-test setup: (1) First shear (horizontal push) the third chip and hold the bottom two chips to obtain the second IMC strength and (2) then push the second chip and hold the base chip to get the first IMC strength. Since the holders are not 100 percent rigid, some of the push forces become relaxed. Also, the larger the holder, the more the push force relaxes.

8.5.8   Electrical Resistance of the InSnAu IMCs

The electrical resistance of the InSnAu IMC interconnects can be measured by Kelvin's four-point measurement method, as shown in Fig. 8.26. Two bumps of 200-µm diameter are formed far away (800 µm) from each other. Figure 8.27 shows the measurement results, and it can be seen that the electrical resistance of the InSnAu IMC joint is approximately 0.12 Ω, and it is not degraded even after three reflows at 260°C.

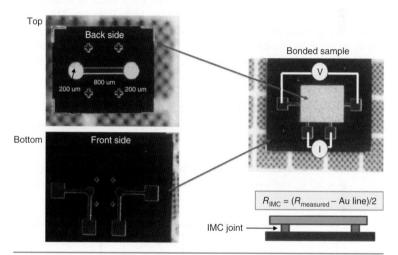

FIGURE 8.26   The test fixtures for the electrical resistance measurement.

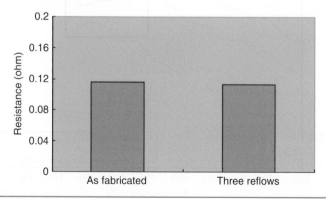

FIGURE 8.27   IMC joint electrical resistance: as fabricated and after three reflows.

8.5.9   When Does the InSnAu IMC Become Unstable?

It has been shown earlier by the DSC method that the remelting temperature of the InSnAu IMC interconnects is greater than 365°C. It would be interesting to find out the temperature when the IMC interconnects of the 3D IC chip stack collapse. However, owing to the size limit of the platinum crucible of the DSC equipment, the whole 3D IC chip stack cannot be put in and measured. Instead, an indirect method is adopted that employs a thermomechanical analyzer (TMA) to measure the expansion of the whole 3D IC chip stack versus temperature, as shown in Fig. 8.28. Before the InSnAu IMC interconnects start to remelt, the curve should be moving upward (very close to a straight line) with temperature. However, when the InSnAu IMC interconnects start to remelt, the expansion curve starts to become unstable until it drops, which means that the 3D IC chip stack is totally collapsed. Thus, based on the three samples tested under TMA and shown in Fig. 8.28, the IMC in the 3D IC chip stack could start remelting at around 380 to 400°C (these are higher than 365°C because the whole 3D IC chip stack is being measured and has absorbed more heat) and collapse at around 450 to 480°C.

8.5.10   Summary and Recommendations

Some important results and recommendations are summarized as follows:

- A low-temperature solder system (InSnAu) has been designed for 3D IC chip-stacking applications. The bonding conditions

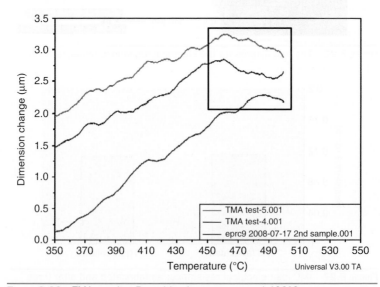

FIGURE 8.28   TMA results: Remelting happens around 400°C.

are 180°C and 6 MPa for 45 seconds. After bonding, the whole assembly is annealed at 120°C for 12 hours.

- SEM, XDR, and TEM images show that uniform bonding has been achieved without any defects, and all the bonding areas have been converted into InSnAu and InAu IMCs. Also, DSC results show that the remelting temperature of the IMCs is greater than 365°C.

- After three reflows (with a maximum temperature of 260°C), there is (1) no visible change of the InSnAu IMCs of the 3D IC chip stacking and (2) no change in the electrical contact resistance of the IMCs.

- Because of the three reflows (with a maximum temperature of 260°C), the shear strength of the InSnAu IMCs is reduced by 15 percent. Even with this reduction, IMC strength is still more than two times higher than the required 20 MPa.

- The Young's modulus of the InSnAu IMCs has been measured on the fracture face by a nanoindentation method and is equal to 81 GPa. The measured hardness of the IMCs is 1.5 GPa.

- The failure modes (fracture surfaces) of the InSnAu IMC interconnects subjected to shear are along the IMC and at the interface between the TiAu UBM and the InSnAu IMC.

- For 3D IC chip stacking, because the chip could be as thin as 20 μm, which makes the ordinary characterization methods very difficult or impossible, new methods and equipment are desperately needed.

8.6   Low-Temperature W2W [TiCuTiAu to TiCuTiAuSnInSnInAu] Bonding

W2W bonding (Fig. 8.29) is one of the most important technologies in high-throughput semiconductor manufacturing. As discussed earlier, (1) W2W bonding is the common assembly technology for 3D Si integration, (2) for 3D IC integration, W2W bonding is not very common except, for example, in MEMS high-yield cap wafers to the MEMS/ASIC (application-specific IC) wafers and high-yield transparent plastic lens wafers to light-emitting diodes (LEDs)/driver wafers, and (3) for microchannel applications, W2W bonding is used for bonding two high-yield TSV microchannnel wafers together. Since the W2W bonding for 3D Si integration is bumpless and far from volume production, only W2W bondings with solder are presented in this chapter. The solder system is the SnIn solder and CuTiAu metallization (≤180°C, W2W bonding for MEMS applications).

Figure 8.29   W2W bonder (EVG).

8.6.1   Test Vehicle

Figure 8.30 [21, 22] shows a schematic view of the top and bottom wafers for bonding and a detailed schematic of metallization deposited on wafers. It can be seen that (1) there are 119 square sealing rings of size of 11 × 11 mm and width of 300 μm, (2) the sealing rings are constructed of SnIn solder (the phase diagram is shown in Fig. 8.31), (3) the UBM is TiCuTiAu, and (4) a cavity exists with an area of 6 × 6 mm and a depth of 250 μm.

8.6.2   Test Vehicle Fabrication

At first, a 300-Å thickness of SiO and a 1500-Å thickness of SiN were formed on a 200-mm silicon wafer by the thermal oxidation and low-pressure chemical-vapor deposition process. They act as a hard mask for the cavity etching. The patterning process was done by photolithography using dry film as a photoresistant material. The cavity inside each bonding ring was formed by using a wet-etching KOH process on both the cap and bottom wafers. The sealing rings were constructed of SnIn solder on Cu-based high-temperature components. TiCuTiAu metallization with thicknesses of 0.05, 2, 0.05, and 0.03 μm was sputtered on the Si/SiO/SiN substrate. Here, the first thin Ti layer acts as the adhesive layer, and the second Ti layer acts as the buffer layer. The thin Au layer was necessary for wetting and preventing metals from oxidizing before solder deposition.

In order to achieve an InSn alloy in a relatively short time, four layers of Sn/In/Sn/In solders were deposited with thicknesses of 1, 1, 0.7, and 0.8 μm, respectively. Based on this design, the atom percentage of Sn in InSn was about 50 percent, which is approximate

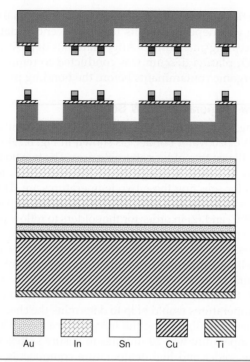

FIGURE 8.30   Schematic view of top and bottom wafers for bonding (*above*) and metallization deposited on the wafers for low-temperature bonding (*below*).

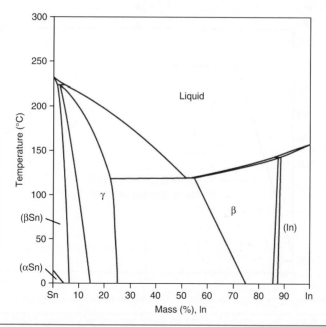

FIGURE 8.31   Phase diagram of InSn alloy.

to the eutectic composition of the InSn alloy. Lastly, a 0.03-μm thin layer of Au was deposited on its top to prevent oxidation during further processing and storage. After materials deposition and dry-film strip, O_2 plasma descum was conducted to remove the oxide layer and organic contaminants before the bonding process.

8.6.3   Low-Temperature W2W Bonding

The bonding was conducted in a controlled nitrogen atmosphere using the EVG520 wafer bonder (as shown in Fig. 8.29) with a pressure of 5.5 MPa on the solder patterning area. The two bonded wafers were heated to a peak temperature of 180 and 150°C, respectively. The reasons for selecting these two temperatures are (1) the melting temperatures of Sn, In, and eutectic InSn alloys are, respectively, 232, 156, and 118°C, and (2) in order for the solders to reflow properly, the temperature must be 30 to 40°C higher than the melting temperature of the solders. Thus the bonding temperature of 180°C is much higher than the melting temperature of both In and eutectic InSn alloys. By way of contrast, the bonding temperature of 150°C is a mere 30°C higher than the eutectic temperature of InSn. Therefore, these two bonding temperatures would help to determine the characteristics of W2W hermetic bonding using the InSn solder layers. Dwell time at the peak temperature was 20 minutes to ensure sufficient diffusion between the solder and high-temperature components. Figure 8.32 shows the W2W bonding temperature profile, and Fig. 8.33 shows some typical C-mode scanning acoustic microscope (C-SAM) graphs

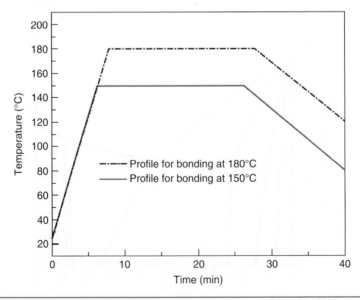

FIGURE 8.32   Temperature profiles for wafer bonding at 180 and 150°C.

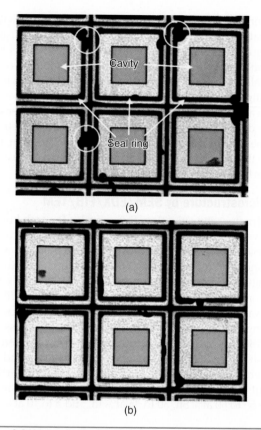

(a)

(b)

FIGURE 8.33   C-SAM graphs for bonded devices: bonding at (a) 180°C and (b) 150°C.

of the bonded modules at 180°C (a) and 150°C (b). After bonding, the bonded wafers were diced into individual dies (13 × 13 mm) with a dicing speed of 2 mm/s.

8.6.4   C-SAM Inspection

After dicing, all the dies were nondestructively inspected by C-SAM for investigation of the W2W bonding interface. Some typical C-SAM images for the bonded modules are shown in Fig. 8.33. A die is labeled as a "good die" if there are no detectable voids and/or cracks. The *bonding yield* is further defined as the percentage of good dies to the gross dies of a bonded wafer. Based on these definitions, 100 percent good seals (no voids and holes) throughout the 200-mm wafers bonded by both the 180 and 150°C bonding temperatures have been obtained, as shown in Fig. 8.33. However, since the resolution of C-SAM is about several micrometers, very tiny voids cannot be detected. The effect of bonding temperature also can be observed by

C-SAM. More solder is squeezed out from the seal joint with a bonding temperature of 180°C than at 150°C. This is so because at higher bonding temperatures, the flowability of liquid solder is better, and under the same pressure, more molten solder would flow over the seal ring, especially at the corner regions, as shown in the circles in Fig. 8.33a. The maximum distance of the squeezed-out solder of the sealing ring is about 2 mm at the corners. For small packages, the flow-over solder may damage the device inside. A couple of simple yet useful methods are (1) to reduce the bonding pressure and/or solder thickness but at the same time ensure the yield and (2) to make grooves around the sealing ring to prevent overflow of the solder.

8.6.5   Microstructure by SEM/EDX/FIB/TEM

For microstructure investigation, five samples for each bonding temperatures were mounted and cross-sectioned for SEM observation. These samples are ground with SiC paper and polished with 1.0-μm diamond and 0.05-μm silica suspensions. The compositions of the bonded ring joints are analyzed by the energy dispersive x-ray (EDX). A focused ion beam (FIB) is employed to prepare the thin film for TEM and EDX examinations of the solder joint. SEM images of sealing-ring joint bonding at 180°C are shown in Fig. 8.34. The thickness of the joint formed at 180°C is about 5.5 μm, and that of the IMC interconnects is 1.5 μm, as shown in Fig. 8.34a, b. The whole joint after bonding is thinner, and there are no detectable voids and cracks. The overflow of solder is about 10 μm, as shown in Fig. 8.34c. The compositions of the solders are 55 to 60 at% Sn, 30 to 35 at% In, and 5 to 10 at% Au.

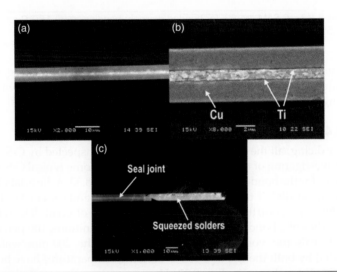

Figure 8.34   Interfacial microstructure of seal joint bonded at 180°C: (a) low magnification; (b) high magnification; (c) solder squeezed out at the edge of the seal joint.

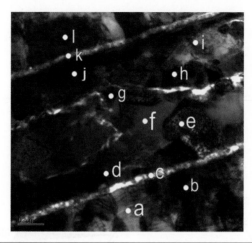

FIGURE 8.35   TEM/EDX analysis for the seal joint bonded at 180°C.

TEM/EDX is used to determine the accurate compositions along the sealing joint. A TEM microstructure is shown in Fig. 8.35, and the EDX results are listed in Table 8.3. It is quite obvious that two parallel Ti lines are embedded in the joint. According to the EDX results for positions **c** and **k**, a small amount of In, Cu, and Sn is found in the Ti layer, which means that these elements can diffuse through the Ti layer. This proves the suitability of using Ti as a buffer layer for

Position	Element (at%)				
	Cu	Ti	Sn	In	Au
a	88.1		7.1	4.7	
b	77		11.8	11.2	
c	5.3	79.6	9.9	5.1	
d	12.9	5.7	12.5	47.7	21.2
e	6.8		3	54.8	35.4
f			33.4	18.8	47.8
g	5.7		2.8	60.2	31.3
h	2.4		6.9	59.2	31.5
i	6.1		7.9	54.5	31.4
j	6	4	21.9	44.4	23.7
k	4.8	90.4	2.6	2.2	
l	50.3		30.7	19.1	

TABLE 8.3   Compositions of Selected Positions in the TEM/EDX Analysis.

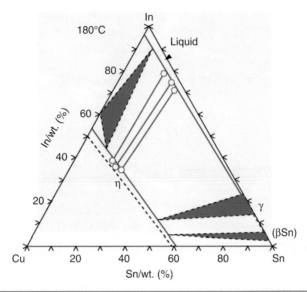

FIGURE 8.36   Isothermal section of CuInSn system at 180°C.

low-temperature W2W bonding. The compositions of positions **a**, **b**, and **l** adjacent to the Cu sides show this phenomenon. For positions **a** and **b**, a small amount of Sn and In is found in Cu, which corresponds to the Cu-rich phase. For position **l**, the ratio between Cu and SnIn is around 1:1. According to the thermodynamic calculation of phase equilibrium in a CuInSn system at 180°C, as shown in Fig. 8.36 [9], an η phase exists in the system. In fact, the η phase exists in both the two binary systems—CuIn and CuSn, that is, η-Cu₂In and η-Cu₆Sn₅, with the same NiAu-type structure [12]. According to the study of Sommadossi et al. and the high solubility of Sn, the compounds should be η-Cu₆(Sn-In)₅ [12].

For positions **e**, **j**, **h**, and **i**, the compositions correspond to the Au(InSn)₂ phases. The small Cu content may be dissolved in the compounds during the liquid/solid diffusion bonding process. For position **d**, the composition of CuSnInAu comes from two phases: AuInSn and CuSnIn. For point **f**, the composition falls in the Au(SnIn) phase, which may be formed owing to the absence of In atoms because a portion of the solder is squeezed out.

SEM photos of the sealing-ring joint bonding at 150°C are shown in Fig. 8.37. The thickness of the joint is about 6.5 μm, in which the thickness of IMCs is around 2 μm. Comparing Fig. 8.34 with Fig. 8.37, we also find that the joint formed at 150°C is thicker than that formed at 180°C. The thickness differences of these two sealing joints are mainly caused by the different bonding temperatures. At higher temperatures, solder material melts faster and has better flowability.

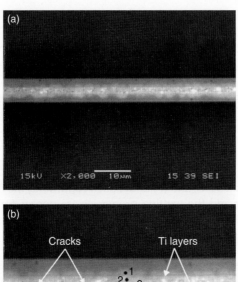

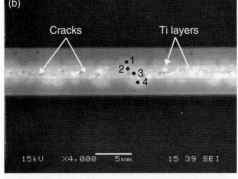

FIGURE **8.37** Interfacial microstructure of seal joint bonded at 150°C: (a) low magnification; (b) high magnification.

Under the same pressure, more solder liquid is squeezed out at higher bonding temperatures. Thus a thin solder joint is formed with a thin IMC layer because more solder alloy is kept at 150°C for diffusion bonding; that is, more solder material is diffused through the Ti layer into the Cu substrates, as shown in Fig. 8.37b. When bonding at 150°C, owing to the low bonding temperature, cracks at the center of the sealing joint are formed that may be harmful to its hermeticity and reliability properties. The reasons for the poor bonding qualities are (1) at the surface of the solder material, after the thin Au layer deposition, one thin AuIn layer may be formed; (2) during bonding, both bonding pairs must break through the AuIn layer and meet with each other; (3) the melting points of In and Sn are 156 and 232°C, respectively; (4) ideally, it would be nice if the InSn layers could form the eutectic alloy at as low as 118°C; (5) unfortunately, they do not, and when the temperature reaches 150°C, it takes a relatively longer time for InSn interdiffusion to form a low-temperature alloy; (6) even if a eutectic alloy is finally obtained, the temperature difference is

Position	Element (at%)				
	Cu	Ti	Sn	In	Au
1	78.01	5.05	3.48	8.94	4.52
2	72.51	5.49	13.60	8.40	0
3	60.17	3.31	22.88	13.64	0
4	51.21	2.02	20.86	18.21	7.70

TABLE 8.4   Compositions of the interface in Fig. 8.37b.

merely around 30°C; and (7) under such conditions, the flowability of liquid solder and the wetting properties are poorer compared with those under higher bonding temperatures. Therefore, less liquid alloy is pushed through the AuIn layer for interdiffusion and joining. As a result, voids along the AuIn layer are formed, and at some points, cracks are generated owing to the continuity of the voids.

EDX analysis shows that Sn and In atoms diffuse through Ti into Cu and near the Cu regions, and $Cu_3(SnIn)$ phases are formed. At the center region, both $Cu_6(SnIn)_5$ and AuIn phases are formed according to the EDX analysis listed in Table 8.4.

8.6.6   Helium Leak-Rate Test and Results

The hermeticity of the ring seals for patterned dies is evaluated by helium leak-rate tests based on MIL-STD-883. Based on the internal cavity volume of the bonded dies (~0.02 cm³), the pressure of helium gas in a bombing chamber should be set at 75 psi, and the exposure time is 2 hours. After bombing, the samples were put into the helium leak detector to measure the leak rates. For the leak-rate test, the sample size was 21. The results show that the leak rate of the devices bonded at 180°C was smaller than 2×10^{-8} (atm · cc)/s, which is smaller than the rejected limit value [5×10^{-8} (atm · cc)/s] of MIL-STD-883E. However, for packages bonded at 150°C, only 86 percent had a leak rate smaller than 5×10^{-8} (atm · cc)/s. This could be due to the voids/cracks in the middle of the sealing ring.

8.6.7   Reliability Tests and Results

Since the hermeticity of the packages bonded at 150°C is not very good, for the reliability study, only packages bonded at 180°C were considered. The reliability tests and results are listed in Table 8.5. The samples were subjected to the pressure-cooker test (121°C at 2 atm for 300 hours), the high-humidity test (85°C/85% relative humidity for 1000 hours), the high-temperature-storage test (125°C for 1000 hours), and the temperature-cycling test (−40 to 125°C for 1000 cycles). For each test, the sample size was 21. Before and after the tests, the samples were examined by C-SAM, helium leak-rate test, and shear test.

Reliability Test Items	Results	
	Leak-Rate Test (5×10^{-8} atm · cc/s)	Shear Strength (MPa)
Pressure-cooker test	$90.5\% < 5 \times 10^{-8}$	27.61
High humidity	$95.2\% < 5 \times 10^{-8}$	24.57
High-temperature storage	$100\% < 5 \times 10^{-8}$	21.27
Temperature cycle	$90.5\% < 5 \times 10^{-8}$	25.62
*After bonding	$100\% < 2 \times 10^{-8}$	46.32

TABLE 8.5 Characterization Results after Reliability Assessment Test

Failure was defined as a leak rate smaller than 5×10^{-8} (atm · cc)/s or a large crack in the sealing ring. The ratios of packages with a leak rate smaller than 5×10^{-8} (atm · cc)/s after the pressure-cooker test, high-humidity storage, high-temperature-storage test, and temperature-cycling test are 90.5, 95.2, 100, and 90.5 percent, respectively. The shear-strength requirement for this package is 6 MPa. Thus it can be seen from Table 8.5 that the sample shearing-stress results for all the tests are larger than the minimum requirement. The interfacial microstructures were analyzed after reliability tests. Figure 8.38 shows that (1) there are no significant changes in the microstructures, (2) with further SEM/EDX analysis, the compositions of the sealing ring also show no detectable changes, which is reasonable because the reliability test temperatures are not very high compared with the melting temperatures of the IMCs, and (3) the Ti layer also can inhibit the diffusion between Cu and IMCs. However, according to the shearing test results, the adhesions between different IMCs decrease after reliability tests. Also, during long-term reliability tests, diffusion occurs between the overflow solder and high-temperature components (e.g., Cu and IMCs), as shown in Fig. 8.39 for the microstructure of the edges of the sealing ring after high-temperature storage. According to EDX analysis, all the overflowed solders became high-temperature compounds, and tens of micrometers of Cu substrates are consumed and turned into $Cu_6(SnIn)_5$. For the other three reliability tests, because the test temperatures were not high, the diffusion phenomena were not as active. For example, for the thermal cycling test, only 10 to 30 μm on each side of the sealing ring Cu substrates was consumed by about 20 μm of solder. Because the diffusion did not create new compounds and cracks, it will not cause the sealing ring to fail during reliability tests.

Samples with poor hermeticity also were analyzed to understand the failure mechanism of the sealing ring. The failure modes are consistent. As shown in Fig. 8.40, a large crack throughout the sealing joint was found. The crack expands along the bonding line between the InAu compounds and the CuSnIn compounds. When different IMCs are formed at the joints, under thermal-cycling and pressure-cooker

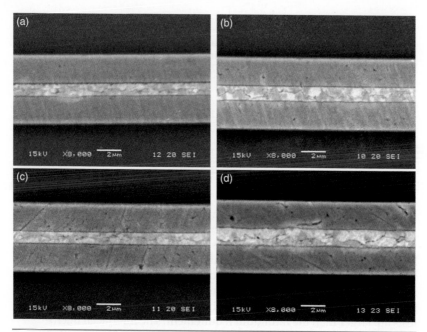

FIGURE 8.38   Interfacial microstructure of the seal joint of good dies after reliability tests: (a) pressure-cooker test, (b) high-humidity storage, (c) high temperature, and (d) temperature cycling.

tests, stresses may be generated. This is the reason why after these two tests, the leak-rate test revealed a relative poorer hermeticity result compared with the other tests. This means that the adhesion between the two phases is the key to determining the reliability of the sealing joint. Therefore, there are at least two ways to enhance the reliability of the present sealing joints: (1) reduce AuIn compounds by a thin Au-layer deposition and (2) reduce the Ti-buffer-layer thickness to allow fast diffusion between the solders and Cu substrate. In this case, less solder will flow out of the joint and thicker IMC joints will be formed with discontinuous AuIn phases.

8.6.8   Summary and Recommendations

Some important results are summarized as follows:

- No detectable voids are found in the sealing rings of all packages after bonding at 150 and 180°C based on C-SAM observation. When bonding at 180°C, all samples obtained a good leak rate [2×10^{-8} (atm · cc)/s]. When bonding at 150°C, only 86 percent of packages can achieve acceptable hermeticity.

- Under the two bonding temperatures (150 and 180°C), In and Sn atoms can diffuse through the Ti layer to form joints with

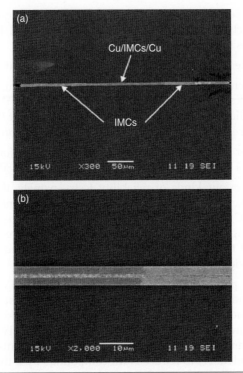

FIGURE 8.39 Diffusion between overflow solder and high-temperature IMC joints after high-temperature storage: (*a*) low magnification; (*b*) high magnification.

FIGURE 8.40 Interfacial microstructure of the failed seal joint after a pressure-cooker test.

Cu. Sealing joints are composed of high-melting-point IMCs, which allow the package for the next level of interconnections to bond at higher temperatures.

- More solders are squeezed out when bonding at 180°C.

- When bonding at lower temperatures, for example, 150°C, a thick sealing joint is obtained owing to the low flowability of the liquid alloy. However, cracks are formed at the center of the sealing rings, which cause the hermeticity problem.

- After bonding, high shear strength is obtained for the sealing rings by both bonding temperatures (150 and 180°C).

- After four kinds of reliability tests—namely, the pressure-cooker test (121°C, 2 atm, 300 hours), high-humidity storage (85°C/85% relative humidity, 1000 hours), high-temperature storage (125°C, 1000 hours), and temperature-cycling test (–40 to 125°C, 1000 hours)—the package bonded at 180°C provides good reliability and hermeticity.

- The propagation of a crack along different IMC layers is found in the failed packages after reliability tests. Therefore, the adhesion between different phases is the key for determining the reliability of the sealing joint. Two ways to enhance the reliability of the present sealing joints are recommended: (1) to reduce AuIn compounds by a thin Au-layer deposition and (2) to reduce the Ti-buffer-layer thickness to allow fast diffusion between the solders and Cu substrate.

- The current material system (Cu/Ti/Au) with the bonding parameters (180°C for 20 minutes under 5.5 MPa of pressure) is recommended to be used for MEMS packaging applications, which require low-temperature bonding and hermetic sealing.

8.7 References

[1] Klumpp, A., "Vertical System Integration by Using Inter-Chip Vias and Solid Liquid Interdiffusion Bonding," *Jpn. J. Appl. Phys.*, Vol. 43, 2004, pp. L829–L830.

[2] Chen, K., "Microstructure Examination of Copper Wafer Bonding," *J. Electron. Mater.*, Vol. 30, 2001, pp. 331–335.

[3] Morrow, P. R., "Three-dimensional Wafer Stacking via Cu-Cu Bonding Integrated with 65-nm Strained-Si/low-*k* CMOS Technology," *IEEE Electron. Device Lett.*, Vol. 27, 2006, pp. 335–337.

[4] Fukushima, T., Y. Yamada, H. Kikuchi, T. Tanaka, and M. Koyanagi, "Self-Assembly Process for Chip-to-Wafer Three-Dimensional Integration," *IEEE/ECTC Proceedings*, Reno, NV, 2007, pp. 836–841.

[5] Sakuma, K., P. Andry, C. Tsang, K. Sueoka, Y. Oyama, C. Patcl, B. Dang, S. Wright, B. Webb, E. Sprogis, R. Polastre, R. Horton, and J. Knickerbocker, "Characterization of Stacked Die using Die-to-Wafer Integration for High Yield and Throughput," *IEEE/ECTC Proceedings*, Lake Beuna Vista, FL, May 2008, pp. 18–23.

[6]   Wakiyama, S., H. Ozaki, Y. Nabe, T. Kume, T. Ezaki, and T. Ogawa, "Novel Low-Temperature CoC Interconnection Technology for Multichip LSI (MCL)," *IEEE/ECTC Proceedings*, Reno, NV, 2007, pp. 610–615.
[7]   Liu, Y. M., and T. H. Chuang, "Interfacial Reactions Between Liquid Indium and Au-Deposited Substrates," *J. Electron. Mater.*, Vol. 29, No. 4, 2000, pp. 405–410.
[8]   Zhang, W., A. Matin, E. Beyne, and W. Ruythooren, "Optimizing Au and In Micro-Bumping for 3D Chip Stacking," *IEEE/ECTC Proceedings*, Lake Beuna Vista, FL, May 2008, pp. 1984–1989.
[9]   Liu, X., H. Liu, I. Ohnuma, R. Kainuma, K. Ishida, S. Itabashi, K. Kameda, and K. Yamaguchi, "Experimental determination and thermodynamic calculation of the phase equilibria in the Cu-In-Sn system," *J. Electron. Mater.*, Vol. 30, no. 9, 2001, pp. 1093–1103.
[10] Choi, W., C. Premachandran, C. Ong, L. Xie, E. Liao, A. Khairyanto, B. Ratmin, K. Chen, P. Thaw, and J. H. Lau, "Development of Novel Intermetallic Joints Using Thin Film Indium Based Solder by Low Temperature Bonding Technology for 3D IC Stacking," *IEEE/ECTC Proceedings*, San Diego, CA, May 2009, pp. 333–338.
[11] Choi, W., D. Yu, C. Lee, L. Yan, A. Yu, S. Yoon, J. H. Lau, M. Cho, Y. Jo, and H. Lee, "Development of Low Temperature Bonding Using In-based Solders," *IEEE/ECTC Proceedings*, Orlando, FL, May 2008, pp. 1294–1299.
[12] Sommadossi, S., and A. F. Guillermet, "Interface Reaction Systematic in the Cu/In-48Sn/Cu System bonded by diffusion soldering," *Intermetallics*, Vol. 15, 2007, pp. 912–917.
[13] Shigetou, A., T. Itoh, K. Sawada, and T. Suga, "Bumpless Interconnect of 6-μm Pitch Cu Electrodes at Room Temperature," *IEEE/ECTC Proceedings*, Lake Buena Vista, FL, May 27–30, 2008, pp. 1405–1409.
[14] Made, R., C. L. Gan, L. Yan, A. Yu, S. U. Yoon, J. H. Lau, and C. Lee, "Study of Low Temperature Thermocompression Bonding in Ag-In Solder for Packaging Applications," *J. Electron. Mater.*, Vol. 38, 2009, pp. 365–371.
[15] Yan, L.-L., C.-K. Lee, D.-Q. Yu, A.-B. Yu, W.-K. Choi, J. H. Lau, and S.-U. Yoon, "A Hermetic Seal Using Composite Thin Solder In/Sn as Intermediate Layer and Its Interdiffusion Reaction with Cu," *J. Electron. Mater.*, Vol. 38, 2009, pp. 200–207.
[16] Yan, L.-L., V. Lee, D. Yu, W. K. Choi, A. Yu, A.-U. Yoon, and J. H. Lau, "A Hermetic Chip to Chip Bonding at Low Temperature with Cu/In/Sn/Cu Joint," *IEEE/ECTC Proceedings*, Orlando, FL, May 2008, pp. 1844–1848.
[17] Yu, A., C. Lee, L. Yan, R. Made, Q. Zhang, S. Yoon, and J. H. Lau, "Development of Wafer Level Packaged Scanning Micromirrors," *Proc. Photon. West*, Vol. 6887, 2008, pp. 1–9.
[18] Lee, C., A. Yu, L. Yan, H. Wang, J. Han, Q. Zhang, and J. H. Lau, "Characterization of Intermediate In/Ag Layers of Low Temperature Fluxless Solder Based Wafer Bonding for MEMS Packaging," *J. Sensors Actuators* (in press).
[19] Yu, D.-Q., C. Lee, L. L. Yan, W. K. Choi, A. Yu, and J. H. Lau, "The Role of Ni Buffer Layer on High Yield Low Temperature Hermetic Wafer Bonding Using In/Sn/Cu Metallization," *Appl. Phys. Lett.* (in press).
[20] Yu, D. Q., L. L. Yan, C. Lee, W. K. Choi, S. U. Yoon, and J. H. Lau, "Study on High Yield Wafer to Wafer Bonding Using In/Sn and Cu Metallization," *Proceedings of the Eurosensors Conference*, Dresden, Germany, 2008, pp. 1242–1245.
[21] Yu, D., C. Lee, and J. H. Lau, "The Role of Ni Buffer Layer Between InSn Solder and Cu Metallization for Hermetic Wafer Bonding," *Proceedings of the International Conference on Electronics Materials and Packaging*, Taipei, Taiwan, October 22–24, 2008, pp. 335–338.
[22] Yu, D., L. Yan, C. Lee, W. Choi, M. Thew, C. Foo, and J. H. Lau, "Wafer Level Hermetic Bonding Using Sn/In and Cu/Ti/Au Metallization," *IEEE Proceeding of Electronics Packaging and Technology Conference*, Singapore, December 2008, pp. 1–6.
[23] Chen, K., C. Premachandran, K. Choi, C. Ong, X. Ling, A. Khairyanto, B. Ratmin, P. Myo, and J. H. Lau, "C2W Bonding Method for MEMS Applications," *IEEE Proceedings of Electronics Packaging Technology Conference*, Singapore, December 2008, pp. 1283–1287.

[24] Premachandran, C. S., J. H. Lau, X. Ling, A. Khairyanto, K. Chen, and Myo Ei Pa Pa, "A Novel, Wafer-Level Stacking Method for Low-Chip Yield and Non-Uniform, Chip-Size Wafers for MEMS and 3D SIP Applications," *IEEE/ECTC Proceedings*, Orlando, FL, May 27–30, 2008, pp. 314–318.

[25] Simic, V., and Z. Marinkovic, "Room Temperature Interactions in Ag-Metals Thin Film Couples," *Thin Solid Films*, Vol. 61, 1979, pp. 149–160.

[26] Lin, J.-C., "Solid-Liquid Interdiffusion Bonding Between In-Coated Silver Thick Films," *Thin Solid Films*, Vol. 61, 1979, pp. 212–221.

[27] Roy, R., and S. K. Sen, "The Kinetics of Formation of Intermetallics in Ag/In Thin Film Couples," *Thin Solid Films*, Vol. 61, 1979, pp. 303–318.

[28] Chuang, R. W., and C. C. Lee, "Silver-Indium Joints Produced at Low Temperature for High-Temperature Devices," *IEEE Trans. Components Packag. Technol. A*, Vol. 25, 2002, pp. 453–458.

[29] Chuang, R. W., and C. C. Lee, "High-Temperature Non-Eutectic Indium-Tin Joints Fabricated by a Fluxless Process," *Thin Solid Films*, Vol. 414, 2002, pp. 175–179.

[30] Lee, C. C., and R. W. Chuang, "Fluxless Non-Eutectic Joints Fabricated Using Gold-Tin Multilayer Composite," *IEEE Trans. Components Packag. Technol. A*, Vol. 26, 2003, pp. 416–422.

[31] Humpston, G., and D. M. Jacobson, *Principles of Soldering*, ASM International, Materials Park, MD, 2004

[32] Chuang, T., H. Lin, and C. Tsao, "Intermetallic Compounds Formed During Diffusion Soldering of $Au/Cu/Al_2O_3$ and Cu/Ti/Si with Sn/In Interlayer," *J. Electron. Mater.*, Vol. 35, 2006, pp. 1566–1570.

[33] Lee, C., and S. Choe, "Fluxless In-Sn Bonding Process at 140°C," *Mater. Sci. Eng.*, Vol. A333, 2002, pp. 45–50.

[34] Lee, C. "Wafer Bonding by Low-Temperature Soldering," *Sensors & Actuators*, Vol. 85, 2000, pp. 330–334.

[35] Vianco, P. T., "Intermetallic Compound Layer Formation Between Copper and Hot-Dipped 100 In, 50In-50Sn, 100Sn, and 63Sn-37Pb Coatings," *J. Electron. Mater.*, Vol. 23, 1994, pp. 583–594.

[36] Frear, D. R., "Intermetallic Growth and Mechanical-Behavior of Low and High Melting Temperature Solder Alloys," *Metall. Mater. Trans. A*, Vol. 25, 1994, pp. 1509–1523.

[37] Morris, J. W., "Microstructure and Mechanical Property of Sn-In and Sn-Bi Solders," *J. Miner. Metals Mater. Soc.*, Vol. 45, 1993, pp. 25–27.

[38] Mei, Z., "Superplastic Creep of Low Melting-Point Solder Joints," *J. Electron. Mater.*, Vol. 21, 1992, pp. 401–407.

[39] Chuang, T. H., "Phase Identification and Growth Kinetics of the Intermetallic Compounds Formed During In-49Sn/Cu Soldering Reactions," *J. Electron. Mater.*, Vol. 31, 2002, pp. 640–645.

[40] http://www.webelements.com/WebElements: The periodic table on the website.

[41] Kazumasa, T., "Micro Cu Bump Interconnection on 3D Chip Stacking Technology," *Jpn J. Appl. Phys.*, Vol. 43, No. 4B, 2004, pp. 2264–2270.

CHAPTER 9

Thermal Management of Three-Dimensional Integrated Circuit Integration

9.1 Introduction

There are many critical issues of three-dimensional (3D) integration [1–33]. For example, electronic design automation (EDA) software is not commonly available; test methods and equipment are lacking; known-good-die (KGD) is required; fast chips are mixed with slow chips; large chips are mixed with small chips; through-silicon vias (TSVs) are required for 3D integration; wafer thinning and thin-wafer handling during processes are necessary; microbumps usually are required; thermal management issues are significant; equipment accuracy for alignments is a problem; there are 3D inspection issues; 3D expertise, infrastructure, and standards are lacking; TSV cost is much higher than that of wire bonding; TSV high-volume production tools are lacking/expensive; TSV design guidelines are not commonly available; TSV design software is lacking; test methodologies and software for TSVs are lacking; copper filling helps on thermal management but increases thermal coefficient of expansion (TCE); voidless copper filling takes a long time (low throughputs); there are tough requirements of TSV wafer yields (>99.9 percent); TSV wafer warpage owing to TCE mismatch is an issue; TSVs with high aspect ratios are difficult to manufacture at high yield; TSV inspection methodology is lacking; and TSV expertise, infrastructure, and standards are lacking.

As mentioned earlier, thermal management is one of the critical issues of 3D integrated circuit (IC) integration. This is so because: (1) the heat flux generated by stacked multifunctional chips in miniature packages is extremely high, (2) 3D circuits increase total power generated per unit surface area, (3) chips in the 3D stack may be overheated if cooling is not properly and adequately provided, (4) the space between the 3D stack may be too small for cooling channels (i.e., no gap for fluid flow), and (5) thin chips may create extreme conditions for on-chip hot spots.

Thermal management is the fourth key-enabling technology for 3D IC integration. However, it should be pointed out that very often thermal management is the show stopper because the system-in-package (SiP) is either too hot to function or the thermal management system is too expensive to build. Thus low-cost and effective thermal management design guidelines and solutions are desperately needed for widespread use of 3D IC integration SiPs.

In this chapter, based on the theory of heat transfer [34–36], the thermal performance of 3D integration SiPs with TSV interposers and memory-chip stacking is studied. The results are plotted in useful design charts for engineering practice, and convenient design guidelines are also provided. The equivalent thermal conductivity developed in Chap. 3 for a TSV interposer/chip with various TSV diameters, pitches, and aspect ratios is used to perform all the simulations reported herein [31]. Finally, a thermal management system with embedded microchannels and TSVs is developed for two chips each with 100 W of heat dissipation [19].

9.2 Effects of TSV Interposer on Thermal Performance of 3D Integration SiPs

9.2.1 Geometry and Thermal Properties of Materials for Package Modeling

Figure 9.1 shows a schematic of a compact model of a 3D SiP with a TSV interposer that is modeled as a block (without TSV) with the determined equivalent thermal conductivities as shown in Sec. 3.4.1 of this book. This model consists of a large chip, a block (for the TSV interposer), a 1-2-1 buildup (BU) substrate, a printed circuit board (PCB), and solder joints. Their dimensions and thermal properties are listed in Tables 9.1 through 9.3.

9.2.2 Effect of TSV Interposer on Package Thermal Resistance

The junction to ambient thermal resistance of a package is often used as an index to describe the thermal performance of the package; the higher the thermal resistance, the poorer is the thermal performance of the package. The junction to ambient thermal

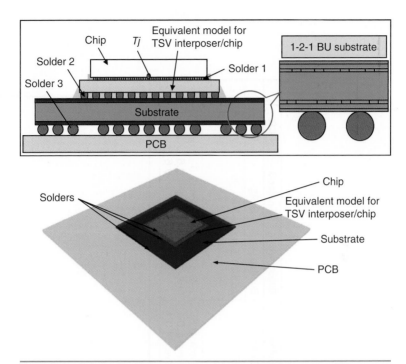

FIGURE 9.1   Schematic diagram and model of a 3D IC integration SiP.

Component	Interposer	TSV	TSV filler			
Material	Silicon	Copper	Polymer with filler particles	Polymer with filler particles	Al	Copper
Thermal conductivity (W/m · °C)	150	390	4.0	25	237	390
Dimension detail (mm)	1.4 × 1.4 × 0.3	Varied	Copper plating thickness for partial filled via (5–25 μm)			
TSV pitch (mm)		0.15–0.6				

TABLE 9.1   Geometry and Thermal Conductivity of Materials for TSV Interposer/Chip

	Chip	TSV	Bumps	Underfill	PCB
Material	Si (TSV)	Cu	SnAg	Polymer	FR4
k (W/m · °C)	Empirical equation	390	57	0.5	// 0.8 ⊥ 0.3
Dimension (mm)	5 × 5	φ 0.05	φ 0.20 Height = 0.15	5 × 5 × 0.15	76 × 114 × 1.6
Power (W)	0.2 W	N.A.	N.A.	N.A.	N.A.

TABLE 9.2   Geometry and Thermal Conductivity of Materials for the 3D IC SiP

Component	Chip	Solder 1	Interposer	Solder 2	Substrate	Solder 3	PCB
Material	Si	SnAg	Si + TSV (Cu)	SnAg	Buildup	SnAg	FR4
Thermal conductivity (W/m · °C)	150	57	K_{eq}	57	$// 100 \perp 0.5$	57	$// 0.8$ $\perp 0.3$
Dimension (mm)	$21 \times 21 \times 0.75$	Pitch: 0.15 Height: 0.08 Ave. D: 0.08	Variable	Pitch: 0.5 Height: 0.1 Ave. D: 0.1	1-2-1 $(45 \times 45 \times 1)$	Pitch: 1.0 Height: 0.6 Ave. D: 0.6	$101 \times$ 114×1.6

TABLE 9.3 Material Properties for Simulations

resistance of the package (R_{ja}) can be expressed as $R_{ja} = (T_j - T_a)/P$, where T_j and T_a, respectively, represent the junction temperature and ambient temperature, and P is total power dissipated from the chip.

9.2.3 Effect of Chip Power

Figure 9.2 shows the junction to ambient thermal resistance of packages with different chip power dissipations. It can be seen that (1) the thermal resistance decreases as the chip power increases (this is so because high chip power induces high package temperature because a high heat-transfer coefficient is induced, and more heat is removed from the package), and (2) the thermal resistance of the TSV package is lower than that of the package without the TSV (this is so because of the TSV interposer's spreading effect). The dash lines in the figure represent the correlations between the thermal resistance and the chip power when the junction temperature equals 85 and 125°C, respectively.

9.2.4 Effect of Interposer Size

Figure 9.3 shows the effect of the TSV interposer area on package thermal performance. It can be seen that the thermal resistance R_{ja} decreases by about 14 percent when the TSV interposer size is increased from 21×21 mm to 45×45 mm. This indicates that package thermal performance can be improved by increasing TSV interposer size.

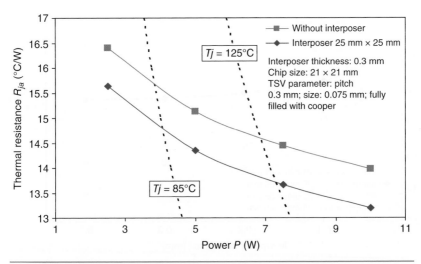

FIGURE 9.2 Junction to ambient thermal resistance for different power dissipation values.

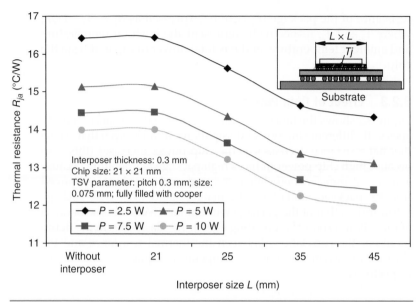

FIGURE 9.3   Junction to ambient thermal resistance for different TSV interposer sizes.

9.2.5   Effect of TSV Interposer Thickness

Figure 9.4 shows the effect of the TSV interposer thickness on the package thermal performance. It can be seen that the thicker the TSV interposer, the lower is the thermal resistance. This is so because of the increased spreading effect with a thicker interposer. However, for small size interposers, this effect is negligible.

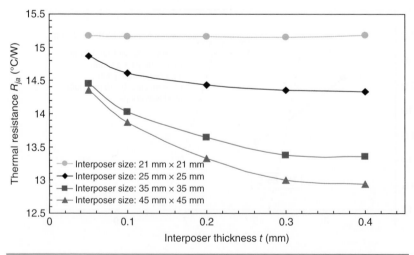

FIGURE 9.4   Junction to ambient thermal resistance for different TSV interposer thicknesses.

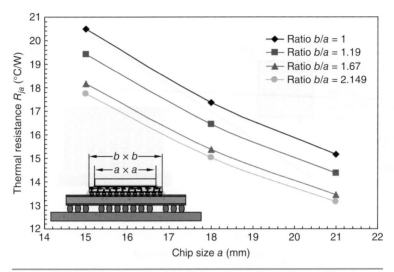

FIGURE 9.5   Junction to ambient thermal resistance for different chip sizes.

9.2.6   Effect of Moore's Law Chip Size

Figure 9.5 shows the effect of the chip size for different size ratios of the TSV interposer to the chip on the thermal performance of the package. It can be seen that for the same size ratio of chip to TSV interposer, the smaller sizes of chips induce higher thermal resistance in the package.

9.3   Thermal Performance of 3D Memory-Chip Stacking

9.3.1   Thermal Performance of 3D Stacked TSV Chips with a Uniform Heat Source

Figure 9.6 shows the maximum junction temperature of a stacked chip package (varying with the number of the chips). In these simulations, all the chips have the same size ($5 \times 5 \times 0.05$ mm), and there are 225 (15×15) copper-filled TSVs with a 0.2-mm pitch on each chip. The power dissipated by each chip is 0.2 W, and it is assumed that the power is uniformly distributed on each chip. The ambient temperature is 25°C. It can be seen from the figure that the maximum junction temperature increases linearly with the number of chips stacked. In addition, it can be seen that if the maximum allowable junction temperature is 85°C, then the maximum number of chips that can be stacked together is seven under the present conditions.

Figure 9.7 shows the maximum junction temperature at each layer of the TSV chip stack. It can be seen that the difference in maximum

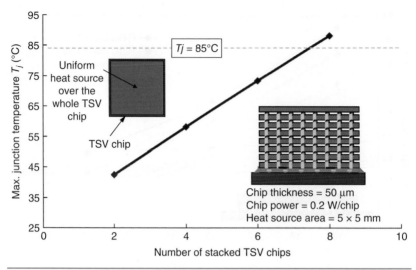

FIGURE 9.6 T_j versus number of stacked TSV chips (uniform heat source).

junction temperature between each layer of the stack is negligible. This means that the temperature distribution for the different layers of chips is uniform because we assumed that the power dissipation is uniformly distributed in each chip. Figure 9.8 shows the maximum temperature contours of a chip.

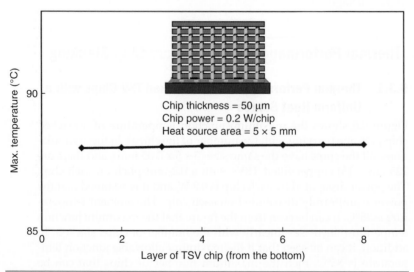

FIGURE 9.7   Maximum temperature at each layer of stacking of TSV chips (uniform heat source).

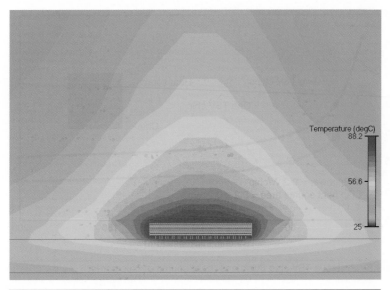

Figure 9.8   Maximum temperature contours of a chip (uniform heat source).

9.3.2   Thermal Performance of 3D Stacked TSV Chips with a Nonuniform Heat Source

The results presented in Figs. 9.6 through 9.8 are based on the assumption that the power is dissipated uniformly over the whole chip. However, in most applications, the power dissipated by each chip is basically nonuniform, and as such, it will induce quite different thermal behaviors of 3D IC chip stacking with TSVs. In addition, it is well known that ordinary silicon chips normally have large parallel conduction of heat (parallel to the chip surface) owing to the large thermal conductivity of the Si material. However, for 3D IC chip stacking, in order to have a low profile, the chip thickness of each layer of the 3D chip stacking must be ground down to 50 μm and less. Thus the parallel spreading effect is suppressed by the very thin chip, and the hot spot will be very intense. Compounding this with the nonuniform heat source, it becomes a challenge in 3D IC integration.

9.3.3   Two TSV Chips (Each with One Distinct Heat Source)

Figures 9.9 and 9.10 show the thermal performance of a 3D integration of two copper-filled TSV chips stacking. All the chips are 5 × 5 mm, and each chip's center is subjected to a distinct heat source (0.2 W) in a tiny area (0.2 × 02 mm). It can be seen from the figures that (for both one- and two-chip stacks) (1) for a nonuniform heat source, the effect of chip thickness on the thermal performance of 3D IC

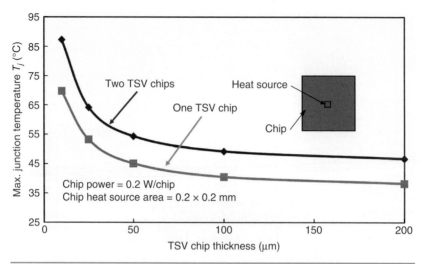

FIGURE 9.9 T_j versus TSV chip thickness (heat source at chip center).

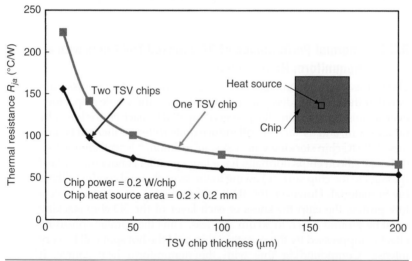

FIGURE 9.10 R_{ja} versus TSV chip thickness (heat source at chip center).

integrations is very important, (2) this thickness effect is even more significant in the application range (≤50 μm) of 3D IC integrations, and (3) the maximum junction temperature and thermal resistance decrease as chip thickness increases.

9.3.4   Two TSV Chips (Each with Two Distinct Heat Sources)

In addition to one heat source per chip (5 × 5 mm), Figs. 9.11 and 9.12 show the effect of two heat sources at a distance apart (gap) on the

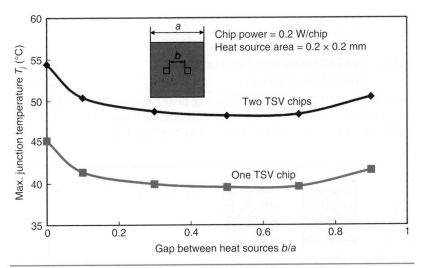

FIGURE 9.11 T_j versus the gap (b) between two heat sources.

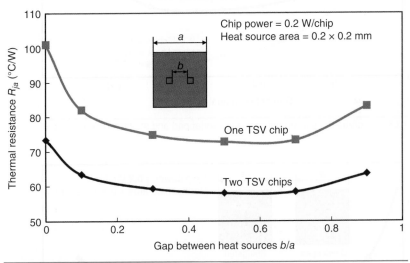

FIGURE 9.12 R_{ja} versus the gap (b) between two heat sources.

thermal performance of 3D IC integrations of copper-filled TSV chips. There are two distinct heat sources (each with 0.1 W and on a 0.2- × 0.2-mm area) on each chip (5 × 5 × 0.05 mm). It can be seen from the figures that (1) the larger the gap ($b/a \leq 0.7$) between the two heat sources, the better is the thermal performance (i.e., lower maximum junction temperature and thermal resistance), and (2) when the gap between the two heat sources is larger than 0.7 (i.e., the heat sources are too close to the edge of the chip), the thermal performance is weaker. This is due to suppressible spreading effects near the edges of the chips.

9.3.5   Two TSV Chips with Two Staggered Distinct Heat Sources

In addition to the case of overlapping heat sources discussed in the preceding section, finally, Figs. 9.13 and 9.14 show the orientation effect (staggered heat sources) of two stacked chips, each

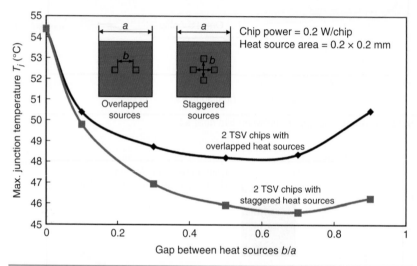

FIGURE 9.13 T_j versus the gap (b) between the overlapped and staggered heat sources.

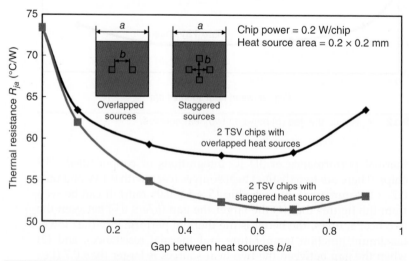

FIGURE 9.14 R_{ja} versus the gap (b) between the overlapped and staggered heat sources.

with two heat sources at a certain distance apart, on the thermal performance of a 3D chip stacking. It can be seen that (1) similar to the case of overlapping heat sources, the larger the gap ($b/a \leq 0.7$) between the two pairs of staggered heat sources, the lower are the maximum junction temperature and thermal resistance, (2) when the gap between the two pairs of staggered heat sources is larger than 0.7 (i.e., the heat sources are too close to the edge of the chip), the thermal performance is weaker, and (3) the maximum junction temperature and thermal resistance of the 3D stacking with two TSV chips subjected to two pairs of staggered heat sources are lower than those with two pairs of overlapping heat sources. This is so because the staggered heat sources avoid the superimposition of heat sources and thus lead to better thermal performance. This result is very useful for the design and layout 3D chip stacking because it permits relocation of the heat sources and/or rotation of the chip.

9.4   Effect of Thickness of the TSV Chip on Its Hot-Spot Temperature

Figure 9.15 shows the temperature maps on a TSV chip with various chip thicknesses (10–200 μm). The chip is 5 × 5 mm, and its center is subjected to a distinct heat source (0.2 W) in a tiny area (0.2 × 0.2 mm).

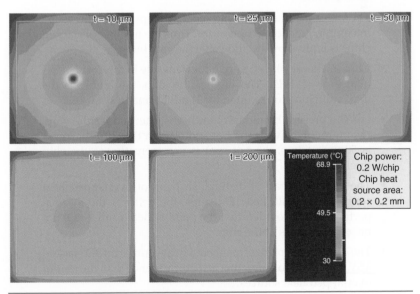

FIGURE 9.15   Hot spots for different TSV chip thickness.

It can be seen that the heat on the chip surface is well dissipated for typical chip thicknesses of 100 to 200 μm subjected to the generated power of 0.2 W. For the 200-μm-thick chip, the temperature distribution is almost uniform and equal to 35°C. However, the hot-spot temperature on the chip increases to 69°C (0.2 W power) if the chip thickness is reduced to 10 μm and the hot-spot area is clearly shown. Thus, for 0.4 W of power (which is very common for DRAM) in a 5- × 5-mm chip with 10-μm thickness, the hot-spot temperature could be 138°C, which far exceeds the maximum allowable junction temperature (usually 85°C) of most Si chips.

9.5   Summary and Recommendations

In this study, modeling and simulation to determine the effects of TSV interposer on the thermal performance of the 3D SiP and chip stacking have been conducted and presented. Some important results are summarized as follows [31]:

- The TSV interposer improves the thermal performance of 3D SiPs owing to spreading effects.

- The junction to ambient thermal resistance of SiPs decreases by 14 percent when the TSV interposer size increased from 21 × 21 mm to 45 × 45 mm.

- Junction to ambient thermal resistance of the SiP decreases by 11 percent when the TSV interposer thickness is increased from 50 to 400 μm.

- The larger the number of stacked TSV chips, the higher is the maximum junction temperature. Thus the number of stacked chips is limited by the allowable thermal budgets.

- Chip thickness plays a very important role in thermal performance of 3D SiPs. The thinner the chip, the higher is the maximum junction temperature and thermal resistance.

- For the present boundary conditions, the thinner the chip, the more intensive is the hot spot. When the chip thickness is 50 μm or less, the hot spot is very sensitive to the chip thickness.

- The larger the gap of distinguish pairs of overlapped heat sources on 3D stacked TSV chips, the better is the thermal performance, provided that the pairs of heat sources are not near to the edges of the chip.

- The distinguish pairs of staggered heat sources on 3D stacked TSV chips lead to better thermal performance than the pair of overlapped heat sources.

9.6 Thermal Management System with TSVs and Microchannels for 3D Integration SiPs

9.6.1 Test Vehicle

Figure 9.16 [19] schematically shows a silicon carrier with microchannels for thermal management and TSVs for electrical feedthrough. This carrier consists of two silicon chips (without any devices), and the difference between these two chips is that the bottom Si chip does not have an outlet (Figs. 9.17 through 9.19). There are 144 TSVs (at a pitch of 0.5 mm), and they are in rows along the periphery of the carrier. After bonding, electrical interconnection through the carrier is made by TSV with on-wall metallization. The fluidic channels are connected through one/two inlets and one/two outlets. There are sealing rings around both the fluidic path and the individual TSV to isolate the fluid from the electrical interconnection. The depth and diameter of the silicon vias are 400 and 150 µm, respectively, whereas both the depth and the width of the fluidic channels are 350 µm. The solder for the sealing ring is Au20Sn, and the under bump metallurgy (UBM) is TiCuNiAu.

9.6.2 Test Vehicle Fabrication

Figure 9.20 shows the fabrication process of the test vehicle on a 200-mm silicon wafer. First, a layer of SiO_2 3-µm thick is deposited by the plasma-enhanced chemical-vapor deposition (PECVD) process in a Novellus PECVD system and then patterned and etched with via and cooling-channel patterns as a hard mask; then the silicon via and the cooling channels are etched with the deep reactive-ion etching (DRIE) process. Since the depths for TSV and cooling-channel structures are

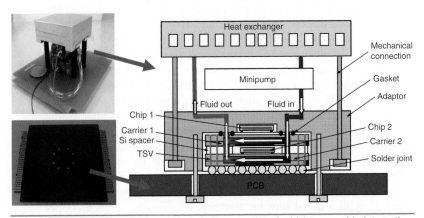

FIGURE 9.16   Illustration of 3D stacking of silicon carriers for high-power chip integration.

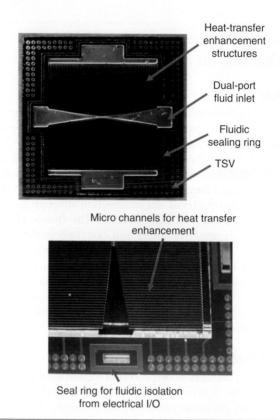

Heat-transfer
enhancement
structures

Dual-port
fluid inlet

Fluidic
sealing ring

TSV

Micro channels for heat transfer
enhancement

Seal ring for fluidic isolation
from electrical I/O

FIGURE 9.17 Design of the inlet/outlet of the microchannel and sealing ring as well as TSVs.

different, the DRIE process is separated into two steps. First, via that is 100-μm deep are etched using photoresist as the etch mask; then the photoresist is stripped, and the via and the channels are etched at the same time at the 350-μm depth. SiO_2 patterned at step 1 is used as the etch mask. Since the DRIE process is restricted to blind silicon vias, in order to achieve a through-silicon via (TSV), the silicon wafer is backgrinded and polished to 450 μm to expose the via after the DRIE process. Then a passivation layer is deposited on both sides of the wafer. The passivation layer is 1 μm of SiO_2 that is deposited by the Plasma-Therm SLR 720 system. After passivation, UBM then is sputtered and patterned around the TSV and the cooling channels on both sides of the wafer. The sputtering is done in the Balzer system. The UBM material used is Ti/Cu/Ni/Au with thicknesses of 0.1, 2, 0.5, and 0.1 μm, respectively. In order to make sure that the sidewall of the TSV is coated with a metal film for electrical connections, a thick copper layer (which is used in the UBM layer) is sputtered from both sides of the wafer. For this special case, the sputtering process is much simpler

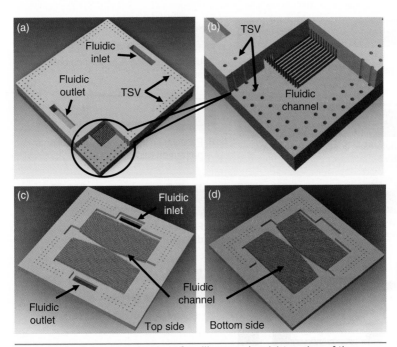

FIGURE 9.18   Schematic drawing of a silicon carrier: (a) top view of the carrier; (b) cross-sectional view of the carrier; (c) top silicon chip; (d) bottom silicon chip.

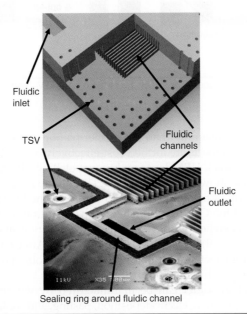

FIGURE 9.19   Zoomed view of the microchannels.

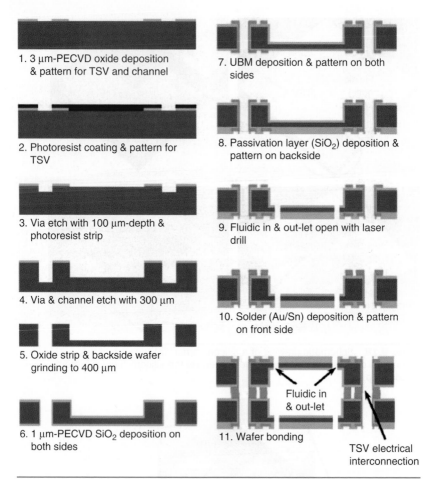

1. 3 µm-PECVD oxide deposition & pattern for TSV and channel

2. Photoresist coating & pattern for TSV

3. Via etch with 100 µm-depth & photoresist strip

4. Via & channel etch with 300 µm

5. Oxide strip & backside wafer grinding to 400 µm

6. 1 µm-PECVD SiO₂ deposition on both sides

7. UBM deposition & pattern on both sides

8. Passivation layer (SiO₂) deposition & pattern on backside

9. Fluidic in & out-let open with laser drill

10. Solder (Au/Sn) deposition & pattern on front side

11. Wafer bonding

Fluidic in & out-let

TSV electrical interconnection

FIGURE 9.20   Fabrication process for silicon carrier with both TSVs and microchannels.

and less time consuming than the copper-plating process. Therefore, electrical interconnects are formed cost-effectively on the TSV.

Usually, a thick UBM layer makes wet etching difficult. A workable process is shown in the following: (1) The top Au layer is etched with a solution of triiodide, (2) the Ni layer is etched with a solution of AC-100, (3) the Cu layer is etched with a solution of A95, and (4) the Ti layer is etched with a 49 percent HF solution that is diluted with deionized (DI) water (49%HF:H₂O = 1:15). After UBM patterning, two or four rectangular holes (for inlets and outlets) are opened in the top silicon chip by laser drilling.

The Au20Sn solder can be manufactured by various methods, including evaporation, electrodeposition, pasting, and solder preforms. Among these methods, evaporated solder allows more accurate control

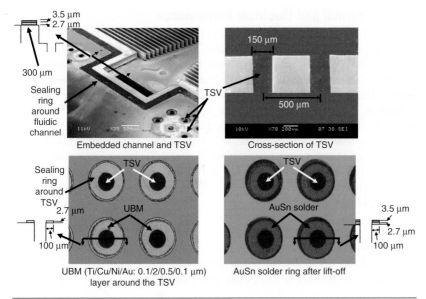

FIGURE 9.21 Detailed geometry and materials of the silicon carrier with both TSVs and microchannels.

of dimensions and position. Therefore, after UBM patterning, a sequence of Sn/Au layer (eight layers of Sn and eight layers of Au in total) is evaporated and simultaneously patterned by a lift-off process. The thickness for each Sn layer and Au layer is 0.2 and 0.24 μm, respectively. The evaporation can be done in a Temescal E-Beam Evaporator (Model VES-2550) from Semicore Equipment, Inc. The evaporation rate is 5 Å/s. During the lift-off process, a 20-μm thickness of dry film is used as the mold.

The fabricated wafer is shown in Fig. 9.21. The AuSn solder is 3.5 μm thick, and the UBM, Ti (0.1 μm)/Cu (2 μm)/Ni (0.5 μm)/Au (0.1 μm), is 2.7 μm thick. The annular sealing ring around the TSV is 100 μm. The width of the sealing ring between the silicon wafers is 300 μm. The TSV is on a 500-μm pitch. The TSV is not totally filled with Cu; instead, the on-wall metallization is done by double-sided metal-film sputtering.

9.6.3   Wafer-to-Wafer Bonding

After solder layer formation, wafer-level bonding is conducted to bond the two wafers together. The bonding is carried out at 350°C with 15 minutes of peak-temperature hold time and 8 kN of compressive force (static pressure of 4.7 MPa). After bonding, the bonded wafer is diced into single-silicon carriers with dimensions of 15.1 × 15.1 × 0.8 mm. The dicing machine used is the DAD 651 from Disco. The spindle speed is 30,000 rpm, and the feed rate is 5 mm/s.

9.6.4   Thermal and Electrical Performance

The patterned UBM layer is shown in Fig. 9.21. It shows that the pattern edge is smooth. Figure 9.22 shows the pressure-drop measurements. It can be seen that the carrier with two inlets and two outlets performs better thermally than the carrier with one inlet and one outlet. The electrical connectivity through the TSVs with different diameters has been tested, and the results are listed in Table 9.4. When the diameter is larger than 150 µm, the resistance through the TSV is less than 0.5 Ω. Therefore, a TSV diameter of 150 µm is used to have higher inputs/outputs (I/Os). Figure 9.23 shows a cross section of a TSV after metallization at a diameter of 150 µm. It can be seen clearly that the sidewall of the TSV is covered with a continuous metal layer.

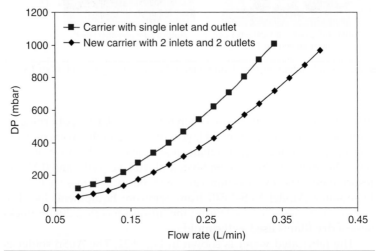

FIGURE 9.22   Pressure-drop measurements. Carrier with two inlets and two outlets performs better than the one with one inlet and one outlet.

TSV Diameter	Resistance	Average Resistance
100 µm	1.5–12 kΩ	6.75 kΩ
125 µm	1.5–6.5 kΩ	4 kΩ
150 µm	0.46–0.52 Ω	0.49 Ω
200 µm	0.38–0.44 Ω	0.41 Ω
300 µm	0.38–0.44 Ω	0.41 Ω

TABLE 9.4   Electrical Resistance for Different Diameters of TSVs

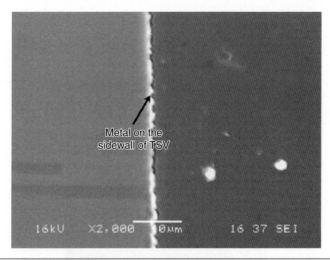

Metal on the
sidewall of TSV

16kU   X2,000   10μm   16 37 SEI

FIGURE 9.23 Metal-thin film on the sidewall of TSV.

9.6.5 Quality and Reliability

Shear-strength measurements of the diced package samples were carried out by a commercially available shear tester (DAGE-SERIES-4000-T, Dage Precision Industries, Ltd., Aylesbury, UK). The shear-strength value measured for 20 different samples ranged from 17.8 to 35.1 MPa, and an average value of 27.2 MPa and a standard deviation of 2.2 were obtained. During the shear test, fracture always occurs at the UBM layer, which indicates that the bonding strength is stronger than the adhesion strength of the UBM layer. Figure 9.24 shows the fracture surface after the shear test, and it can be seen that the metal is peeled from the surface, showing good bonding strength. Figure 9.25 shows the cross section of the bonding interface, and it can be seen that a uniform cross section has been achieved.

During AuSn bonding, pressure is one of the critical parameters to obtain good bonding results. Figure 9.26 shows the cross section of a poor AuSn solder ring after bonding at 350°C for 15 minutes of peak-temperature hold time and 3 kN of compression force (static pressure of 1.7 MPa). There is an obvious crack in the middle of the interface owing to not enough bonding pressure. Higher pressure can improve the contact area between the two bonding surfaces from both macro and micro points of view. Normally, the silicon wafer is not ideally flat, especially after back-grinding and polishing and some microfabrication processes. Therefore, high pressure can push (force) the two bonding surfaces against each other intimately. At the same time, the root-mean-square (rms) roughness of the AuSn surface after evaporation is about 24 nm, as shown in Fig. 9.27. The rough surface will decrease the contact area between the two bonding surfaces.

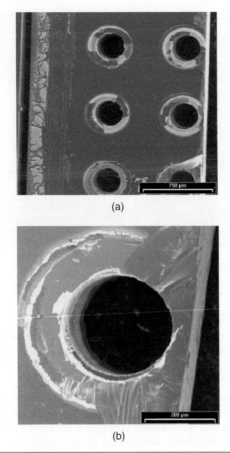

(a)

(b)

FIGURE 9.24   (a) Scanning electron microscopic (SEM) images of the solder surface after shear test. (b) Zoomed view.

Therefore, high pressure is required to overcome the surface roughness to improve the contact area between the two bonding surfaces.

Figure 9.28 shows a bonded pair of silicon wafers (the top and bottom surfaces of a diced silicon carrier, respectively). The metal patterns on both sides provide electrical interconnection and rerouting. The yield after dicing is higher than 98 percent, which indicates that the bonding strength is high enough to pass through the dicing step.

A thermal cycling test between –40°C (15 minutes) and 125°C (15 minutes) for 1000 cycles has been conducted. The temperature ramping rate is 15°C per minute. After the test, the average shear strength for 20 samples was 26 MPa, which is almost the same as that before the test.

Hermetic sealing of the bonded carrier is a key requirement of the project. A test setup has been developed to circulate water under high

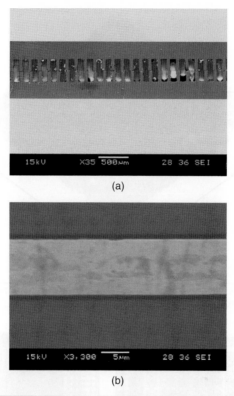

(a)

(b)

FIGURE 9.25   Cross section of the bonding interface: (a) cross section of embedded fluidic channels; (b) cross section of AuSn solder sealing ring after bonding.

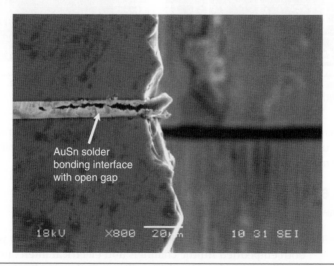

FIGURE 9.26   Crack in the bonding interface with bonding pressure of 3 kN.

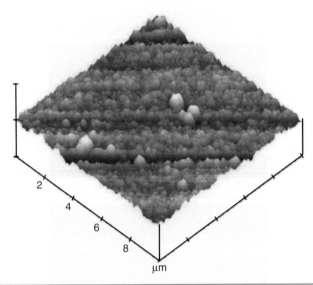

FIGURE **9.27**   Surface roughness of AuSn solder after evaporation on UBM layer. The rms roughness is 25 nm.

FIGURE **9.28**   Optical images of a diced silicon carrier. (*Left*) Top surface. (*Right*) Bottom surface

pressure through the carrier. The water flow rate is increased gradually up to 350 mL/min. No leakage is observed at a high flow rate. The maximum pressure drop in the carrier is 6×10^4 Pa.

9.6.6   Summary and Recommendations

Some important results are summarized as follows:

1.   An integration process for fabricating silicon carriers with both embedded fluidic channels and electrical interconnections has been developed. The silicon carriers can be stacked

on top of each other, with silicon interposers in between, to make up a stacked cooling module for high-power heat dissipation.

2. After on-wall metallization of TSVs with a diameter larger than 150 μm and thickness less than 400 μm, the electrical resistance through the TSVs is 0.49 Ω. After bonding, the electrical resistance between the top and bottom surfaces of the silicon carrier is 1.16 Ω. It can be reduced by using a thicker metal layer and/or a larger TSV diameter.

3. For a single carrier, no leakage is observed at a high flow rate (350 mL/min). The maximum pressure drop in the carrier is 6×10^4 Pa.

4. The thermal management system with two inlets and two outlets performs better thermally than the one with one inlet and one outlet.

5. High bonding pressure can improve the bonding quality of evaporated AuSn solder. With a bonding pressure of 4.7 MPa, the die shear-test strength is higher than 27 MPa, and it is more than adequate for dicing/handling the silicon carrier and to provide a leakage-free fluidic path. After 1,000 cycles of the thermal-cycling test (−40 to 125°C), the average die shear strength is 26 MPa and shows insignificant degradation after the thermal-cycling test.

9.7 References

[1] Andry, P. S., C. K. Tsang, B. C. Webb, E. J. Sprogis, S. L. Wright, B. Bang, and D. G. Manzer, "Fabrication and Characterization of Robust Through-Silicon Vias for Silicon-Carrier Applications," *IBM Journal of Research and Development*, Vol. 52, No. 6, 2008, pp. 571–581.

[2] Knickerbocker, J. U., P. S. Andry, B. Dang, R. R. Horton, C. S. Patel, R. J. Polastre, K. Sakuma, E. S. Sprogis, C. K. Tsang, B. C. Webb, and S. L. Wright, "3D Silicon Integration," *IEEE/ECTC Proceedings*, Orlando, FL, May 2008, pp. 538–543.

[3] Kumagai, K., Y. Yoneda, H. Izumino, H. Shimojo, M. Sunohara, and T. Kurihara, "A Silicon Interposer BGA Package with Cu-Filled TSVs and Multilayer Cu-Plating Interconnection," *IEEE/ECTC Proceedings*, Orlando, FL, May 2008, pp. 571–576.

[4] Sunohara, M., T. Tokunaga, T. Kurihara, and M. Higashi, "Silicon Interposer with TSVs (Through-Silicon Vias) and Fine Multilayer Wiring," *IEEE/ECTC Proceedings*, Orlando, FL, May 2008, pp. 847–852.

[5] Lee, H. S., Y.-S. Choi, E. Song, K. Choi, T. Cho, and S. Kang, "Power Delivery Network Design for 3D SIP Integrated over Silicon Interposer Platform," *IEEE/ECTC Proceedings*, Reno, NV, May 2007, pp. 1193–1198.

[6] Matsuo, M., N. Hayasaka, and K. Okumura, "Silicon Interposer Technology for High-Density Package," *IEEE/ECTC Proceedings*, Las Vegas, NV, May 2000, pp. 1455–1459.

[7] Selvanayagam, C., J. H. Lau, X. Zhang, S. Seah, K. Vaidyanathan, and T. Chai, "Nonlinear Thermal Stress/Strain Analysis of Copper Filled TSVs (Through-Silicon Vias) and Their Flip-Chip Microbumps," *IEEE/ECTC Proceedings*, Orlando, FL, May 2008, pp. 1073–1081.

[8] Wong, E., J. Minz, and S. K. Lim, "Effective Thermal Via and Decoupling Capacitor Insertion for 3D System-on-Package," *IEEE/ECTC Proceedings*, San Siego, CA, May 2006, pp. 1795–1801.

[9] Lau, J. H., C. Lee, C. Premachandran, and A. Yu, *Advanced MEMS Packaging*, McGraw-Hill, New York, 2010.

[10] Lau, J. H., *Reliability of RoHS-Compliant 2D and 3D IC Interconnects*, McGraw-Hill, New York, 2011.

[11] Lau, J. H., "Critical Issues of TSV and 3D IC Integration," *IMAPS Transactions, Journal of Microelectronics and Electronic Packaging*, First Quarter Issue, 2010, pp. 35–43.

[12] Lau, J. H., "Design and Process of 3D MEMS System-in-Package (SiP)," *IMAPS Transactions, Journal of Microelectronics and Electronic Packaging*, First Quarter Issue, 2010, pp. 10–15.

[13] Lau, J. H., R. Lee, M. Yuen, and P. Chan, "3D LED and IC Wafer-Level Packaging," *Journal of Microelectronics International*, Vol. 27, No. 2, 2010, pp. 98–105.

[14] Lau, J. H., "State-of-the-Art and Trends in 3D Integration," *Chip Scale Review*, March–April 2010, pp. 22–28.

[15] Lau, J. H., Y. S. Chan, and R. S. W. Lee, "3D IC Integration with TSV Interposers for High-Performance Applications," *Chip Scale Review*, September–October 2010, pp. 26–29.

[16] Lau, J. H., "Overview and Outlook of Through-Silicon Via (TSV) and 3D Integrations," *Journal of Microelectronics International* Vol. 28, No. 2, 2011, pp. 8–22.

[17] Yu, A., J. H. Lau, S. Ho, A. Kumar, H. Yin, J. Ching, V. Kripesh, D. Pinjala, S. Chen, C. Chan, C. Chao, C. Chiu, M. Huang, and C. Chen, "Three-Dimensional Interconnects with High Aspect Ratio TSVs and Fine-Pitch Solder Microbumps," *IEEE/ECTC Proceedings*, San Diego, CA, May 2009, pp. 350–354. Also, accepted for publication in *IEEE Transactions in Advanced Packaging*.

[18] Yu, A., A. Kumar, S. Ho, H. Yin, J. H. Lau, J. Ching, V. Kripesh, D. Pinjala, S. Chen, C. Chan, C. Chao, C. Chiu, M. Huang, and C. Chen, "Development of Fine-Pitch Solder Microbumps for 3D Chip Stacking," *IEEE/EPTC Proceedings*, Singapore, December 2008, pp. 387–392. Also, accepted for publication in *IEEE Transactions in Advanced Packaging*.

[19] Yu, A., N. Khan, G. Archit, D. Pinjalal, K. Toh, V. Kripesh, S. Yoon, and J. H. Lau, "Fabrication of Silicon Carriers with TSV Electrical Interconnects and Embedded Thermal Solutions for High Power 3D Packages," *IEEE Transactions on Components and Packaging Technology*, Vol. 32, No. 3, September 2009, pp. 566–571.

[20] Tang, G., O. Navas, D. Pinjala, J. H. Lau, A. Yu, and V. Kripesh, "Integrated Liquid Cooling Systems for 3D Stacked TSV Modules," *IEEE Transactions on Components and Packaging Technology*, Vol. 33, No. 1, 2010, pp. 184–195.

[21] Zhang, X., T. Chai, J. H. Lau, C. Selvanayagam, K. Biswas, S. Liu, D. Pinjala, G. Tang, Y. Ong, S. Vempati, E. Wai, H. Li, B. Liao, N. Ranganathan, V. Kripesh, J. Sun, J. Doricko, and C. Vath, "Development of Through Silicon Via (TSV) Interposer Technology for Large Die (21- × 21-mm) Fine-Pitch Cu/Low-*k* FCBGA Package," *IEEE/ECTC Proceedings*, May 2009, pp. 305–312. Also, *IEEE Transactions in Advanced Packaging* (in press).

[22] Kumar, A., X. Zhang, Q. Zhang, M. Jong, G. Huang, V. Kripesh, C. Lee, J. H. Lau, D. Kwong, V. Sundaram, R. Tummala, and M. Georg, "Evaluation of Stresses in Thin Device Wafers Using Piezoresistive Stress Sensors," *IEEE Proceedings of EPTC*, December, pp. 1270–1276. Also, *IEEE Transactions in Components and Packaging Technology*, Vol. 1, No. 6, June 2011, pp. 841-850.

[23] Vempati1, S. R., S. Nandar, C. Khong, Y. Lim, K. Vaidyanathan, J. H. Lau, B. P. Liew, K. Y. Au, S. Tanary, A. Fenne, R. Erich, and J. Milla, "Development of 3D Silicon Die Stacked Package Using Flip-Chip Technology with Microbump Interconnects," *IEEE/ECTC Proceedings*, San Diego, CA, May 2009, pp. 980–987.

[24] Lim, S., V. Rao, H. Yin, W. Ching, V. Kripesh, C. Lee, J. H. Lau, J. Milla, and A. Fenner, "Process Development and Reliability of Microbumps," *IEEE/ECTC*

Proceedings, December 2008, pp. 367–372. Also, *IEEE Transactions in Components and Packaging Technology*, Vol. 33, No. 4, December 2010, pp. 747 –753.

[25] Selvanayagam, C., J. H. Lau, X. Zhang, S. Seah, K. Vaidyanathan, and T. Chai, "Nonlinear Thermal Stress/Strain Analysis of Copper-*Filled TSVs (Through-Silicon Vias) and Their Flip-Chip Microbump*s," *IEEE Transactions in Advanced Packaging*, Vol. 32, No. 4, Nov. 2009, pp. 720–728.

[26] Made, R., C. L. Gan, L. Yan, A. Yu, S. U. Yoon, J. H. Lau, and C. Lee, "Study of Low Temperature Thermocompression Bonding in Ag-In Solder for Packaging Applications," *Journal of Electronic Materials*, Vol. 38, 2009, pp. 365–371.

[27] Yan, L.-L., C.-K. Lee, D.-Q. Yu, A.-B. Yu, W.-K. Choi, J. H. Lau, and S.-U. Yoon, "A Hermetic Seal Using Composite Thin Solder In/Sn as Intermediate Layer and Its Interdiffusion Reaction with Cu," *Journal of Electronic Materials*, Vol. 38, 2009, pp. 200–207.

[28] Lee, C., A. Yu, L. Yan, H. Wang, J. Han, Q. Zhang, and J. H.Lau, "Characterization of Intermediate In/Ag Layers of Low-Temperature Fluxless Solder–Based Wafer Bonding for MEMS Packaging," *Journal of Sensors & Actuators* (in press).

[29] Yu, D.-Q., C. Lee, L. L. Yan, W. K. Choi, A. Yu, and J. H. Lau, "The Role of Ni Buffer Layer on High-Yield Low-Temperature Hermetic Wafer Bonding Using In/Sn/Cu Metallization," *Applied Physics Letters*, Vol. 94, No. 3, January 2009, pp. 34105–34105-3.

[30] Zhang, X., A. Kumar, Q. X. Zhang, Y. Y. Ong, S. W. Ho, C. H. Khong, V. Kripesh, J. H. Lau, D.-L. Kwong, V. Sundaram, Rao R. Tummula, and Georg Meyer-Berg, "Application of Piezoresistive Stress Sensors in Ultrathin Device Handling and Characterization," *Journal of Sensors & Actuators: A. Physical*, Vol. 156, November 2009, pp. 2–7.

[31] Lau, J. H., and G. Tang, "Thermal Management of 3D IC Integration with TSV (Through Silicon Via)", *IEEE/ECTC Proceedings*, San Diego, May 2009, pp. 635–640.

[32] Yamaji, Y., "Thermal Characterization of Baredie Stacked Modules with Cu Through-Vias," *IEEE/ECTC Proceedings*, 2001, pp. 730–732.

[33] Wong, E., "Effective Thermal Via and Decoupling Capacitor Insertion for 3D System-on-Package," *IEEE/ECTC Proceedings*, 2006, pp. 1795–1801.

[34] Bar-Cohen, A., and A. Kraus, *Advances in Thermal Modeling of Electronic Components and Systems*, Vol. 2, ASME, New York, 1988.

[35] Kraus, A., and A. Bar-Cohen, *Thermal Analysis and Control of Electronic Equipments*, Hemisphere, New York, 1983.

[36] Simons, R. E., "Simple Formulas for Estimating Thermal Spreading Resistance," *Electronics Cooling*, Vol. 10, 2004, pp. 6–8.

CHAPTER 10

Three-Dimensional Integrated Circuit Packaging

10.1 Introduction

As mentioned in Chap. 1, three-dimensional (3D) integration consists of 3D integrated circuit (IC) packaging, 3D IC integration, and 3D silicon (Si) integration. The latter two have been discussed in Chaps. 1 through 9 because they are very important technologies for the future. However, as shown in Fig. 10.1, 3D IC packaging is now mature for production, and some of the latest advances have been keeping 3D IC/Si integrations away from volume production. In this chapter, (1) wire bonding of stack dies on Cu–low-k chips, (2) bare chip-to-chip (face-to-face) stacking, and (3) fan-out embedded wafer-level-package (WLP)–to–chip (face-to-face) assembly will be discussed. Finally, a note on wire-bonding reliability is provided. The cost issue of through-silicon-via (TSV) technology will be briefly mentioned first.

10.2 Cost: TSV Technology versus Wire-Bonding Technology

As mentioned earlier, TSV is a disruptive technology, and it is trying to displace wire-bonding technology, which is the most mature, high-yield, and low-cost technology. As shown in Fig. 10.2 and Chap. 2, there are six key steps in making TSVs. For 300-mm wafers, each piece of automatic equipment and its accessories for each step usually cost more than a million dollars and have to be placed in cleanrooms (class 10 or lower). The depreciation of the equipments and accessories is very high, and their lives are very short (e.g., owing to upgrades and advances in technology requirements). In addition, the cost of building and maintaining cleanrooms

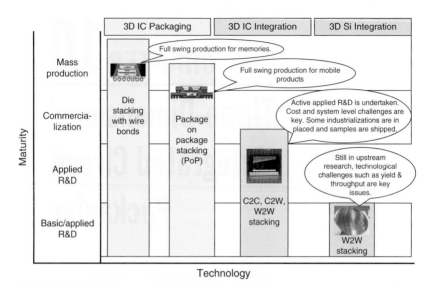

FIGURE 10.1   Maturity of 3D integration technology.

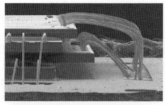

Automatic wire bonding is a mature technology with 99.99% yield and very low cost and is reliable.

There are six key steps in making a TSV:
♦ Via forming (DRIE)
♦ Dielectric layer (PECVD)
♦ Barrier/seed layers (PVD)
♦ Via filling (Cu plating)
♦ CMP
♦ TSV Cu reveal
   The order of costs is
PVD > PECVD > CMP > Plating > Etching

FIGURE 10.2   Wire bonds are replaced by TSVs. There are six key steps in making a TSV.

(running 24 hours a day) is very expensive. On the other hand, the wire bonders are placed in class 1000 (some even in class 10,000) cleanrooms. Thus, compared with wire-bonding technology, TSV technology is very expensive, even taking cost of ownership into consideration!

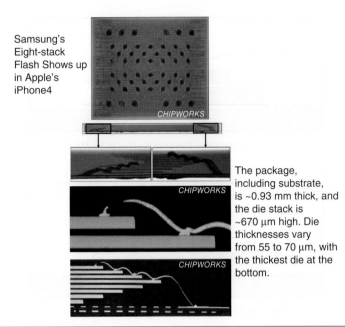

Samsung's
Eight-stack
Flash Shows up
in Apple's
iPhone4

The package,
including substrate,
is ~0.93 mm thick, and
the die stack is
~670 µm high. Die
thicknesses vary
from 55 to 70 µm, with
the thickest die at the
bottom.

FIGURE 10.3   Samsung's eight-stack flash shows up in Apple's iPhone4.

Today, more than 85 percent of the semiconductor chips use wire-bonding technology. For example, Fig. 10.3 illustrates Samsung's eight-stack flash chips with wire-bonding technology that show up in Apple's iPhone4. The package, including substrate, is approximately 0.93 mm thick, and the die stack is approximately 670 µm high. The die thicknesses vary from 55 to 70 µm, with the thickest die at the bottom.

However, just like the solder-bumped flip-chip technology [1, 2] developed by IBM more than 40 years ago, because of their unique advantages, TSVs will be here to stay and for a very long time for low-power, wide-bandwidth, high-performance, and high-density applications. Comparing wire-bonding technology with TSV technology, the latter provides the opportunity for the shortest chip-to-chip interconnects and smallest pad sizes and pitches of interconnects, which lead to (1) better electrical performance, (2) lower power consumption, (3) wider data width and thus bandwidth, (4) higher density, (5) smaller form factor, and (6) lighter weight.

10.3   Wire Bonding of Stack Dies on Cu–Low-*k* Chips

10.3.1   Test Vehicles

Figure 10.4 shows the two test vehicles. For both test vehicles, the bottom chip is a 65-nm Cu–low-*k* die that is die-attached to an organic substrate. This Cu–low-*k* die supports (1) two daughter chips with a

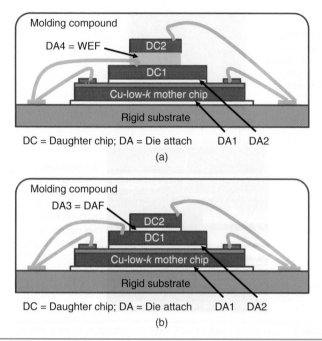

Figure 10.4 Schematic diagram of a 65-nm Cu–low-k stack die package: (a) test vehicle 1 (TV1) using wire-embedded film (WEF); (b) test vehicle 2 (TV2) using die-attach film (DAF) for two daughter dies that are placed at right angles to each other.

wire-embedded film (WEF) between them, as shown in Fig. 10.4a, test vehicle-1 (TV1), and (2) two daughter chips with a die-attach film (DAF) between them, as shown in Fig. 10.4b, test vehicle-2 (TV2). For both test vehicles, (1) the DAF is used between the bottom daughter die and the Cu–low-k die, and (2) a die-attach paste is used between the Cu–low-k die and the substrate.

The 65-nm Cu–low-k die has six metal layers with 408 input/output pins (I/Os) and is 7×7 mm in size. Its thickness is 0.3 mm. The daughter die dimension is $3 \times 5 \times 0.075$ mm. The organic substrate used is two metal layers with a plastic ball-grid-array (PBGA) substrate. The final package dimension is $17 \times 17 \times 1.1$ mm. The substrate and mold compounds are of green materials in line with environmental legislation.

10.3.2   Stresses at the Cu–Low-k Pads

The inferior mechanical properties of the Cu–low-k material complicate the packaging process in terms of low-k peeling during the dicing process, bond-pad cratering after wire bonding, and delamination of the passivation layer during reliability tests [3–5]. Common failure patterns encountered when dicing low-k wafers include metal/interlayer dielectric (ILD) delamination, metal/ILD peeling, and discoloration of the metal layer [6–13]. Under the Cu–low-k packaging program, the Institute of Microelectronics, Singapore, has developed polymer-encapsulated dicing-lane (PEDL) technology [6, 14] to address the dicing challenges of Cu–low-k packaging and simultaneously improve the reliability of the packages by reducing the corner stress of the Cu–low-k die compared with the conventional dicing method.

The local maximum shear-stress concentration is located at the outermost corner of the silicon die. This is due to thermal coefficient of expansion (TCE) mismatch among different materials plus the stress singularity at the corners of the die. The shear stress around the die corner is one of the major contributors to delamination failure between the low-k layer and the fluorinated silicate glass (FSG) layer [14]. Results have shown that by chamfering the edge of the silicon die at a 45° angle for 35 μm, the maximum stress can be reduced by 34 percent or more [14].

The package configuration analyzed is shown in Fig. 10.4b and the materials properties are shown in Table 10.1. PEDL technology, i.e., bevel-cut and benzocyclobutene (BCB) coating was adopted here to reduce the stress in the low-k layer as shown in Fig. 10.5. Figure 10.6 shows the finite-element analysis (FEA) results with a boundary condition of temperature change of the whole structure from 175 to 25°C [15]. Figure 10.6a shows the shear-stress distributions for the low-k layer near die attach 2, whereas Fig. 10.6b shows the shear-stress distributions in low-k corner. It is observed that the maximum stresses are almost at the same location for four cases.

Materials	CTE (ppm/°C)	E (GPa)	Poisson's Ratio
BCB	52	3	0.35
Low-k (BD)	10	8	0.3
FTEOS	2	105.44	0.3
TEOS	0.57	66	0.18
Silicon	2.7	131	0.28
Cu	17	110	0.34
Mold compound	8.9/40.4 ($T_g = 147.8$°C)	20.7 (25°C) 0.296 (250°C)	0.3
Die attach paste	50/100 ($T_g = 150$°C)	5.9 (25°C) 0.56 (260°C)	0.3
DAF	80/170 ($T_g = 128$°C)	1.66 (25°C) 0.037 (200°C)	0.3
Substrate	15	29	0.3
Solder mask	52	4	0.4

TABLE 10.1 Material Properties Used in Finite-Element Modeling and Analysis

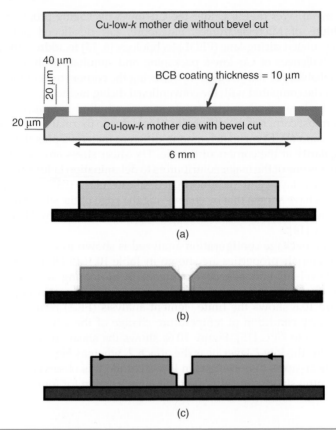

FIGURE 10.5 (Top-1) Schematic of a stack die without BCB coating and without a bevel cut. (Top-2) Schematic of a stack die with BCB coating and with a bevel cut. Different types of cuts: (a) straight; (b) bevel; (c) two-step.

Figure 10.7 shows the shear stresses in the low-k layer, die-attach layers, and epoxy-mold compound for the three case studies with or without using bevel cut and BCB coating, namely (1) straight cut without BCB; (2) bevel cut without BCB; (3) BCB without bevel cut; and (4) bevel cut with BCB. Since the low-k layer is easy to delaminate, the shear stress in the low-k layer at two critical locations is closely monitored—one at the low-k corner and another near die attach 2. It is found that a bevel cut is effective in reducing the shear stresses at the low-k corner. When the base case (i.e., without a bevel cut and without BCB coating) is replaced by the case with a bevel cut and without BCB coating, the shear stress at the low-k corner is reduced from 100.90 to 57.68 MPa. When the base case is replaced by the case with a bevel cut and BCB coating, the shear stress at the low-k corner is reduced from 100.90 to 23.75 MPa. On the other hand, it is found that the BCB coating is effective in reducing the shear stress in the low-k layer near die attach 2. When the base case is replaced by the

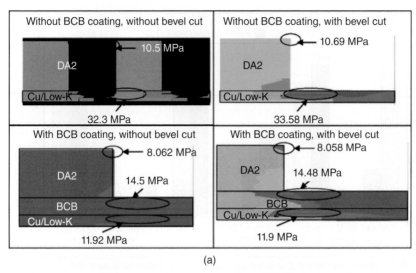

(a)

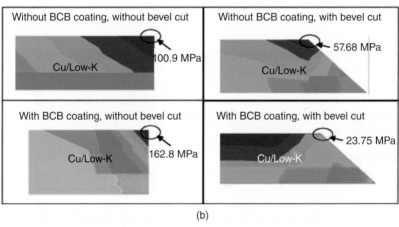

(b)

Figure 10.6 Shear stress distributions (*a*) for low-*k* layer near die attach 2 (DA2) (*b*) in low-*k* corner.

case with BCB coating and with a bevel cut, the shear stress in the low-*k* layer near die attach 2 is reduced from 32.30 to 11.90 MPa.

In summary, as shown in Fig. 10.7, both a bevel cut and BCB coating have to be presented to reduce the stresses in the overall low-*k* layer. Thus PEDL technology implemented in the package not only reduces the stress at the die corner but also reduces the stress at the low-*k* layer directly under the second-stack die (i.e., daughter die 1).

10.3.3 Assembly and Process

As mentioned in Chaps. 4 and 5, all 3D integration requires thin wafers and chips. In this study, the Cu–low-*k*-pad wafers were thinned down to 300 μm, and the Al-pad wafers (for the daughter dies) were

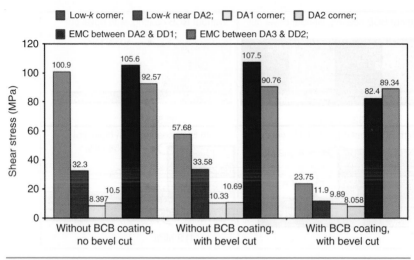

FIGURE **10.7**   Sheer stresses in the low-*k* layer at the low-*k* corner and near die attach 2 for different cases using BCB coating and a bevel cut.

thinned to 75 µm. All wafers were first thinned by means of coarse mechanical grinding, followed by fine mechanical grinding. In order to remove the microcracks on the backside of the thinned wafers, an additional process (namely, dry polishing) also was needed. The intra-to-inter wafer thickness variation of all the Cu–low-*k*-pad wafers was 2.21-µm total thickness variation (TTV). The TTV for the Al-pad wafers was 2.28 µm.

Evaluation of Dicing Methods

Die singulation is a major process that every individual IC chip goes through. With the trend toward 3D SiP and thinner and low-*k* wafers, this singulation process becomes critical. In this study, comparison among (1) straight-cut (Fig. 10.5*a*), (2) bevel-cut (Fig. 10.5*b*), and (3) two-step-cut (Fig. 10.5*c*) wafers is performed in terms of die strength and chipping results. The dicing parameters for straight-cut, bevel-cut, and two-step-cut wafers are shown in Table 10.2.

Dicing Method	Parameter	Blade Selection	Blade Speed (rpm)	Feed Speed (mm/s)	Blade Height (mm)
Straight cut		27 HCAA	30000	15	0.06
Bevel cut	Blade 1	2050D-T1	30000	30	0.115
	Blade 2	HDCC-B	30000	30	0.06
Two-step cut	Blade 1	27 HCCC	30000	20	0.1
	Blade 2	27 HAAA	30000	20	0.06

TABLE **10.2**   The Dicing Parameters for the Three Dicing Methods

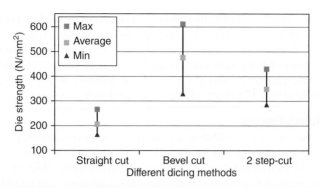

FIGURE 10.8   Die strength for 300-µm-thick Cu–low-*k* chips using different dicing methods.

To ascertain the die strength, a bending test has been shown to be a reliable measurement method [16]. In this study, a three-point bending test was used to determine the die strength, and the results are shown in Fig. 10.8 and Table 10.3. The sample size was 33 for each case. It was found that (1) a bevel cut results in higher die strength

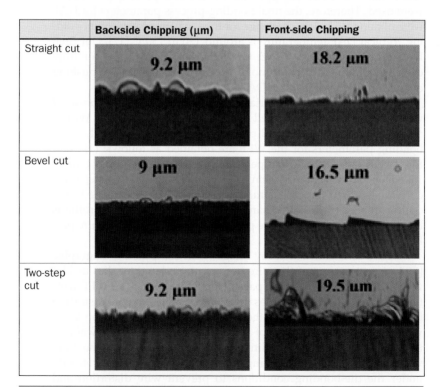

TABLE 10.3   Chipping Values for the Three Dicing Methods

than the other two dicing methods, (2) a bevel cut instead of a straight cut would lead to a 130 percent increase in die strength, (3) a bevel cut induced less chipping than a straight cut or a two-step cut, (4) a bevel cut instead of a straight cut would lead to a 10 percent decrease in front-side chipping, and (5) a bevel cut instead of a two-step cut would lead to a 15 percent decrease in front-side chipping.

Die-Attach Process

Thin dies are very flexible and require full support during pick and place to avoid die bending and to ensure uniform bond-line thickness, both of which are critical to packaging reliability. Thin die-attach layers are easier to control with film-type materials. Thus DAF was used for the two top daughter dies in TV2, whereas WEF was used for the two top daughter dies in TV1 (see Fig. 10.4). For both TV1 and TV2, the die attach used between daughter die 1 and the Cu–low-k mother die also was DAF. Die-attach paste was used for the Cu–low-k mother die attached onto the substrate, which usually has an uneven surface and requires material flow for good coverage without voids or delamination.

Table 10.4a shows the bonding-process setup for DAF. It was found that the voids disappeared when no. 4 process parameters were used. Therefore, the final bonding-process parameters for DAF were obtained. The key challenge for the DAF process was film voiding. Typical DAF temperature is around 120 to 150°C [17]. Our final DAF temperature was 150°C. Once the DAF is void-free or fully cured, employing a transfer-mold process is less effective at pushing out any trapped voids in the DAF.

WEF is a die-attach-material technology that allows same-sized wire-bonded dies to be stacked directly on top of each other. WEF is applicable to both center and peripheral wire-bond pads. The final bonding parameters for the WEF are obtained as follows: (1) bond force = 0.3 kgf, (2) bond temperature = 150°C, and (3) hold time at 150°C for 1 second. Comparison between one-step and two-step postcure for the WEF process is made in Table 10.4b. According to the optical images of die-shear samples, a two-step postcure profile is better than a one-step postcure profile. Thus a two-step postcure profile was used for the WEF.

Figure 10.9 shows an x-ray image (with computed tomographic [CT] reconstruction) of the wire between two daughter dies in TV1 when the final WEF process parameters were used. The WEF wets the wire and fully encapsulates the wire within it. No wire distortion is observed. In this example, the WEF has two layers. The top layer, which is attached to the back of a top die, has very limited flowability under the die-bonding conditions. The bottom layer of the film, which is facing the wire bonding of the bottom die, has low viscosity under the die-bonding conditions to prevent wire distortion and achieve complete gap filling of the bottom-die topography.

No.	Process parameters	CSAM after post bond cure (1-step post cure: 160°C for 60 min)
1	Bond force: 1.0 kgf; Bond temp: 150°C; Hold time: 1 sec	Void
2	Bond force: 1.0 kgf; Bond temp: 150°C; Hold time: 2 sec	Voids
3	Bond force: 2.0 kgf; Bond temp: 150°C; Hold time: 2 sec	Voids
4	Bond force: 2.5 kgf; Bond temp: 150°C; Hold time: 2 sec	OK

(a)

Postbond cure parameters	Optical images after different postbond cure and die shear testing	Observation
1-step post cure: 160°C for 60 min		Light region after die shear testiong
2-step post cure: 120°C for 60 min, followed by 140°C for 60 min		Uniform contrast on the film after die shear testiong

(b)

TABLE 10.4   (a) Bonding process setup for Die-Attach Film (DAF). (b) Postbond Cure Process Setup for Wire-Embedded Film (WEF)

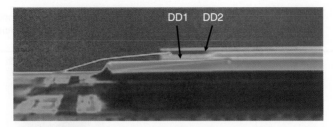

FIGURE 10.9 X-ray image showing the wire between the two daughter dies in TV1.

Wire-Bonding Process

Wire bonding is one of the key interconnection methods in 3D SiPs. State-of-the-art wire bonders with software and capabilities for complex looping, long and thin wires, and high bond-placement accuracy are required for the stacking of dies with different sizes, bonding on thin overhanging dies, forward and reverse bonding with very low loop heights, bonding on pads with low-k dielectrics, and bonding over active circuits under pad structures.

The upper part of Fig. 10.10 shows the ultralow-loop-height (50-μm) wire bonds on the first daughter die in TV1 [15]. It uses a new forward-motion method with the loop profile (as shown in the bottom part of Fig. 10.10) specially designed for the 50-μm loop height. The wire used is 0.8-mil (20-μm) NL4 gold wire. The wire bonder used is an ASM Eagle 60. The wire pitch is 53 μm, and the passivation opening is 43×100 μm. The key wire-bonding parameters are shown in Table 10.5. The new forward motion provides several benefits for stack-die packages. In addition to offering a lower loop height and less neck damage than traditional forward bonding, it provides for reduced pad damage, higher throughput, and better mold sweep than reverse bonding. It also offers a finer pitch capability than reverse bonding because of the low deformation on the first bond.

After wire bonding, the samples underwent the wire-pull test and the ball-shear test. It was found that (1) all ball-shear readings were above the minimum requirement (8 gf) with a desired shear failure mode, (2) no peeling was observed for all test samples during the ball-shear test, and (3) wire-pull readings were above the minimum requirement (3 gf) with a desired wire-pull failure mode—neck break (Fig. 10.11). Finally, some of the wire-bonded test vehicles (TV1, TV2) were built as shown in Fig. 10.12.

Molding Process

Increased wire density and wire length in 3D die stacking makes the mold process more difficult compared with conventional single dies. Different layers of wire-bond loops that are subjected to varying amounts of drag force can result in wire-sweep differences. This increases the possibility of wire shorts. Therefore, a careful selection

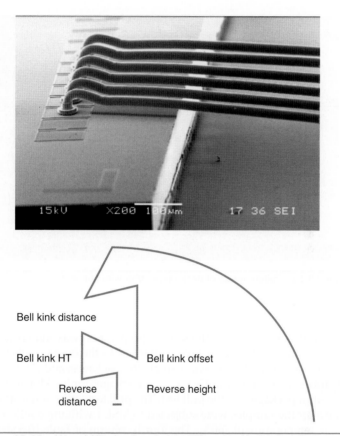

FIGURE 10.10   (*Above*) Ultralow-loop-height (50 µm) gold wire–bonding process established. (*Below*) Ultralow-loop motion profile: bell loop.

	Setting	
Bonding Parameters	**First bond**	**Second bond**
Bond time (ms)	10	8
Bond power (Dac)	36	30
Bond force (g)	8	27
EFO current (mA)	3800	
EFO time (ms)	223	
Bond temperature (°C)	170	

TABLE 10.5   Key Wire-Bonding Parameters

of molding-compound (MC) materials is needed for a reliable 3D die-stacking package.

For the selection of MC materials, an adhesion test (pull test) and moisture-sensitive-level (MSL) tests (MSL 2 and MSL 3) are performed

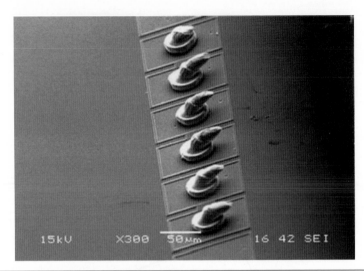

FIGURE 10.11   Failure mode of neck break after wire-pull test.

on four MCs (Table 10.6). The sample preparation was shown as follows: (1) dummy low-k dies were attached to the organic substrate,
(2) four MCs were used to encapsulate the packages, and (3) molded
substrates were diced into 17- × 17-mm packages for the MSL test and
8- × 8-mm packages for the pull test. The pull test (adhesion test) was
done after the samples were subjected to MSL 1 with three reflows at
a peak temperature of 260°C. The fourth column of Table 10.6 shows
that there were no significant differences in the pull values for all four
MC [15].

For MSL test evaluation, the samples are subjected to MSL 3 and
MSL 2, respectively, and three reflows at a peak temperature of 260°C.
Both CSAM and Thru-Scan analyses are carried out to check the performance of the assembled packages with four MCs. A failure criterion based on the Joint Electron Device Engineering Council (JEDEC)
recommendation of 10 percent delamination is employed. The second
column of Table 10.6 shows that all four MCs passed MSL 3 without
any voids and delamination. As to MSL 2, MC2 passed MSL 2 without
any failure in seven samples (see the third column of Table 10.6). However, one sample out of six of MC1 failed at MSL 2. Five samples out of
seven of MC3 failed at MSL 2. Three samples out of seven of MC4
failed at MSL 2. Therefore, MC2 was selected for use in subsequent
assembly builds.

After selection of the MC material was done, molding was performed on TV1 and TV2. Wire sweeps were measured for TV1 and
TV2, as shown in Fig. 10.13. All maximum wire sweeps were less than
10 percent, which is within acceptance limits.

(a)

(b)

Figure 10.12   Wire-bonded test vehicles: (a) TV1; (b) TV2.

Reliability Testing and Results

The final TV1 and TV2 were subjected to reliability tests (Table 10.7) according to JEDEC standards for components. Both CSAM and Thru-Scan were performed to examine the occurrence of delamination. No delamination had been observed on all samples up to 1000 cycles in thermal cycling (TC) tests and 1000 hours in high-temperature storage (HTS) tests (Fig. 10.14).

Mc Code	MSL 3 (Failed Samples/ Total Samples)	MSL 2 (Failed Samples/ Total Samples)	Pull Value in the Pull Test (kgf)	Warpage (μm)
MC1	0/7	1/6	2.33	377
MC2	0/7	0/7	2	504
MC3	0/7	5/7	2.33	692
MC4	0/7	3/7	1.33	620

TABLE 10.6   Molding Compound (MC) Selection

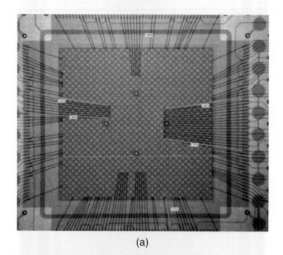

(a)

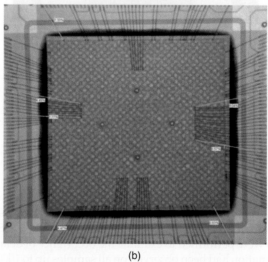

(b)

FIGURE 10.13   (a) Wire-sweep measurement for TV1 with a maximum wire sweep of 9.51 percent. (b) Wire-sweep measurement for TV2 with a maximum wire sweep of 9.04 percent.

Reliability Test	Group	Sample Size	Electrical Test Fail	CSAM Delamination
Thermal cycling (−40°C/125°C)	TV1	22	0	0
	TV2	21	0	0
High-temperature storage (150°C)	TV1	22	0	0
	TV2	21	2	0

Table 10.7 Reliability Results for 65-nm Cu–Low-*k* Stack-Die Package during Thermal Cycling and High-Temperature Storage

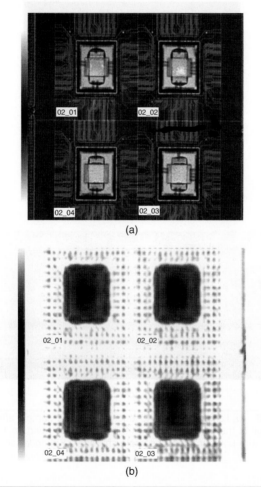

(a)

(b)

Figure 10.14 (*a*) Typical CSAM result showing no delamination after 1000 cycles in a TC test (−40 to 125°C). (*b*) Typical Thru-Scan result showing no delamination after 1000 cycles in a TC test (−40 to 125°C).

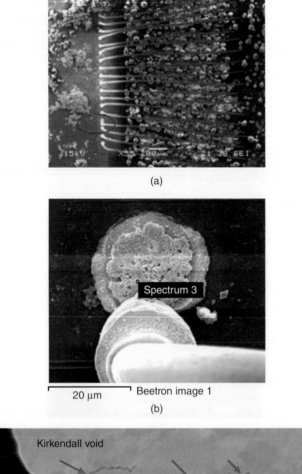

(a)

Spectrum 3

20 µm   Beetron image 1

(b)

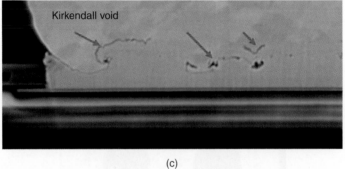

Kirkendall void

(c)

FIGURE 10.15   (a) Failed wire-bond location of one TV2 sample after decapsulation of the MC. (b) Top view of the failed wire bond. (c) Cross section of a wire bond of the TV2 sample showing a Kirkendall void.

An electrical test also was performed to monitor electrical resistance along the daisy-chain routing during reliability TC and HTS testing. No electrical failure was encountered on all samples up to 1000 cycles in TC testing (−40 to 125°C) and 1000 hours of HTS testing (150°C) except for two TV2 samples in HTS testing (Table 10.7). The two samples were shown to be open circuit in the daisy chain. Failure analysis was performed by decapsulation of the MC followed by pull testing and shear testing on the wire bonds. Both tests showed the wire bonds with lower bond strength. In the worst case, the ball bond was detached from the pad, indicating a typical Kirkendall void failure (Fig. 10.15). It was found that the failed wire bond occurred at the right side of the top die (i.e., daughter die 2). It should be noted that there was a 1-mm overhang between daughter die 2 and daughter die 1 in TV2 samples (see Fig. 10.12*b*). However, TV1 does not have any overhang issues. It might be the overhang of 1 mm that caused the poor wire-bond strength at time zero, although all ball-shear readings and wire-pull readings of the TV2 samples were above the minimum requirement.

10.3.4 Summary and Recommendations

Some important results and recommendations are summarized as follows [15]:

- PEDL technology has been adopted to reduce delamination of the Cu–low-*k* layer by reducing the stress in the layer. The stress in the Cu–low-*k* layer has been reduced by a factor of 4.2.

- Based on the dicing evaluation, the bevel-cut dicing methodology is better than the single-step dicing methodology in terms of chipping and die-strength results: The front-side chipping has been reduced by 15 percent, and the die strength has been increased by a factor of 2.3.

- An ultralow-loop (50-μm) wire-bonding process has been established.

- The wire-embedded-film process has been established, and it allows the special die-attach film to penetrate between gold wires such that the dummy die (or spacer die) can be removed.

- The 65-nm Cu–low-*k* stack-die PGBA package (i.e., TV1) has been developed, and it has successfully passed JEDEC component-level tests such as TC for 1000 cycles (−40 to 125°C) and HTS (150°C) for 1000 hours.

10.4 Bare Chip-to-Chip and Face-to-Face Interconnects

As mentioned in Chap. 1, face-to-face silicon chip stacking is one of the package-on-package (PoP) 3D IC packaging technologies [18–56]. In this section, the interconnect reliability of face-to-face silicon chips stacked with AuSn lead-free solder material is presented. Wafer bumping and the chip-to-wafer (C2W) bonding assembly process of the PoP are also discussed.

10.4.1   3D IC Packaging with AuSn Interconnects

AuSn solders are especially suited for optoelectronics and medical packaging. Most medical and optoelectronic devices have to be soldered in a fluxless process, which is possible using AuSn solders, as shown in the phase diagram in Fig. 10.16. Also, 80wt%Au20wt%Sn solder is used widely because of its advantages: (1) high strength, (2) high corrosion resistance, and (3) high fatigue resistance. Thus the Au20Sn solder system usually is selected as the interconnect material for fluxless flip-chip applications. Besides AuSn solder, SnAg is also considered.

10.4.2   Test Vehicle and Fabrication

Test Vehicle

The test vehicle was a silicon-stacked module, as shown in Fig. 10.17 [49]. The assembled stacked module consisted of a top silicon chip known as the *daughter die* and a bottom silicon chip known as the *mother die*, which were fabricated in wafers. On the daughter die, there were bumps made of Cu stud coated with the AuSn solder. On the mother die, there were electroless NiAu under bump metallurgy (UBM) for

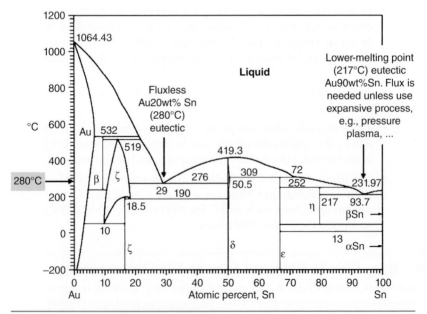

FIGURE 10.16   AuSn solder phase diagram.

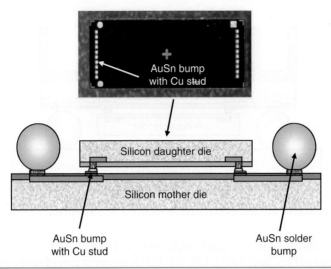

FIGURE 10.17 Stacked silicon module (daughter die and mother die) and a close look at the daughter die.

the bumps from the daughter die and AuSn solder bumps for the next level of interconnect, such as substrates. For both dies used in this test, the pads were interconnected in an alternating pattern so as to provide a daisy-chain connection when they were soldered together. There were 20 pads on the daughter die, and they were arranged in two rows, as shown in Fig. 10.17. The materials and dimensions of the key elements of the Si-stacked module are listed in Table 10.8. This PoP module eventually was attached to either a rigid or flexible substrate, as shown in Fig. 10.18.

The PoP (in a broad sense) shown in Fig. 10.18 could be a very low-cost package for high-performance, high-density, low-power-consumption, and wide-bandwidth applications. The heat spreader/sink is optional. The chip-to-chip interconnects are face-to-face (the shortest distance), and the signals, powers, grounds, etc. easily can go to the next

Package Type	Silicon Stacked module	
Test Die	Daughter	Mother
Die Size (mm)	3.405 × 1.34 × 0.06	4.793 × 1.34 × 0.13
Pad opening	FC pad 30 μm	FC pad 30 μm
Pad pitch	100 μm	100 μm
Bump type, (Height)	Au20Sn on Cu stud (25 μm)	Au20Sn (125 μm)
Under bump metallization	Al	Electroless NiAu

TABLE 10.8 Materials and Geometry of Key Elements of the PoP Module

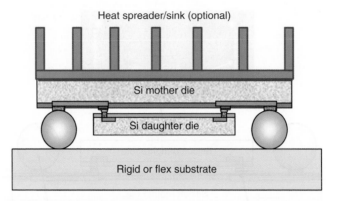

Figure 10.18   Stacked silicon module attached on a substrate.

layer of interconnect through solder bumps. For the wide I/O memory mentioned in Chap. 1 (Fig. 1.9), this PoP is the most cost-effective package (memory as the daughter die and logic as the mother die). It should be emphasized that there are no TSVs in the System-on-Chip (SoC)/Logic and memory chip.

Test Vehicle Fabrication

AuSn solder can be manufactured by various methods, such as electron-beam evaporation, electron deposition, and paste and solder perform. Among these methods, evaporated solder allows more accurate control of dimensions and positions. Therefore, after UBM patterning, a sequence of alternating layers of Au and Sn are deposited to form an AuSn layered structure. Some of the advantages of using the electron-beam evaporation process for AuSn deposition include [49]

1. The amount of oxide formed after deposition will be reduced.

2. The thickness and position of the solder can be controlled precisely.

3. The Au deposition rate is high.

4. Uniform AuSn thickness can be achieved across a 200-mm wafer.

p-Type silicon (100) wafers of 200-mm diameter with a thin electrical insulating film of 5000 Å of SiO_2 were used for fabricating the test vehicle. Figure 10.19 provides a schematic of the process for making the AuSn solder bumps on the daughter die. The first step was to deposit an AlCu layer on the Si wafer and then pattern the metal pads with daisy chains and etch the AlCu layer. Second, passivation layers of 5000 Å of SiO_2/5000 Å of SiN were deposited, and then the passivation layer was patterned using dry etching to open the metal pad. The Ti/Cu seed layer was sputtered as a conductive

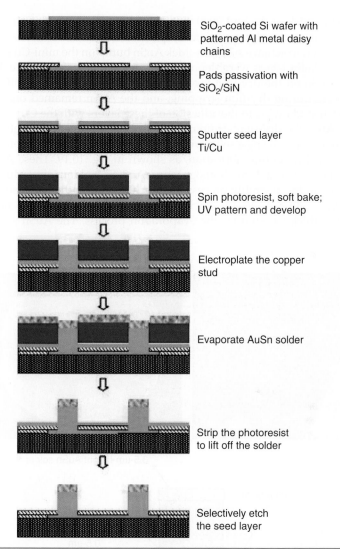

SiO$_2$-coated Si wafer with patterned Al metal daisy chains

Pads passivation with SiO$_2$/SiN

Sputter seed layer Ti/Cu

Spin photoresist, soft bake; UV pattern and develop

Electroplate the copper stud

Evaporate AuSn solder

Strip the photoresist to lift off the solder

Selectively etch the seed layer

Figure 10.19 Fabrication process of the AuSn solder bump on the daughter-die wafer.

layer for a further mini-Cu-stud electroplating process. Then a 20-µm-thick dry-film photoresist was laminated using a hot-roll laminator and ultraviolet (UV) light employing an EVG 640 contact-mask aligner followed by developing the photoresist to define the UBM pads. These patterned vias were filled with copper up to 18 to 20 µm using the RENA 200-mm wafer electroplating tool, which is a fountain-type (cup-type) tool in which Sperolyte CuSO$_4$-based solution (Atotech Pte, Ltd.) was used as the electrolyte solution. Then layers

2200 Å thick of Au and 2000 Å thick of Sn were deposited alternatively using the electron-beam evaporator. A total of 16 layers were deposited to achieve a 3.5-μm-thick AuSn bump on the mini-Cu stud. These AuSn layers were deposited on the whole wafer surface, but the AuSn on the dry-film layer could be removed using the lift-off process during dry-film stripping, and the AuSn remained only on the Cu stud owing to the adhesion of these layers with the Cu.

After the dry-film stripping, the Ti/Cu seed layers were etched off one by one using a selective wet-etching chemical to complete the AuSn solder-bump fabrication, as shown in Fig. 10.19. These AuSn layers melt and form homogeneous Au20Sn solder joints during reflow in the assembly process. Figure 10.20 shows the AuSn solder bump, where a 3.5-μm thickness of AuSn was deposited on a 20-μm-thick Cu stud by the evaporation process. The AuSn solder

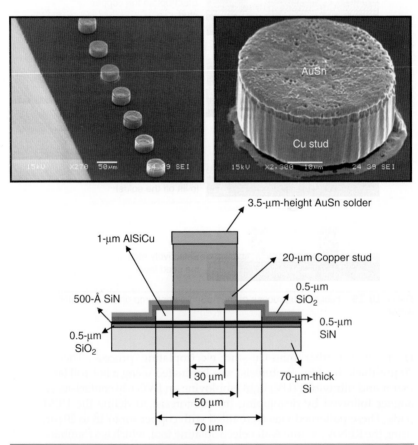

FIGURE 10.20 Dimensions and uniformity of the AuSn solder bump on the daughter-die wafer by the electron-beam evaporation process.

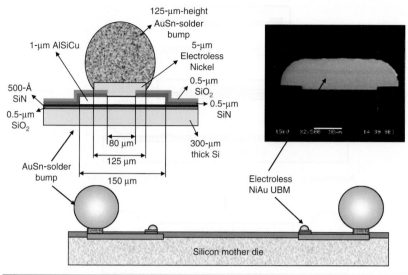

FIGURE 10.21 Dimensions of and close-up look at the electronless NiAu UBM on the mother die.

bumps were distributed at a 100-μm pitch along the two sides of the daughter die, and they were quite uniform, as the figure shows.

Figure 10.21 shows the bumps on the mother die—namely, the AuSn solder bump (which will not be discussed in this section) and the electroless NiAu UBM. The fabrication process for the mother die was the same as for the daughter die up to the AlCu metal pads being patterned for the daisy chain and metal pads' passivation opening. Then the electroless NiAu layers were deposited as the UBM, as shown in Fig. 10.21.

10.4.3 Chip-to-Wafer Assembly

The overall assembly process of the AuSn solder-bumped daughter die on the mother die is by a chip-to-wafer (C2W) bonding method; that is, first dice the daughter-die wafer into individual dies and then pick up the good daughter dies and bond them on good mother dies of the mother-die wafer, as shown in Fig. 10.22.

The bonding process is done with the flip-chip bonder FC150 from Karl Suss, as shown in Fig. 10.23. The assembly is carried out using nonreflowed bumps with N_2 purging during bonding. Argon sputter cleaning was performed on the daughter die and mother die (on a wafer) prior to the bonding process. The main challenge in AuSn flip-chip bonding is to achieve a good AuSn wetting on the electroless NiAu UBM bond pad with minimum AuSn solder squeezed between the bump and the UBM after bonding. In addition to bump-height coplanarity, the assembly process optimization in

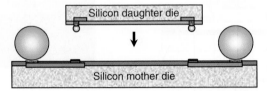

AuSn bumps formed by evaporation method

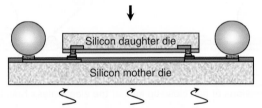

Alignment of daughter die to mother die by flip-chip bonder

↓

Heat applied during bonding to form AuSn interconnection

FIGURE 10.22   Overall PoP module assembly process.

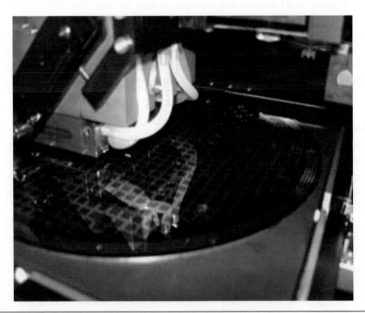

FIGURE 10.23   FC150 chip-to-wafer (C2W) bonder.

terms of alignment accuracy, bond force, bond temperature, and bond time on contact resistance, AuSn wetting, and AuSn solder-joint shape was investigated.

Bump-Height Coplanarity

It is important to characterize the Au bump height before the flip-chip attachment process because a nonuniform bump height will result in weak joints and even open continuity. Figure 10.24 shows the bump-height measurements in an acceptable range of a maximum 2.5 µm at the die level and 1.5 µm at the wafer-to-wafer level.

Alignment Accuracy

In the flip-chip conventional soldering process, solder bumps are dipped into flux and attached onto the substrate pads before being subjected to the reflow process. During the oven reflow process, the solder bump on the die is able to self-align to the substrate pads and form good solder interconnections. However, with thermal compression bonding, the solder is unable to self-align during assembly. Hence bonding alignment accuracy becomes critical, especially when using large dies with solder bumps. Precise bonding equipment with good alignment and bonding accuracy is required to form a good AuSn interconnect with stable contact resistance and good AuSn wetting on the substrate pad.

The contact resistance of the interconnects is characterized at different bonding alignment accuracies, as shown in Fig. 10.25. A bond force of 0.8 kg is applied during the flip-chip bonding process. There is a significant difference in the contact resistance and the AuSn

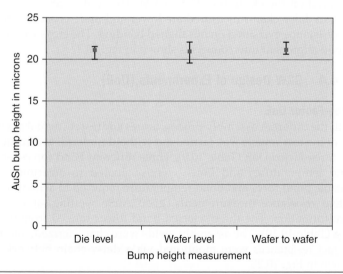

FIGURE 10.24 AuSn bump-height measurement on various wafers.

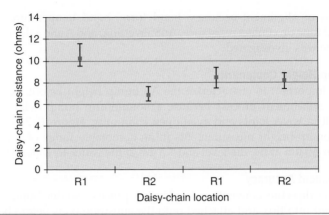

FIGURE 10.25 C2W bonding-alignment accuracy versus contact-resistance measurements.

wetting on the electroless NiAu bond pads on misaligned AuSn bumps to electroless NiAu pads, compared with 100 percent aligned AuSn bumps to electroless NiAu pads, as shown in Fig. 10.26. This mode of joint configuration results in poor AuSn wetting on the bond pad and higher resistance readings. One of the reasons for this is that there is no self-alignment of the AuSn solder bumps to the electroless NiAu UBM after bonding. The AuSn solder is unable to wet the electroless NiAu UBM if there is a great difference in the misalignment offset. On the other hand, for the 100 percent aligned bonding, stable contact resistance and good AuSn wetting are observed on the electroless NiAu UBM. The contact resistances on the two daisy chains measured are lower and more consistent than the resistance measured on misaligned bonding. The cross section of a good alignment interconnect is shown in Fig. 10.27.

10.4.4 C2W Design of Experiments (DoE)

Three-Factor DoE

With the different assembly process issues addressed, actual assembly of the test vehicle was carried out without underfill according to the three-factor DoE (Table 10.9). Three different bond forces—two bond temperatures and time—were evaluated to determine the suitable bond force and bond temperature required to achieve consistent resistance measurements, good AuSn wetting, and AuSn joint formation. The three selected bond forces ranged from 0.4 to 1 kg (Table 10.10). The assembly yields were characterized, and the contact resistances were monitored via a daisy-chain network, as shown in Fig. 10.28.

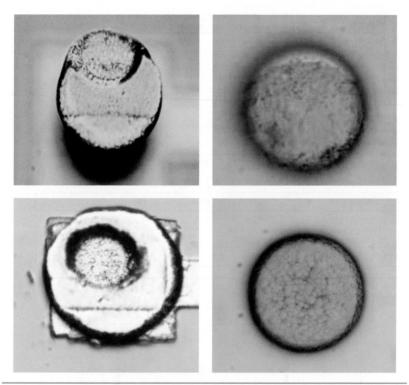

FIGURE 10.26   AuSn wetting failure modes. (*Above*) Misaligned bonding; (*below*) 100 percent bonding alignment. Also, (*left*) daughter dies; (*right*) mother dies.

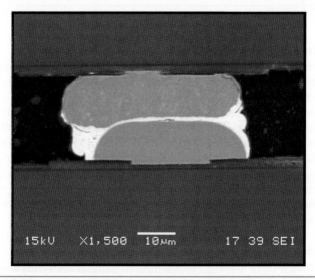

FIGURE 10.27   A cross section of a good AuSn solder joint.

Batch	Bond force (kg)	Bond Temp. (°C)	Bond Time (s)
A	0.4	290	15
B		315	15
C	0.8	290	15
D		315	15
E	1	290	15
F		315	15

TABLE 10.9   Three-Factor DoE of C2W Bonding Evaluation

Factors	Variations			Response
Bond force (kg)	0.4	0.8	1	• Contact resistance
Bond temperature (°C)	290	315		• AuSn joint shape
Bond time (s)	15			• AuSn wetting

TABLE 10.10   C2W Bonding with Various Combinations of Force, Temperature, and Speed

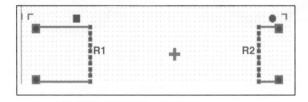

FIGURE 10.28   Resistance measurements through the chain on the mother die.

DoE Results

The bond force has a significant influence on the interconnection resistance and AuSn joint formation. Figure 10.29 shows that stable contact resistance can be achieved at 0.8 and 1 kg of bond forces. However, 1 kg is undesirable because it results in chip passivation cracking at the Al pad–UBM interface, as shown in Fig. 10.30. Furthermore, a high bonding force tends to squeeze out the AuSn eutectic solder between the bump and bond pad and results in poor AuSn wetting on the bond pad. Figure 10.31 shows a cross-sectional view of the AuSn assembly at 1 kg of bond force. The findings also suggested that the contact resistance was more sensitive to bond force than to bond temperature and bond time.

10.4.5   Reliability Tests and Results

A set (batch D) of test-vehicle assemblies (without underfill) was subjected to temperature cycle testing (–40 to 125°C) to access the thermal-fatigue characteristics of the AuSn interconnects [49]. The electrical contact resistance was measured prior to reliability testing and at

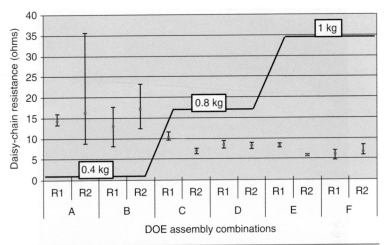

FIGURE 10.29   Contact-resistance measurements from the DoE assemblies.

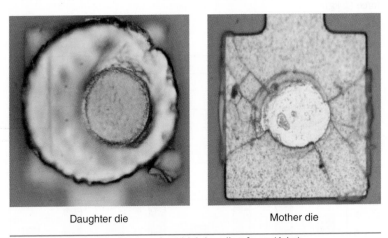

Daughter die   Mother die

FIGURE 10.30   Passivation crack at high bonding force (1 kg).

every 25 cycles during the testing for 100 cycles. The criteria for failure would be an open continuity in the resistance value (∞) of the daisy chain. This set of samples passed the TC test without open continuity up to 100 cycles, as shown in Fig. 10.32. A minimum change of less than 10 percent in the contact resistance was observed for bond force at 0.8 kg. This suggests that the interconnect configuration of a homogeneous AuSn eutectic solder with direct contact to the electroless NiAu UBM bond pad is robust and able to withstand the TC conditions. The results indicate that stability in contact resistances and good reliability performance can be achieved in combination D (bonding temperature = 315°C; bonding time = 15 seconds) at 0.8 kg of bond force.

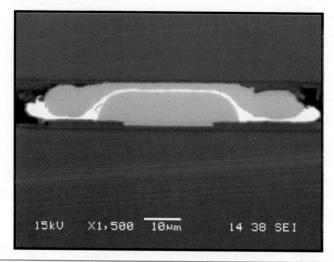

FIGURE 10.31 A cross section of an AuSn assembly made by high bonding force (1 kg).

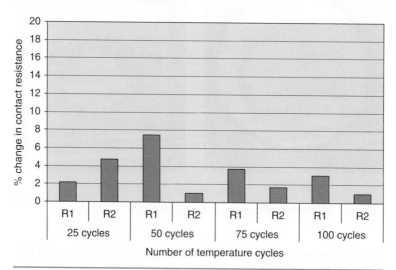

FIGURE 10.32 Percentage changes in contact resistance during temperature cycles (−40 to +125°C).

10.4.6   3D IC Packaging with SnAg Interconnects

Figure 10.33 shows the same daughter die and mother die but with a different solder system. Instead of the AuSn solder, the Sn3wt%Ag + Cu stud is considered on the daughter die, as shown in Fig. 10.34. Also, the solder ball of the chip-scale package (CSP) on the mother die is replaced by Sn37wt%Pb. The focus, of course, is on the reliability of

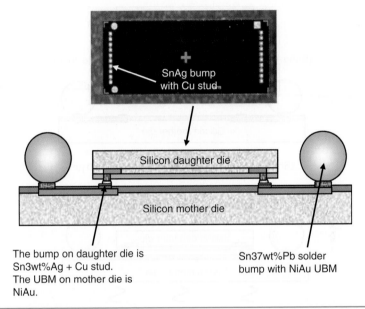

FIGURE 10.33   Stacked silicon module (daughter die and mother die) with SnAg lead-free solder.

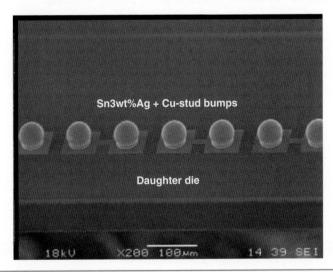

FIGURE 10.34   Lead-free Sn3wt%Ag + Cu-stud (~40-µm) bumps on daughter die.

the SnAg solder joints between the daughter die and mother die [50]. Prior to assembly, the daughter die wafer with a 40-µm height of SnAg + Cu-stud microbumps was thinned down to 70 µm, and the mother die wafer with a 200-µm height of CSP SnPb + UBM microbumps was thinned down to 300 µm, as shown in Fig. 10.35. Figure 10.36 shows a

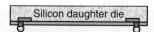

Daughter die wafer with 40 μm height
(SnAg + Cu-Stud) micro bumps is thinned down to 70 μm

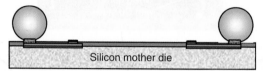

Mother die wafer with 200 μm height CSP
(SnPb + UBM) bumps is thinned down to 300 μm

Alignment of daughter die to mother die by flip-chip bonder

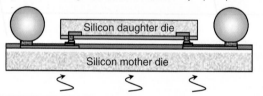

Heat applied during bonding to form SnAgCu interconnection

FIGURE 10.35 Overall PoP module assembly process with SnAg.

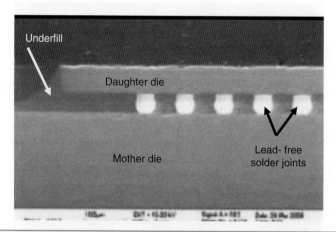

FIGURE 10.36 Daughter die assembled on mother die with SnAg solder joints.

cross section of the assembly where the underfill was applied. It should be pointed out that the solder joints are under natural reflow, and because of the surface tension of the molten solder, the shape of the solder joints is nice and smooth, which is very different from the C2W bonding of AuSn solder joints.

After assembly, the samples were subjected to unbiased highly accelerated stress testing (uHAST): +130°C, 85% relative humidity

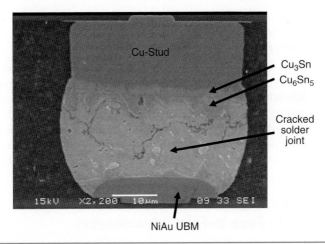

FIGURE 10.37 Failed sample after high-temperature storage test, moisture-sensitivity test levels 1 and 3, and 1000-thermal-cycles test.

(RH), 96 hours; MST L3: 30°C, 60%RH, 192 hours, three reflows at 260°C; MST L1: 85°C, 85%RH, 168 hours, three reflows at 260°C; HTS: +125°C, 1000 hours; and TC: +125/−55°C, 15-minute dwell, 15°C/min ramp, 1000 cycles. A typical image of failed samples after 1000 cycles is shown in Fig. 10.37. The crack initiates near the corner of the interface between the intermetallic compounds (IMCs; Cu_6Sn_5) and the bulk solder and then propagates through the solder joint.

10.4.7 Summary and Recommendations

Some of the important results are summarized as follows [49, 50]:

- Uniform bump coplanarity at the die level (±3.5 μm) is required to ensure that uniform pressure can be applied on the test die and form good AuSn interconnects.

- Stable contact resistance (±2 Ω) and good AuSn wetting on the electroless NiAu pads are achieved on 100 percent aligned bonding at 0.8 kg of bond force.

- Assembly results show that samples assembled at high bonding force (1 kg) are undesirable because this force results in chip passivation cracking at the Al pad–UBM interface. Furthermore, a high bonding force tends to squeeze out the AuSn eutectic solder between the bump and bond pad and results in poor AuSn wetting on the bond pad.

- Temperature-cycling results show that stable contact resistances (±9 Ω) and good reliability performance (>100 temperature cycles) can be achieved without underfill at 0.8 kg of bond force.

- More important, the results of this work will serve as a reference for industries establishing flip-chip assemblies using AuSn fluxless solder-bump technology.

- Chip to wafer flip chip natural reflow assembly process was demonstrated (with good alignments) for SnAg micro bumps of 40 μm height at 100 μm pitch. Underfill process also was demonstrated (without voids) for micro bumps at 100 μm pitch. The stacked packages have passed the JEDEC standard package level reliability tests, in terms of thermal cycling, high temperature storage, moisture sensitivity level 1 and level 3, and uHAST.

10.5   Low-Cost, High-Performance, and High-Density SiPs with Face-to-Face Interconnects

10.5.1   Cu Wire-Interconnect Technology for Moore's Law Chips with Ultrafine-Pitch Cu–Low-k Pads

In Secs. 10.3 and 10.4, 3D IC packaging with (1) Au wire bonds made by a wire bonder, (2) AuSn interconnects made by C2W (thermocompressure) bonding, and (3) SnAg interconnects made by natural reflow were presented and discussed. For Moore's law chips with ultrafine-pitch Cu–low-k pads, perhaps copper wire-interconnect technology (WIT) is more appropriate. WIT is from a patent written by Love et.al., "Wire Interconnect Structures for Connecting an Integrated Circuit to a Substrate" (U.S. Patent 5,334,804, granted on August. 2, 1994). It also can be found in Chap. 14 of Ref. 2. Today it is also called *copper column*, *copper stud*, *copper post*, or *copper pillar*. Figure 10.38 shows a schematic of the pad of a Moore's law chip with Cu–low-k metal layers. Usually, a redistribution layer (RDL) is needed to redefine the pad for the copper column and solder, as shown in the figure.

10.5.2   Reliability Assessment of Ultrafine-Pitch Cu–Low-k Pads with Cu WIT

Figure 10.39a, b [56] show the 110-μm-thick photoresist film JSR151N provided by the JSR Company and the openings (100 and 160 μm in diameter) for making the Cu wires. Figure 10.39c, d show, respectively, the as-plated Cu wire plus solder before and after reflow. Figure 10.39e, f show the TC test results. It can be seen that the failure mode is cracking of the 1-μm-thick RDL, and thus thicker RDLs are needed for TC reliability. For more information, please see Ref. 56.

10.5.3   A Few New Design Proposals

Figure 10.40 shows a few new design examples for low-cost, high-density, and high-performance chip-to-chip and face-to-face interconnects; for example, (1) the Cu–low-k SoC can support the wide I/O memory chip with either solder bumps or Cu wire with solder; (2) the

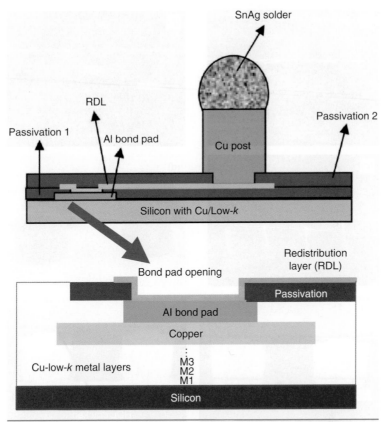

FIGURE 10.38 Schematic of the Cu–low-*k* layer stack structure with emphasis near the bonding pads and RDL.

SoC can simultaneously support the memory and graphic chips; (3) the SoC can simultaneously support the memory-chip stack and other chips; (4) the power, grounds, signals, etc. can connect to the next level of interconnect through either solder bumps or Cu wire with solder; and (5) the heat spreader/sink is optional. All these are conventional and proven packaging technologies, and there are no TSVs in any of these active-device chips.

10.6   Fan-Out-Embedded WLP-to-Chip (Face-to-Face) Interconnects

10.6.1   2D eWLP/RCP

The 2D fan-out-embedded wafer-level package (eWLP) has been very popular in the past few years. It was invented by Infineon and Freescale and is in high-volume production by STMicroelectronics, ASE, StacChippPack, etc. Freescale called it the *redistributed chip package* (RCP) [57–59].

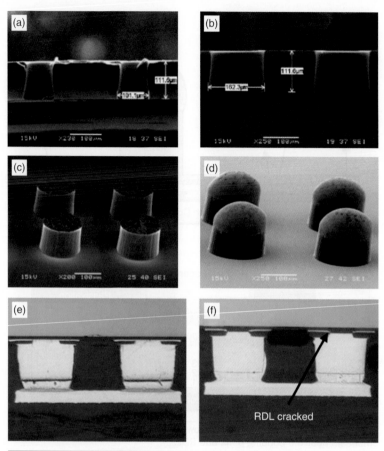

Figure 10.39   Scanning electron microscopic (SEM) and optical images of the Cu-post (column or wire) interconnects: (*a*) patterns of 100-μm diameter in the JSR151N photoresist film at a thickness of 110 μm; (*b*) patterns of 160-μm diameter in the JSR151N photoresist film at a thickness of 110 μm; (*c*) electroplated copper column; (*d*) copper column with SnAg solder cap; (*e*) Before test samples; (*f*) failure mode after TC testing.

10.6.2   3D eWLP/RCP

The 2D eWLP/RCP can be extended into 3D IC packaging by using through-package vias (TPVs). Unlike the TSVs mentioned in Chaps. 1 through 9, TPV allows a package to be designed with 3D connections internal to the package body, as well as creating eWLP/RCPs with double-sided redistribution and TPVs, which enable traditional PoP configurations and additional assembly opportunities [57].

Figure 10.41 shows top and bottom views of Freescale's 3D RCP [57–59] SiP with the core 3D RCP elements as a foundation. To add to this foundation, contained within the RCP system base package, is an embedded MEMS device and its paired controller die. The portion of

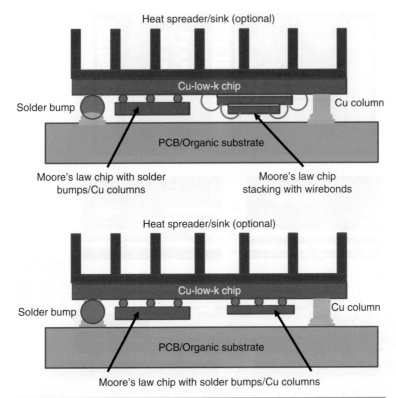

FIGURE 10.40 Proposed low-cost and high-performance 3D IC packaging. Face-to-face and chip-to-chip interconnects and solder bumps (or columns) to the next level of interconnect.

the RCP SiP containing the functional MEMS and ASIC devices is approximately 600 µm in thickness. Also, included is a near face-to-face chip-to-chip assembly with faces separated by only C4 bumps and a single layer for RCP routing.

10.6.3   Summary and Recommendation

Some of the important results are summarized as follows:

- A few low-cost examples of chip-to-chip (with face-to-face) interconnects for 3D IC packaging with conventional and proven technologies have been presented and discussed.

- For cost and reliability reasons, the industry should develop more 3D IC packaging techniques such as face-to-face for chip-to-chip interconnects without using TSVs.

- Of course, for very high-performance and high-density, low power and wide bandwidth applications, TSV technology is needed.

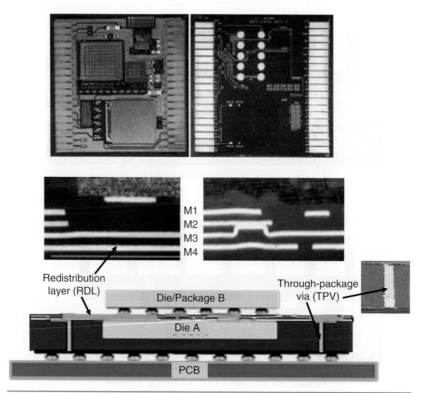

M1
M2
M3
M4

Redistribution
layer (RDL)

Die/Package B

Through-package
via (TPV)

Die A

PCB

FIGURE 10.41   Freescale's low-cost and high-performance chip-to-chip and fact-to-face 3D IC packaging. The vertical interconnects (which are similar to TSVs) are made by through-package vias (TPVs) without using any semiconductor equipment.

10.7   A Note on Wire-Bonding Reliability

A quick and dirty method to judge the quality and reliability of wire bonds is by the pull test [60].The pull force and displacement of gold (Au) and copper (Cu) wires in microelectronics are investigated in this section. Emphasis is placed on (1) the development of a set of equations for determining the internal forces and deformations of wires when subjected to an external pull force, (2) determination of the maximum pull force (when the wire breaks) based on some failure criteria of the wires, and (3) experimental verification of the equations presented.

10.7.1   Common Chip-Level Interconnects

There are three major chip-level interconnects [61, 62], namely, (1) wire bonds [63], (2) solder bumps [1,2], and (3) tape-automated bonding [64]. More than 85 percent of the chips in the world use wire-bonding technology. Since 2006, the annual number of semiconductor chips using wire-bonding technology to transmit their signals to the next level of interconnects has been exceeding the world population (7 billion). One of the simplest qualification and reliability test methods for a wire bond is the pull test, which is the focus of this study.

The internal forces in a wire, when it is subjected to an external pull force, have been given in Ref. 62. However, it only considers the equilibrium of (external and internal) forces acting on the un-deformed wire. In this study, the equilibrium of forces is applied at the deformed shape of the wire, as shown in Figs. 10.42 and 10.43, and the material properties of wires are considered. Thus the results presented herein are a more general and accurate pull relation that can be applied to wires composed of different materials, such as Au and Cu.

Most of the chips to be wire-bonded today are with Au wires. However, because of the surge in Au prices and research and development progress in Cu wire-bonding technology [65–103], many companies have been looking for low-cost solutions, and the shift from Au to Cu wire bonding is genuinely picking up. Useful results for both Au and Cu wires are given in this study and are verified by experiments.

10.7.2 Boundary-Value Problem

Figure 10.42 shows a wire subjected to an external force F at point Q that displaces Q to position Q'. The statement of the current problem is to determine the internal forces (f_d and f_t) in CQ' and BQ' and the new position (displacement) of point Q (i.e., point Q').

In Fig. 10.42, l_d and l_t are the original wire lengths, θ_d and θ_t are the original wire angles, d is the distance between the wire bonds (supports), H is the height different between the wire supports, h is the height of wire loop, and ϕ is the angle of the applied force (F) from the vertical axis.

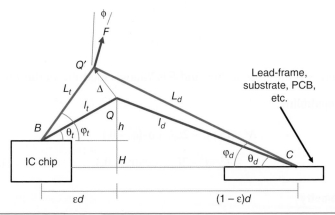

FIGURE 10.42   Wire BQC is subjected to a pull force F at point Q and is displaced Δ to point Q'.

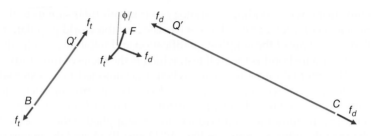

Figure 10.43 Free-body diagrams for equilibrium equations of the pull force F and internal forces f_d and f_t.

The equilibrium equations can be obtained from the free-body diagram with the external force F and internal forces f_t and f_d as shown in Fig. 10.43. There are six unknowns: f_d = internal force in CQ'; f_t = internal force in BQ'; L_d = final length of CQ (CQ'); L_t = final angle of BQ (BQ'); φ_d = final angle of CQ'; and φ_t = final angle of BQ', as shown in Figs. 10.42 and 10.43. The six equations that are necessary and sufficient for determination of these six unknowns ($f_d, f_t, L_d, L_t, \varphi_d$, and φ_t) are shown in Eqs. (10.1) through (10.6).

Equilibrium Equations

$$\frac{f_t}{F} = \frac{\cos\phi \, \cos\,\varphi_d + \sin\phi \, \sin\,\varphi_d}{\sin\varphi_t \, \cos\,\varphi_d + \cos\,\varphi_t \, \sin\varphi_d} \tag{10.1}$$

$$\frac{f_d}{F} = \frac{\cos\phi \, \cos\varphi_t - \sin\phi \, \sin\varphi_t}{\sin\varphi_t \, \cos\varphi_d + \cos\varphi_t \, \sin\varphi_d} \tag{10.2}$$

Strength of Materials

$$L_t - l_t = \frac{f_t l_t}{AE} \tag{10.3}$$

$$L_d - l_d = \frac{f_d l_d}{AE} \tag{10.4}$$

where A is the cross section and E is Young's modulus of the wire.

Compatibility

$$\Delta^2 = L_t^2 + l_t^2 - 2L_t l_t \, \cos\left(\varphi_t - \theta_t\right)$$

$$= L_d^2 + l_d^2 - 2L_d l_d \, \cos\left(\varphi_d - \theta_d\right) \tag{10.5}$$

Geometry

$$L_t = \frac{-g \, \cos(X + Y)}{\sin(90 + \varphi_t)} \tag{10.6}$$

where

$$g = L_t \cos(X-Y) + d\cos(X+Y)$$

$$X = 90 - \frac{\varphi_d}{2}$$

$$Y = \tan^{-1}\left(\frac{L_d-d}{L_d+d}\right)\cot\frac{\varphi_d}{2}$$

The closed-form solutions of Eqs. (10.1) through (10.6) are very difficult to obtain, if not impossible. However, with a computer and software, for a given set of F, h, H, d, ε, ϕ, l_d, l_t, θ_d, and θ_t, the values of L_d, L_t, φ_d, φ_t, f_d, and f_t can be determined. The displacement Δ of point Q also can be determined by Eq. (10.5).

If $L_d = l_d$, $L_t = l_t$, $\varphi_d = \theta_d$, and $\varphi_t = \theta_t$ (i.e., there is no deformation of the wire under the external force), then Eqs. (10.3) through (10.6) do not exist, and Eqs. (10.1) and (10.2) degenerate to the following equations, as given in Ref. 62:

$$\frac{f_t}{F} = \frac{\sqrt{h^2 + \varepsilon^2 d^2}}{(h+\varepsilon H)}\left[(1-\varepsilon)\cos\phi + \frac{(h+H)}{d}\sin\phi\right]$$

$$\frac{f_d}{F} = \frac{\sqrt{1 + \dfrac{(1-\varepsilon)^2 d^2}{(h+H)^2}}}{(h+\varepsilon H)}\left[(h+H)\left(\varepsilon\cos\phi - \frac{h}{d}\sin\phi\right)\right]$$

10.7.3   Numerical Results

It should be reemphasized that the internal forces f_d and f_t and the displacement Δ at point Q can be determined by a computer and software for solving the nonlinear Eqs. (10.1) through (10.6) for any external force F applied at point Q with an angle ϕ from the vertical axis. However, for engineering practice, it is more useful to predict the maximum pull force when the wire breaks. In order to do that, we have to apply a failure criterion (e.g., tensile strength and elongation) to the wire. If the tensile strength σ_s of the wire is chosen as the failure criterion, then the internal force is equal to $A\sigma_s$ when the wire breaks. For example, if the internal force (f_t) in the leg reaches the tensile strength first, then Eq. (10.3) becomes

$$L_t - l_t = \frac{\sigma_s l_t}{E} \tag{10.3'}$$

Otherwise, Eq. (10.4) becomes

$$L_d - l_d = \frac{\sigma_s l_d}{E} \tag{10.4'}$$

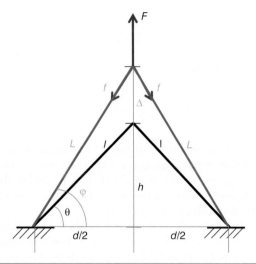

FIGURE 10.44   A special case: $H = 0$, $\varepsilon = 0.5$, and $\phi = 0$.

Now, let's consider a special case ($H=0$, $\varepsilon=0.5$, $\phi=0$, and $d=1.5$ mm), as shown in Fig. 10.44, and assume that the wire is a Cu hard wire with a tensile strength of 25 gf (gram force) and a Young's modulus of 130 GPa. By solving Eqs. (10.1) through (10.6), the maximum pull force F when the wire breaks (with tensile strength as the failure criterion) of the Cu hard wire with various wire loop height h is shown in Fig. 10.45. It can be seen that the maximum pull force is larger for larger wire loops.

10.7.4   Experimental Results

The wire-pull experiments were performed in a uniaxial pull tester (Electro Force) with extremely small force and displacement scales, as shown in Fig. 10.46. The feed (strain) rate was 0.025 mm/s. The distance between the supports d was 1.5 mm. Some of the pull-test results are shown in Figs. 10.45 and 10.46. It can be seen that (1) most of the wires were broken between the pull force and support and (2) the experimental results compared very well with the results predicted by the equations presented earlier.

In Fig. 10.45, the results from the equations provided by Ref. 62 for this special case are also presented. It can be seen that they are different, and the reasons are (1) in Ref. 62, the author does not consider the strength of materials and (2) in Ref. 62, the equilibrium equations are not applied at the deformed shape of the wire.

10.7.5   More Results on Cu Wires

Figure 10.47 shows the maximum pull force (gf) versus wire-loop height for different Cu wires (Cu hard wire and Cu annealed wire).

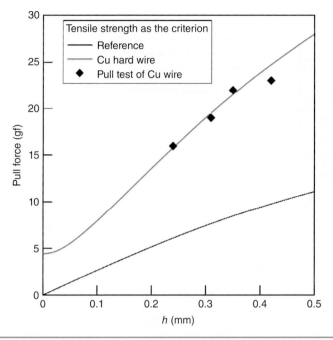

FIGURE 10.45 Maximum pull force (gf = gram force) for Cu hard wires with tensile strength = 25 gf and Young's modulus = 130 GPa versus wire-loop height h. The results from the equations of Ref. 62 are also plotted.

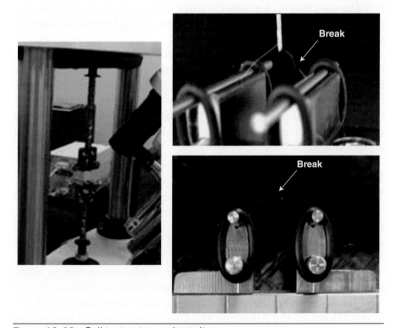

FIGURE 10.46 Pull-test setup and results.

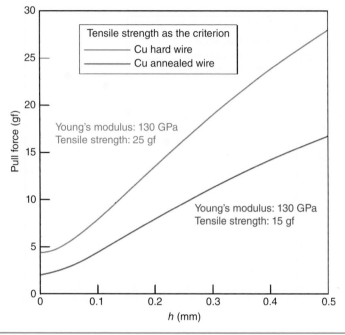

FIGURE 10.47   Maximum pull force versus loop height for different Cu wires.

It can be seen that (1) for both Cu wires, the larger the loop height, the larger is the maximum pull force and (2) the larger the tensile strength, the larger is the maximum pull force.

The maximum displacement of the Cu wires is shown in Fig. 10.48. It can be seen that (1) for both wires, the larger the loop height, the smaller is the displacement, (2) for both wires, the variation in maximum displacement is more significant at smaller ($h <$ 0.3 mm) wire-loop heights, and (3) for all the wire loop heights, the displacement of the Cu hard wire is larger than that of the Cu annealed wire.

10.7.6   Results on Au Wires

The maximum pull force (gf) versus wire-loop height for various Au wires is shown in Fig. 10.49. It can be seen that (1) for all the Au wires, the larger the wire-loop height, the larger is the maximum pull force and (2) the larger the tensile strength and Young's modulus, the larger is the maximum pull force.

Figure 10.50 shows the maximum displacement versus wire-loop height for various Au wires (Au hard wire, Au stress-relieved wire, and Au annealed wire). It can be seen that just as with Cu wires, (1) for all the Au wires, the larger the wire-loop height, the

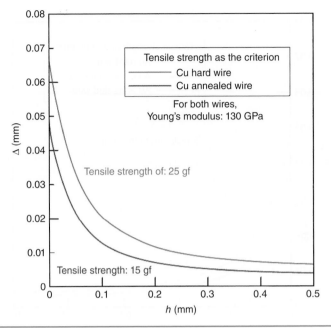

FIGURE 10.48 Maximum displacement Δ versus Cu wire-loop height h.

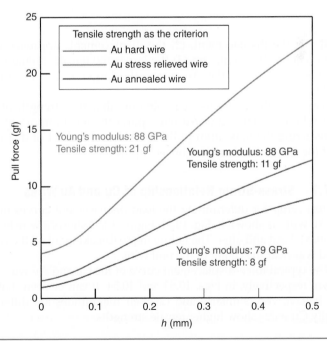

FIGURE 10.49 Maximum pull force (gf) versus loop height for various Au wires.

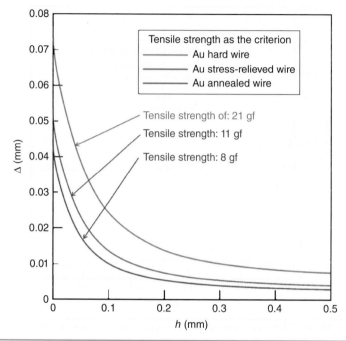

FIGURE 10.50   Maximum displacement versus loop height for various Au wires.

smaller is the displacement, (2) the displacements approach constants when $h > 0.3$ mm, and (3) the larger the tensile strength and Young's modulus of the Au wires, the larger is the maximum displacement.

Based on these results, it is obvious that the strength of the bond-wire material (Cu or Au wire) plays the most important role in predicting the maximum pull force! Also, the effect of displacement for small wire-loop heights is important.

10.7.7   Stress-Strain Relationship of Cu and Au Wires

The test setup for determining the load-displacement curves of Cu and Au wires is shown in Fig. 10.51. Figure 10.52 shows a wire before the unaxial test and after. The failure mode (breaking) of all the wires tested was near the center of the pull wire.

The typical force-displacement curve of the Cu and Au wires are shown, respectively, in Figs. 10.53 and 10.54. It can be seen that (1) they behave very similarly and (2) their magnitudes are different; usually Cu wires show higher tensile strengths.

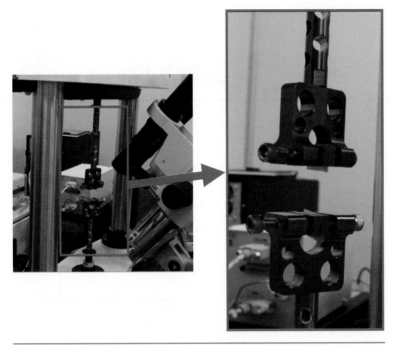

FIGURE 10.51   Uniaxial test setup of a straight wire.

Before pulling   **After pulling**

FIGURE 10.52   A typical wire before and after uniaxial testing.

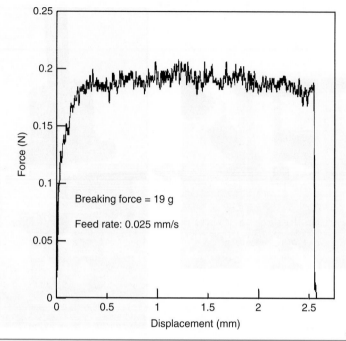

FIGURE 10.53   Typical uniaxial force-displacement curve of Cu wire.

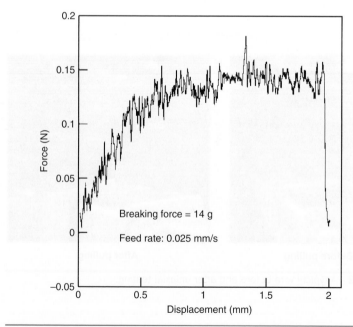

FIGURE 10.54   A typical uniaxial force-displacement curve of Au wire.

10.7.8 Summary and Recommendations

The pull force and displacement of gold (Au) and copper (Cu) wires in microelectronics have been investigated in this study. Some important results and recommendation are summarized as follows:

- Six necessary and sufficient equations have been developed for determining the internal forces and displacement of a wire subjected to an external pull force.

- Based on a failure criterion (i.e., tensile strength) of the wire, the maximum pull force and the maximum displacement of the wire can be determined/predicted.

- An experiment has been set up for measuring the maximum pull force. The test results compared very well with the results predicted by the equations developed herein.

- For both Cu and Au wires, the larger the wire-loop height, the larger is the maximum pull force and the smaller is the maximum displacement.

- For both Cu and Au wires, the larger the tensile strength of the wire, the larger are the maximum pull force and displacement.

- For both Cu and Au wires, their strength of materials plays the most important role in maximum pull force and displacement.

- For both Cu and Au wires, the deformation of the wires plays the next important role in determining the maximum pull force and displacement.

- In general, the tensile strength and Young's modulus (stiffness) of Cu wire are larger than those of Au wire.

- More accurate equations can be obtained by considering the large deformation, change of cross-sectional area (necking), and nonlinear stress-strain relationship of the wire.

- Additional tests should be performed for Au wire and even more so for Cu wire.

10.8 References

[1] Lau, J. H., *Low-Cost Flip Chip Technology*, McGraw-Hill, New York, 2000.

[2] Lau, J. H., *Flip Chip Technology*, McGraw-Hill, New York, 1995.

[3] van Driel, W. D., G. Wisse, A. Y. L. Chang, J. H. J. Jansen, X. Fan, G. Q. Zhang, and L. J. Ernst, "Influence of Material Combinations on Delamination Failures in a Cavity-Down TBGA Package," *IEEE Transactions of Components and Packaging Technology*, Vol. 27, No. 4, December 2004, pp. 651–658.

[4] Harman, G. G., and C. E. Johnson, "Wire Bonding to Advanced Copper, Low-*k* Integrated Circuits, the Metal/Dielectric Stacks, and Materials Considerations," *IEEE Transactions of Components and Packaging Technology*, Vol. 25, No. 4, December 2002, pp. 677–683.

[5] Zhang, J., and J. Huneke, "Stress Analysis of Spacer Paste Replacing Dummy Die in a Stacked CSP Package", *Proceedings of IEEE/ICEPT*, 2003, pp. 82–85.

[6] Yoon, S. W., D. Wirtasa, S. Lim, V. Ganesh, A. Viswanath, V. Kripesh, and M. K. Iyer, "PEDL Technology for Copper/Low-*k* Dielectrics Interconnect," *IEEE Proceedings of EPTC*, Decemeber 2005, pp. 711–715.

[7] Hartfield, C. D., E. T. Ogawa, Y.-J. Park, T.-C. Chiu, and H. Guo, "Interface Reliability Assessments for Copper/Low-*k* Products," *IEEE Transactions of Device Materials Reliability*, Vol. 4, No. 2, June 2004, pp. 129–141.

[8] Alers, G. B., K. Jow, R. Shaviv, G. Kooi, and G. W. Ray, "Interlevel Dielectric Failures in Copper/Low-*k* Structures," *IEEE Transactions of Device Materials Reliability*, Vol. 4, No. 2, June 2004, pp. 148–152.

[9] Chiu, C. C., H. H. Chang, C. C. Lee, C. C. Hsia, and K. N. Chiang, "Reliability of Interfacial Adhesion in a multilevel copper/low-*k* interconnect structure," *Microelectronic Reliability*, Vol. 47, Nos. 9–11, 2007, pp. 1506–1511.

[10] Wang, Z., J. H. Wang, S. Lee, S. Y. Yao, R. Han, and Y. Q. Su, "300-mm Low-*k* Wafer Dicing Saw Development," *IEEE Transactions of Electronic Packaging & Manufacturing*, Vol. 30, No. 4, October 2007, pp. 313–319.

[11] Chungpaiboonpatana, S., and F. G. Shi, "Packaging of Copper/Low-*k* IC Devices: A Novel Direct Fine-Pitch Gold Wire-Bond Ball Interconnects onto Copper/Low-*k* Terminal Pads," *IEEE Transactions of Advanced Packaging*, Vol. 27, No. 3, August 2004, pp. 476–489.

[12] Wang, T. H., Y.-S. Lai, and M.-J. Wang, "Underfill Selection for Reducing Cu/Low-*k* Delamination Risk of Flip-Chip Assembly," *IEEE Proceedings of EPTC*, December 2006, pp. 233–236.

[13] Zhai, C. J., U. Ozkan, A. Dubey, S. Sidharth, R. Blish II, and R. Master, "Investigation of Cu/Low-*k* Film Delaminaiton in Flip Chip Packages," *IEEE/ECTC Proceedings*, June 2006, pp. 709–717.

[14] Ong, J. M. G., A. A. O. Tay, X. Zhang, V. Kripesh, Y. K. Lim, D. Yeo, K. C. Chan, J. B. Tan, L. C. Hsia, and D. K. Sohn, "Optimization of the Thermo-Mechanical Reliability of a 65-nm Cu/Low-*k* Large-Die Flip Chip Package," *IEEE Transactions of Components and Packaging Technology*, Vol. 32, No. 4, December 2009, pp. 838–848.

[15] Zhang, X., J. H. Lau, C. Premachandran, S. Chong, L. Wai, V. Lee, T. Chai, V. Kripesh, V. Sekhar, D. Pinjala, and F. Che, "Development of a Cu/Low-k Stack Die Fine Pitch Ball Grid Array (FBGA) Package for System in Package Applications," *IEEE Transactions on CPMT*, Vol. 1, No. 3, March 2011, pp. 299–309.

[16] Chen, S., C. Z. Tsai, E. Wu, I. G. Shih, and Y. N. Chen, "Study on the Effects of Wafer Thinning and Dicing on Chip Strength," *IEEE Transactions of Advanced Packaging*, Vol. 29, No. 1, February 2006, pp. 149–157.

[17] Toh, C., M. Gaurav, H. Tan, and P. Ong, "Die Attached Adhesives for 3D Same-Size Dies Stacked Packages," *IEEE/ECTC Proceedings*, May 2008, pp. 1538–1543.

[18] Carson, F., K. Ishibashi, and Y. Kim, "Three-Tier PoP Configuration Utilizing Flip Chip Fan-In PoP Bottom Package," *IEEE/ECTC Proceedings*, San Diego, CA, 2009, pp. 313–318.

[19] Xie, D., D. Shangguan, D. Geiger, D. Gill, V. Vellppan, and K. Chinniah, "Head in Pillow (HIP) and Yield Study on SIP and PoP Assembly," *IEEE/ECTC Proceedings*, San Diego, CA, 2009, pp. 752–758.

[20] Sze, T., M. Giere, B. Guenin, N. Nettleton, D. Popovic, J. Shi, R. Ho, R. Drost, D. Douglas, and S. Bezuk, "Proximity Communication Flip-Chip Package with Micron Chip-to-Chip Alignment Tolerances," *IEEE/ECTC Proceedings*, San Diego, CA, 2009, pp. 966–971.

[21] Kumar, A., V. Sekhar, S. Lim, C. Keng, G. Sharma, S. Vempati, V. Kripesh, J. H. Lau, D. Kwong, and X. Dingwei, "Wafer Level Embedding Technology for 3D Wafer Level Embedded Package," *IEEE/ECTC Proceedings*, San Diego, CA, 2009, pp. 1289–1296.

[22] Kim, D., Y. Kim, K. Seong, J. Song, B. Kim, C. Hwang, and C. Lee, "Evaluation for UV Laser Dicing Process and Its Reliability for Various Designs of Stack Chip Scale Package," *IEEE/ECTC Proceedings*, San Diego, CA, 2009, pp. 1531–1536.

[23] Sharma, G., S. Vempati, A. Kumar, S. Nandar, Y. Lim, K. Houe, S. Lim, N. Vasarla, R. Ranjan, V. Kripesh, and J. H. Lau, "Embedded Wafer Level Packages with Laterally Placed and Vertically Stacked Thin Dies," *IEEE/ECTC Proceedings*, San Diego, CA, 2009, pp. 1537–1543.

[24] Pendse, R., "Flip Chip Package-in-Package (fcPiP): A New 3D Packaging Solution for Mobile Platforms," *IEEE/ECTC Proceedings*, Reno, NV, May 2007, pp. 1425–1430.

[25] Carson, F., S. M. Lee, and L. S. Yoon, "The Development of the Fan-in Package-on-Package," *IEEE/ECTC Proceedings*, Lake Buena Vista, FL, May 2008, pp. 956–963.

[26] Kazuo, I., "PoP (Package-on-Package) Stacking Yield Loss Study," *IEEE/ECTC Proceedings*, Reno, NV, May 2007, pp. 1403–1408.

[27] Carson, F., S. Lee, and N. Vijayaragavan, "Controlling Top Package Warpage," *IEEE/ECTC Proceedings*, Reno, NV, May 2007, pp. 737–742.

[28] Vijayaragavan, N., F. Carson, and A. Mistri, "Package on Package Warpage—Impact on Surface Mount Yields and Board Level Reliability," *IEEE/ECTC Proceedings*, Lake Buena Vista, FL, May 2008, pp. 389–396.

[29] IPC/JEDEC, "Moisture/Reflow Sensitivity Classification for Nonhermetic Solid-State Surface Mounted Devices," JSTD-020D, June 2007.

[30] Geiger, D., D. Shangguan, S. Tam, and D. Rooney, "Package Stacking in SMT for 3D PCB Assembly," *IEEE Proceedings of IEMT*, San Jose, CA, July 2003, p. S207P6.

[31] Carson, F., "Package-on-Package Variations on the Horizon," *Semiconductor International*, May 1, 2008.

[32] Wei, L., A. Yoshida, and M. Dreiza, "Control of the Warpage for Package-on-Package (PoP) Design," *Proceedings of SMTAI*, Chicago, IL, 2007, pp. 320–327.

[33] Dongji, X., D. Geiger, D. Shangguan, B. Hu, and J. Sjoberg, "Yield Study of Inline Package on Package (PoP) Assembly," *IEEE Proceedings of Electronic Packaging Technology Conference*, Singapore, Lake Buena Vista, FL, May 2008, pp. 1202–1208.

[34] Kazuo, I., "PoP (Package-on-Package) Stacking Yield Loss Study," *IEEE/ECTC Proceedings*, Reno, NV, 2007, pp. 1043–1048.

[35] Jonas, S., D. Geiger, and D. Shangguan, "Package on Package Process Development and Reliability Evaluation," *IEEE/ECTC Proceedings*, Lake Beuna Vista, FL, May 2008, pp. 2005–2010.

[36] Souriau, J., O. Lignier, M. Charrier, and G. Poupon, "Wafer Level Processing of 3D System in Package for RF and Data Applications," *IEEE/ECTC Proceedings*, Lake Beuna Vista, FL, 2005, pp. 356–361.

[37] Brunnbauer, M., E. Furgut, G. Beer, T. Meyer, H. Hedler, J. Belonio, E. Nomura, K. Kiuchi, K. Kobayashi, "An Embedded Device Technology Based on a Molded Reconfigured Wafer," *IEEE/ECTC Proceedings*, San Diego, CA, 2006, pp. 547–551.

[38] Keser, B., C. Amrine, T. Duong, Owen Fay, S. Hayes, G. Leal, W. Lytle, D. Mitchell, R. Wenzel "The Redistributed Chip Package: A Breakthrough for Advanced Packaging," *IEEE/ECTC Proceedings*, Reno, NV, 2007, pp. 286–290.

[39] Fillion, R., C. Woychik, T. Zhang, and D. Bitting, "Embedded Chip Build-Up Using Fine Line Interconnect," *IEEE/ECTC Proceedings*, Reno, NV, 2007, pp. 49–53.

[40] Kripesh, V., V. S. Rao, A. Kumar, G. Sharma, K. C. Houe, Z. Xiaowu, K. Y. Mong, N. Khan, and J. H. Lau, "Design and Development of a Multi-Die Embedded Micro Wafer Level Package," *IEEE/ECTC Proceedings*, Lake Beuna Vista, FL, May 2008, pp. 1544–1549.

[41] Hamano, T., T. Kawahara, and J. I. Kasai, "Super CSPTM: WLCSP Solution for Memory and System LSI," *Proceedings of International Symposium on Advanced Packaging Materials*, 1999, pp. 221–225.

[42] Braun, T., K. Becker, M. Koch, V. Bader, U. Oestermann, D. Manessis, R. Aschenbrenner, and H. Reichl, "Wafer Level Encapsulation—A Transfer

Molding Approach to System in Package Generation," *IEEE/ECTC Proceedings*, San Diego, CA, 2002, pp. 235–244.

[43] Ostmann, A., A. Neumann, S. Weser, E. Jung, L. Bottcher, and H. Reichl, "Realization of a Stackable Package Using Chip in Polymer Technology," *IEEE Proceedings of Polytronic*, 2002, pp. 160–164.

[44] Ko, C., S. Chen, C. W. Chiang, T.-Y. Kuo, Y. C. Shih, and Y.-H. Chen, "Embedded Active Device Packaging Technology for Next-Generation Chip-in-Substrate Packages, CiSP," *IEEE/ECTC Proceedings*, San Diego, CA, 2006, pp. 322–329.

[45] Takyu, S., T. Kurosawa, N. Shimizu, and S. Harada, "Novel Wafer Dicing and Chip Thinning Technologies Realizing High Chip Strength," *IEEE/ECTC Proceedings*, San Diego, CA, 2006, pp. 1623–1627.

[46] Werner K., and F., Mariani, "Thinning and Singulation of Silicon: Root Causes of the Damage in Thin Chips," *IEEE/ECTC Proceedings*, San Diego, CA, 2006, pp. 1317–1322.

[47] Steen, M., *Laser Material Processing*, 3rd ed., Springer, New York, pp. 107–125.

[48] Li, J., H. Hwang, E. C. Ahn, Q. Chen, P. W. Kim, T. H. Lee, M. K Chung, and T. G. Chung, "Laser Dicing and Subsequent Die Strength Enhancement Technologies for Ultra-thin Wafer," *IEEE/ECTC Proceedings*, Reno, NV, 2007, pp. 761–766.

[49] Lim, S., V. Rao, H. Yin, W. Ching, V. Kripesh, C. Lee, J. H. Lau, J. Milla, and A. Fenner, "Process Development and Reliability of Microbumps," *IEEE/ECTC Proceedings*, December 2008, pp. 367–372. Also, *IEEE Transactions on Components and Packaging Technology*, Vol. 33, No. 4, 2010, pp. 747–753.

[50] Vempati, S., S. Nandar, C. Khong, Y. Lim, V. Kripesh, J. H. Lau, B. P. Liew, K. Y. Au, S. Tamary, A. Fenner, R. Erich, and J. Milla, "Development of 3D Silicon Die Stacked Package Using Flip-Chip Technology with Micro Bump Interconnects," *IEEE/ECTC Proceedings*, San Diego, CA, 2009, pp. 980–987.

[51] Zhang, F., M. Li, W. T. Chen, and K. S. Chan, "An Investigation into the Effects of Flux Residues on Properties of Underfill Materials for Flip-Chip Packages," *IEEE Transactions of Components and Packaging Technology*, Vol. 26, No. 1, June 2003, pp. 233–237.

[52] Kloeser, J., E. Zakel, F. Bechtold, and H. Reichl, "Reliability Investigation of Fluxless Flip-Chip Interconnections on Green Tape Ceramic Substrates," *IEEE Transactions of Components and Packaging Technology A*, Vol. 19, No. 1, March 1996, pp. 24–33.

[53] Hutter, M., H. Oppermann, G. Engelmann, L. Dietricj, and H. Reichl, "Precise Flip-Chip Assembly Using Electroplated $AuSn_{20}$ and $SnAg_{3.5}$ Solder," in *Proceedings of the 57th Electronic Components and Technology Conference*, 2006, pp. 1087–1094.

[54] Hutter, M., F. Hohnke, H. Oppermann, M. Klein, and G. Engelmann, "Assembly and Reliability of Flip-Chip Solder Joints Using Miniaturized Au/Sn Bumps," in *Proceedings of the 54th Electronic. Components and Technology Conference*, 2004, pp. 49–56.

[55] Baggerman, A. F. J., and M. J. Batenburg, "Reliable Au-Sn Flip-Chip Bonding on Flexible Prints," *IEEE Transactions of Components and Packaging Technology B*, Vol. 18, No. 2, May 1995, pp. 257–263.

[56] Rao, V., X. Zhang, S. Ho, R. Rajoo, C. Premachandran, V. Kripesh, S. Yoon, and J. H. Lau, "Design and Development of Fine Pitch Copper/Low-k Wafer Level Package", *IEEE Transactions on Advanced Packaging*, Vol. 33, No. 2, May 2010, pp. 377–388.

[57] Hayes, S., N. Chhabra, T. Duong, Z. Gong, D. Mitchfell, and J. Wright, "System-in-Package Opportunities with the Redistributed Chip Package (RCP)," *Proceedings of International Wafer Level Packaging Conference*, San Jose, CA, October 2011, pp. 1–7.

[58] Keser, B., C. Amrine, T. Duong, S. Hayes, G. Leal, W. Lytle, D. Mitchell, and R. Wenzel, "Advanced Packaging: The Redistributed Chip Package," *IEEE Transactions on Advanced Packaging*, Vol. 31, No. 1, February 2008, pp. 39–43.

[59] Trotta, S., M. Wintermantel, J. Dixon, U. Moeller, R. Jammers, T. Hauck, T., Samulak, B. Dehlink, K. Shun-Meen, H. Li, A. Ghazinour, Y. Yin, S. Pacheco, R. Euter, S. Majied, D. Moline, T. Aaron, V. P. Trivedi, D. J. Morgan, and J. John, "An RCP Packaged Transceiver Chipset for Automotive LRR and SRR Systems in SiGe BiCMOS Technology," *IEEE Transactions on Microwave Theory and Techniques*, Vol. PP, No. 99, 2012, pp. 1–17.

[60] Lau, J. H., *Reliability of RoHS-Compliant 2D and 3D IC Interconnects*, McGraw-Hill, New York, 2011.

[61] Lau, J. H., C. K. Lee, C. S. Premachandran, and A. Yu, *Advanced MEMS Packaging*, McGraw-Hill, New York, 2010.

[62] Harman, G., *Wire Bonding in Microelectronics*, McGraw-Hill, New York, 2010.

[63] Prasad, S., *Advanced Wirebond Interconnection Technology*, Springer, New York, 2004.

[64] Lau, J. H., *Hankbook of Tape Automated Bonding*, Van Nostrand Reinhold, New York, 1992.

[65] Onuki, J., M., Koizumi, and I. Araki, "Investigation of the Reliability of Copper Ball Bonds to Aluminum Electrodes," *IEEE Transactions on Components, Hybrid, and Manufacturing Technology*, Vol. 12, No. 4, 1987, pp. 550–555.

[66] Toyozawa, K., K. Fujita, S. Minamide, and T. Maeda, "Development of Copper Wire Bonding Application Technology," *IEEE Transactions on Components, Hybrid, and Manufacturing Technology*, Vol. 13, No. 4, 1990, pp. 667–672.

[67] Khoury, S., D. Burkhard, D. Galloway, and T. Scharr, "A Comparison of Copper and Gold Wire Bonding on Integrated Circuit Devices," *IEEE Transactions on Components, Hybrid, and Manufacturing Technology*, Vol. 13, No. 4, 1990, pp. 673–681.

[68] Caers, J., A. Bischoff, J. Falk, and J. Roggen, "Conditions for Reliable Ball-Wedge Copper Wire Bonding," *Proceedings of 1993 Japan International Electronic Manufacturing Technology Symposium*, Kanazawa, Japan,1993, pp. 312–315.

[69] Nguyen, L., D. McDonald, A. Danker, and P. Ng, "Optimization of Copper Wire Bonding on Al-Cu Metallization," *IEEE Transactions on Components, Hybrid, and Manufacturing Technology A*, Vol. 18, No. 2, 1995, pp. 423–429.

[70] Tan, J., B. Toh, and H. Ho, "Modelling of Free Air Ball for Copper Wire Bonding," *Proceedings of the 6th IEEE Electronics Packaging and Technology Conference*, Singapore, 2004, pp. 711–717.

[71] Xu, H., C. Liu, V. Silberschmidt, and H. Wang, "Effects of Process Parameters on Bondability in Thermosonic Copper Ball Bonding," *IEEE/ECTC Proceedings*, Lake Buena Vista, FL, 2008. 1424-1430.

[72] Uno, T., S. Terashima, and T. Yamada, "Surface-Enhanced Copper Bonding Wire for LSI," *IEEE/ECTC Proceedings*, San Diego, CA, 2009, pp. 1486–1495.

[73] Deley, M., and L. Levine, "Copper Ball Bonding Advances for Leading Edge Packaging," *Proceedings of Semicon Singapore 2005*, Singapore, 2005, pp. 1–6.

[74] Shah, A., M. Mayer, Y. Zhou, S. Hong, and J. Moon, "In Situ Ultrasonic Force Signals During Low-temperature Thermosonic Copper Wire Bonding," *Microelectronics Engineering*, Vol. 85, No. 9, 2008, pp. 1851–1857.

[75] Hang, C. J., I. Lum, J. Lee, M. Mayer, C. Q. Wang, Y. Zhou, S. J. Hong, and S. M. Lee, "Bonding Wire Characterization Using Automatic Deformability Measurement," *Microelectronics Engineering*, Vol. 85, No. 8, 2008, pp. 1795-1803.

[76] Kim, H., J. Lee, K. Paik, K. Koh, J. Won, S. Choe, J. Lee, J. Moon, and Y. Park, "Effects of Cu/Al Intermetallic Compound (IMC) on Copper Wire and Aluminum Pad Bondability," *IEEE Transactions on Component Packaging Technology*, Vol. 26, No. 2, 2003, pp. 367–374.

[77] Murali, S. et al., "An Analysis of Intermetallics Formation of Gold and Copper Ball Bonding on Thermal Aging," *Material Research Bulletant*, Vol. 38, No. 4, 2003, pp. 637–646.

[78] Murali, S. et al., "Effect of Wire Size on the Formation of Intermetallics and Kirkendall Voids on Thermal Aging of Thermosonic Wire Bonds," *Materials Letters*, Vol. 58, No. 25, 2004, pp. 3096–3101.

[79] Wulff, F., C. Breach, D. Stephan, A. Saraswati, and K. Dittmer, "Characterization of Intermetallic Growth in Copper and Gold Ball Bonds on Aluminum Metallization," *IEEE/ECTC Proceedings*, Singapore, 2004, pp. 348–353.

[80] Ratchev, P., S. Stoukatch, and B. Swinnen, "Mechanical Reliability of Au and Cu Wire Bonds to Al, Ni/Au and Ni/Pd/Au Capped Cu Bond Pads," *Microelectronic Reliability*, Vol. 46, No. 8, 2006, pp. 1315–1325.

[81] Murali, S., N. Srikanth, and C. Vath, "An Evaluation of Gold and Copper Wire Bonds on Shear and Pull Testing," *ASME Transactions, Journal of Electronic Packaging*, Vol. 128, No. 3, 2006, pp. 192–201.

[82] Ibrahim, M. R., Y. Choi, L. Lim, J. Lu, L. Poh, and P. Ai, "The Challenges of Fine Pitch Copper Wire Bonding in BGA Packages," *IEEE Proceedings of 31st International Electronic Manufacturing Technology Symposium*, Kuala Lumpur, Malaysia, 2006, pp. 347–353.

[83] Hang, C. J., C. Wang, M. Mayer, Y. Tian, Y. Zhou, and H. Wang, "Growth Behavior of Cu/Al Intermetallic Compounds and Cracks in Copper Ball Bonds During Isothermal Aging," *Microelectronic Reliability*, Vol. 48, No. 3, 2008, pp. 416–424.

[84] Xu, H., C. Liu, V. Silberschmidt, S. Pramana, T. White, and Z. Chen, "A Re-examination of the Mechanism of Thermosonic Copper Ball Bonding on Aluminium Metallization Pads," *Scripta Materialia*, Vol. 61, No. 2, 2009, pp. 165–168.

[85] Zhang, S., C. Chen, R. Lee, A. Lau, P. Tsang, L. Mohamed, C. Chan, and M. Dirkzwager, "Characterization of Intermetallic Compound Formation and Copper Diffusion of Copper Wire Bonding," *IEEE/ECTC Proceedings*, San Diego, CA, 2006, pp. 1821–1826.

[86] Yeoh, L. S., "Characterization of Intermetallic Growth for Gold Bonding and Copper Bonding on Aluminum Metallization in Power Transistors," *IEEE/ECTC Proceedings*, Singapore, 2007, pp. 731–736.

[87] Murali, S., N. Srikanth, and C. Vath, "Effect of Wire Diameter on the Thermosonic Bond Reliability," *Microelectronic Reliability*, Vol. 46, Nos. 2–4, 2006, pp. 467–475.

[88] Tan, C. W., A. Daud, and M. Yarmo, "Corrosion Study at Cu-Al Interface in Microelectronics Packaging," *Appied Surface Science*, Vol. 191, Nos. 1–4, 2002, pp. 67–73.

[89] England, L., and T. Jiang, "Reliability of Cu Wire Bonding to Al Metallization," *IEEE/ECTC Proceedings*, Reno, NV, 2007, pp. 1604–1613.

[90] Kaimori, S., T. Nonaka, and A. Mizoguchi, "The Development of Cu Bonding Wire with Oxidation-Resistant Metal Coating," *IEEE Transactions on Advanced Packagine*, Vol. 29, No. 2, 2006, pp. 227–231.

[91] Murali, S., N. Srikanth, and C. Vath, "Grains, Deformation Substructures, and Slip Bands Observed in Thermosonic Copper Ball Bonding," *Materials Characterization*, Vol. 50, No. 1, 2003, pp. 39–50.

[92] Murali, S., and N. Srikanth, "Acid Decapsulation of Epoxy Molded IC Packages with Copper Wire Bonds," *IEEE Transactions on Electronic Packaging Manufacturing*, Vol. 29, No. 3, 2006, pp. 179–183.

[93] Chylak, R., "Developments in Fine Pitch Copper Wire Bonding Production," *IEEE/ECTC Proceedings*, Singapore, 2009, pp. 1–6.

[94] Appelt, B., A. Tseng, and Y. Lai, "Fine Pitch Copper Wire Bonding—Why Now?" *IEEE/ECTC Proceedings*, Singapore, 2009, pp. 469–472.

[95] Yauw, O., H. Clauberg, K. Lee, L. Shen, and B. Chylak, "Wire Bonding Optimization with Fine Copper Wire for Volume Production," *IEEE/EPTC Proceedings*, Singapore, December 2010, pp. 467–472.

[96] Leng, E., C. Siong, L. Seong, P. Leong, K. Gunase, J. Song, K. Mock, C. Siew, and K. Siva, "Ultra Fine Pitch Cu Wire Bonding on C45 Ultra Low k Wafer Technology," *IEEE/EPTC Proceedings*, Singapore, December 2010, pp. 484–488.

[97] Teo, J., "17.5-μm Thin Cu Wire Bonding for Fragile Low-*k* Wafer Technology," *IEEE/EPTC Proceedings*, Singapore, December 2010, pp. 355–358.

[98]   Kim, S., J. Park, S. Hong, and J. Moon, "The Interface Behavior of the Cu-Al Bond System in High Humidity Conditions," *IEEE/EPTC Proceedings*, Singapore, December 2010, pp. 545–549.

[99]   Song, M., G. Gong, J. Yao, S. Xu, S. Lee, M. Han, and B. Yan, "Study of Optimum Bond Pad Metallization Thickness for Copper Wire Bond Process," *IEEE/EPTC Proceedings*, Singapore, December 2010, pp. 597–602.

[100] Boettcher, T., M. Rother, S. Liedtke, M. Ullrich, M. Bollmann, A. Pinkernelle, D. Gruber, H. Funke, M. Kaiser, K. Lee, M. Li, K. Leung, T. Li, M. Farrugia, and O. O'Halloran, "On the Intermetallic Corrosion of Cu-Al Wire Bonds," *IEEE/EPTC Proceedings*, Singapore, December 2010, pp. 585–590.

[101] Tang, L., H. Ho, Y. Zhang, Y. Lee, and C. Lee, "Investigation of Palladium Distribution on the Free Air Ball of Pd-Coated Cu Wire," *IEEE/EPTC Proceedings*, Singapore, December 2010, pp. 777–782.

[102] Kumar, B., M. Sivakumar, R. Malliah, M. Li, and S. Yew, "Process Characterization of Cu & Pd Coated Cu Wire Bonding on Overhang Die: Challenges and Solution," *IEEE/EPTC Proceedings*, Singapore, December 2010, pp. 859–867.

[103] Breach, C., N. Shen, T. Mun, T. Lee, and R. Holliday, "Effects of Moisture on Reliability of Gold and Copper Ball Bonds," *IEEE/EPTC Proceedings*, Singapore, December 2010, pp. 44–51.

CHAPTER **11**

Future Trends of 3D Integration

11.1 Introduction

The trends of 3D integration by the end of this decade will be presented in this chapter. Figure 11.1 shows the roadmap [1, 2]. It can be seen that the TSVs for passive interposers will be used the most by 2020.

11.2 The Trend of 3D Si Integration

As mentioned in Chap. 1, there are many issues such as technology, electronic design automation (EDA), and an ecosystem of 3D Si integration. However, it is the right way to go and compete with Moore's law. The industry should immediately build an ecosystem incorporating standards and infrastructures so that EDA vendors can create and qualify the software for design, simulation, analysis and verification, manufacturing preparation, and test for 3D Si integration and *strive* to make it happen!

Hopefully, by 2020 (Fig. 11.1), the memory chips stacking (memory cube) could be manufactured at lower costs, higher throughputs, and lower profile with the 3D Si integration technology by *bumpless* W2W bonding as shown in Fig. 11.2. This will be performed by semiconductor foundries. The packaging assembly and test houses receive the stacked-wafer (without any bumps) and will go through the routine tasks such as wafer bumping, dicing, packaging, and testing.

11.3 The Trend of 3D IC Integration

The potential applications of 3D IC integration have been shown in Fig. 1.6. By 2020, most TSVs will be fabricated on passive interposers (the 2.5D IC integration) as shown in Fig. 11.1. For active interposers (with TSVs on device chips), we have to wait for the ecosystem and

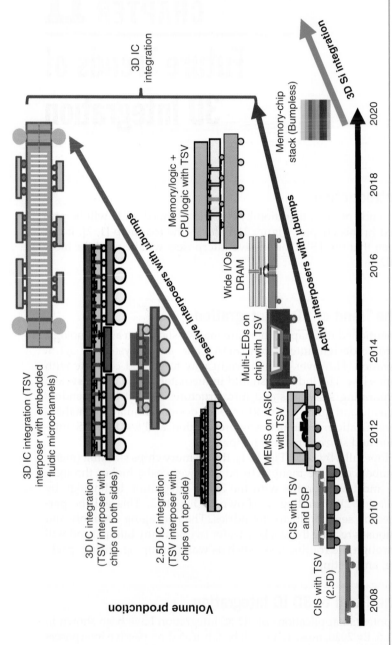

FIGURE 11.1 Roadmap of 3D Si integration and 3D IC integration.

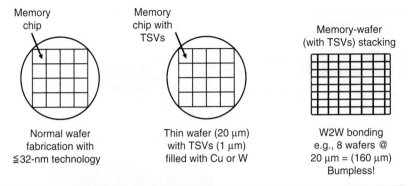

FIGURE 11.2 Memory cube fabricated by semiconductor foundries.

EDA, except for niche applications such as 3D MEMS [3] and 3D LED [4]. Meantime, the wide I/O memory and DRAM are replaced by the 2.5D IC integration as shown in Fig. 11.3: (1) the TSV SoC/logic of the wide I/O memory shown in Figs. 1.6 and 1.9 can be placed side by side with the memory and is TSV-less, and (2) similarly the TSV logic controller of the wide I/O DRAM shown in Figs. 1.6 and 1.11 can be placed side by side with the memory cube and is TSV-less.

Figure 11.4 shows 3D IC integration with passive interposers. It can be seen that the 2.5D IC integration can be a real 3D IC integration by placing the Moore's law chips (without TSVs) on both sides of the interposer in a face-to-face format. In this case, the chip-to-chip interconnect is shorter and the interposer size is smaller. For examples, please see Figs. 1.23 through 1.33. Also, the wide I/O memory (Figs. 1.6 and 1.9) and wide I/O DRAM (Figs. 1.6 and 1.11) can be arranged in this format.

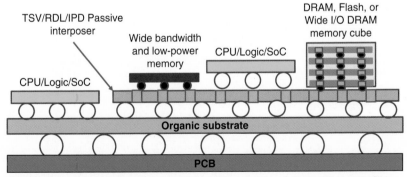

Underfill is needed between the interposer and the organic substrate. Also, underfill is needed between the interposer and the Moore's law chips and the memory cube

FIGURE 11.3 Potential applications of 2.5D IC integration.

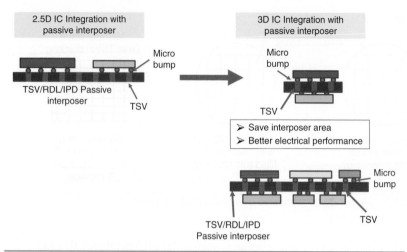

Figure 11.4 3D IC integration with a passive interposer.

11.4 References

[1] Lau, J. H., "Evolution, Challenges, and Outlook of 3D IC/Si Integration," *Plenary Keynote at IEEE ICEP*, Nara, Japan, April 2011, pp. 1–16.
[2] Lau, J. H., "3D IC Integration and 3D Si Integration," *Plenary Keynote at IWLPC*, San Jose, CA, October 2011, pp. 1–18.
[3] Lau, J. H., C. K. Lee, C. S. Premachandran, and A. Yu, *Advanced MEMS Packaging*, McGraw-Hill, New York, 2010.
[4] Lau, J. H., R. Lee, M. Yuen, and P. Chan, "3D LED and IC Wafer Level Packaging," *Journal of Microelectronics International*, Vol. 27, Issue 2, 2010, pp. 98–105.

Index

Note: Page numbers followed by *f* denote figures; page numbers followed by *t* denote tables.